# Conservation Research, Policy and Practice

Conservation research is essential for advancing knowledge, but to make an impact, scientific evidence must influence conservation policies, decision-making and practice. This raises a multitude of challenges. How should evidence be collated and presented to policy-makers to maximise its impact? How can effective collaboration between conservation scientists and decision-makers be established? How can the resulting messages be communicated to bring about change?

Emerging from a successful international symposium organised by the British Ecological Society and the Cambridge Conservation Initiative, this is the first book to practically address these questions across a wide range of conservation topics. Well-renowned experts guide readers through global case studies and their own experiences.

This is a must-read for practitioners, researchers, graduate students and policy-makers wishing to enhance the prospect of their work 'making a difference'. This title is also available as Open Access on Cambridge Core.

WILLIAM J. SUTHERLAND is Miriam Rothschild Chair in Conservation Biology, University of Cambridge, UK. He is an Honorary Member and previous President of the British Ecological Society. He is involved in horizon-scanning, agenda-setting and improving practice.

PETER N. M. BROTHERTON is a Director at Natural England, the official nature conservation agency for England. He has a particular interest in the interface between conservation science, policy and practice. He was lead advisor to the UK government on the England Biodiversity Strategy (2011) and co-authored the influential 'Making Space for Nature' report.

ZOE G. DAVIES is Professor of Biodiversity Conservation, University of Kent, UK. She has worked on applied projects in the UK, across Europe, Chile, Guyana, Kenya, Madagascar, Indonesia, Malaysia and Singapore. Her research involves integrating natural and social science disciplines to answer important questions regarding how we should conserve the natural environment.

NANCY OCKENDON is the Science Coordinator of the Endangered Landscapes Programme, Cambridge Conservation Initiative, UK, where she aims to ensure the more effective use and generation of scientific evidence in conservation projects. She is also interested in improving communication between scientists, practitioners and policy-makers.

NATHALIE PETTORELLI is Senior Research Fellow, Zoological Society of London, UK. She has published four books and over 150 articles on the topic of biodiversity monitoring and wildlife management. She is a senior editor for *Journal of Applied Ecology*, the UK representative to GEO Programme Board, and a member of the British Ecological Society Policy Committee.

JULIET A. VICKERY is Head of International Research, RSPB Centre for Conservation Science, Bedfordshire, UK. She is an Honorary Research Fellow in the Conservation Science Group, University of Cambridge; Chair of the Policy Committee of the British Ecological Society; and President of the British Ornithologists' Union.

# Ecological Reviews

Ecological Reviews publishes books at the cutting edge of modern ecology, providing a forum for volumes that discuss topics that are focal points of current activity and likely long-term importance to the progress of the field. The series is an invaluable source of ideas and inspiration for ecologists at all levels from graduate students to more-established researchers and professionals. The series has been developed jointly by the British Ecological Society and Cambridge University Press and encompasses the Society's Symposia as appropriate.

*Biotic Interactions in the Tropics: Their Role in the Maintenance of Species Diversity* Edited by David F. R. P. Burslem, Michelle A. Pinard and Sue E. Hartley

*Biological Diversity and Function in Soils* Edited by Richard Bardgett, Michael Usher and David Hopkins

*Island Colonization: The Origin and Development of Island Communities* By Ian Thornton Edited by Tim New

*Scaling Biodiversity* Edited by David Storch, Pablo Margnet and James Brown

*Body Size: The Structure and Function of Aquatic Ecosystems* Edited by Alan G. Hildrew, David G. Raffaelli and Ronni Edmonds-Brown

*Speciation and Patterns of Diversity* Edited by Roger Butlin, Jon Bridle and Dolph Schluter

*Ecology of Industrial Pollution* Edited by Lesley C. Batty and Kevin B. Hallberg

*Ecosystem Ecology: A New Synthesis* Edited by David G. Raffaelli and Christopher L. J. Frid

*Urban Ecology* Edited by Kevin J. Gaston

*The Ecology of Plant Secondary Metabolites: From Genes to Global Processes* Edited by Glenn R. Iason, Marcel Dicke and Susan E. Hartley

*Birds and Habitat: Relationships in Changing Landscapes* Edited by Robert J. Fuller

*Trait-Mediated Indirect Interactions: Ecological and Evolutionary Perspectives* Edited by Takayuki Ohgushi, Oswald Schmitz and Robert D. Holt

*Forests and Global Change* Edited by David A. Coomes, David F. R. P. Burslem and William D. Simonson

*Trophic Ecology: Bottom-Up and Top-Down Interactions across Aquatic and Terrestrial Systems* Edited by Torrance C. Hanley and Kimberly J. La Pierre

*Conflicts in Conservation: Navigating Towards Solutions* Edited by Stephen M. Redpath, R. J Gutiérrez, Kevin A. Wood and Juliette C. Young

*Peatland Restoration and Ecosystem Services* Edited by Aletta Bonn, Tim Allott, Martin Evans, Hans Joosten and Rob Stoneman

*Rewilding* Edited by Nathalie Pettorelli, Sarah M. Durant and Johan T. du Toit

*Grasslands and Climate Change* Edited by David J. Gibson and Jonathan A. Newman

*Agricultural Resilience: Perspectives from Ecology and Economics* Edited by Sarah M. Gardner, Stephen J. Ramsden and Rosemary S. Hails

*Wildlife Disease Ecology: Linking Theory to Data and Application* Edited by Kenneth Wilson, Andy Fenton and Dan Tompkins

*Microbiomes of Soils, Plants and Animals: An Integrated Approach* Edited by Rachael E. Antwis, Xavier A. Harrison and Michael J. Cox

# Conservation Research, Policy and Practice

Edited by

WILLIAM J. SUTHERLAND
*University of Cambridge*

PETER N. M. BROTHERTON
*Natural England*

ZOE G. DAVIES
*Durrell Institute of Conservation and Ecology (DICE), University of Kent*

NANCY OCKENDON
*University of Cambridge*

NATHALIE PETTORELLI
*Zoological Society of London*

JULIET A. VICKERY
*Royal Society for the Protection of Birds*

CAMBRIDGE
UNIVERSITY PRESS

# CAMBRIDGE
## UNIVERSITY PRESS

University Printing House, Cambridge CB2 8BS, United Kingdom

One Liberty Plaza, 20th Floor, New York, NY 10006, USA

477 Williamstown Road, Port Melbourne, VIC 3207, Australia

314–321, 3rd Floor, Plot 3, Splendor Forum, Jasola District Centre, New Delhi – 110025, India

79 Anson Road, #06–04/06, Singapore 079906

Cambridge University Press is part of the University of Cambridge.

It furthers the University's mission by disseminating knowledge in the pursuit of education, learning and research at the highest international levels of excellence.

www.cambridge.org
Information on this title: www.cambridge.org/9781108714587
DOI: 10.1017/9781108638210

First published 2020

*A catalogue record for this publication is available from the British Library.*

ISBN 978-1-108-71458-7 Paperback

Cambridge University Press has no responsibility for the persistence or accuracy of URLs for external or third-party internet websites referred to in this publication and does not guarantee that any content on such websites is, or will remain, accurate or appropriate.

# Contents

*Colour plates can be found between pages 144 and 145.*

# Contributors

JOHN D. ALTRINGHAM
School of Biology
University of Leeds
Leeds
UK
j.d.altringham@leeds.ac.uk

EDITH ARNDT
Centre of Excellence for Biosecurity
Risk Analysis
University of Melbourne
Parkville
Australia
edith.arndt@unimelb.edu.au

MALCOLM AUSDEN
Reserves Ecology, Royal Society
for the Protection of Birds
Sandy
UK
malcolm.ausden@rspb.org.uk

BEAU AUSTIN
Research Institute for the
Environment and Livelihoods
Charles Darwin University
Casuarina
Australia
Land & Water Flagship
CSIRO
Berrimah
Australia

Beau.Austin@cdu.edu.au

IAN BATEMAN
Land, Environment, Economics and
Policy Institute (LEEP)
University of Exeter
Exeter
UK
i.bateman@exeter.ac.uk

ANNA BERTHINUSSEN
Conservation First
UK
anna@conservationfirst.co.uk

AMY BINNER
Land, Environment, Economics and
Policy Institute (LEEP)
University of Exeter
Exeter
UK
A.R.Binner@exeter.ac.uk

ALETTA BONN
Helmholtz Centre for
Environmental Research – UFZ
Leipzig
Germany
Institute of Biodiversity
Friedrich Schiller University Jena
Jena
Germany

German Centre for Integrative
Biodiversity Research
Leipzig
Germany
aletta.bonn@idiv.de

IAN BOYD
University of St Andrews
Scottish Oceans Institute
St Andrews
UK
ilb@st-andrews.ac.uk

PETER N. M. BROTHERTON
Natural England
Peterborough
UK
Peter.Brotherton@naturalengland.
org.uk

MARK BURGMAN
Centre for Environmental Policy
Imperial College
London
UK
m.burgman@imperial.ac.uk

SAMANTHA H. CHENG
Center for Biodiversity Outcomes
Arizona State University
Tempe
USA
Center for Biodiversity and
Conservation
American Museum of Natural
History
New York
USA
scheng@amnh.org

STEVEN J. COOKE
Canadian Centre for
Evidence-Based
Conservation

Carleton University
Ottawa
Canada
Steven.Cooke@carleton.ca

MARIANNE DARBI
Helmholtz Centre for
Environmental Research
– UFZ
Leipzig
Germany
marianne.darbi@ufz.de

ZOE G. DAVIES
Durrell Institute of Conservation
and Ecology (DICE)
School of Anthropology and
Conservation
University of Kent
Canterbury
UK
z.g.davies@kent.ac.uk

BRETT DAY
Land, Environment,
Economics
and Policy Institute (LEEP)
University of Exeter
Exeter
UK
Brett.Day@exeter.ac.uk

CATHY DEAN
Save the Rhino International
London
UK
cathy@savetherhino.org

LYNN V. DICKS
Department of Zoology
University of Cambridge
Cambridge
UK
lynn.dicks@zoo.cam.ac.uk

MEGAN C. EVANS
School of Earth and Environmental
Sciences
University of Queensland
Brisbane
Australia
Centre for Policy Futures
University of Queensland
Brisbane
Australia
megan.evans@uq.edu.au

MICHELA FACCIOLI
Land, Environment, Economics and
Policy Institute (LEEP)
University of Exeter
Exeter
UK
M.Faccioli@exeter.ac.uk

CARLO FEZZI
Land, Environment, Economics and
Policy Institute (LEEP)
University of Exeter
Exeter
UK
Department of Economics and
Management
University of Trento
Trento, Italy
C.Fezzi@exeter.ac.uk

KERRIE FOXWELL-NORTON
School of Humanities, Languages
and Social Sciences
Griffith University
Gold Coast
Australia
k.foxwell@griffith.edu.au

ROBERT P. FRECKLETON
Department of Animal & Plant
Sciences

University of Sheffield
Sheffield
UK
r.freckleton@sheffield.ac.uk

NEAL R. HADDAWAY
Stockholm Environment Institute
Stockholm
Sweden
Africa Centre for Evidence
University of Johannesburg
Johannesburg
South Africa
neal.haddaway@sei.org

AMY HINSLEY
Department of Zoology
University of Oxford
Oxford
UK
amy.hinsley@zoo.ox.ac.uk

HARRIET IBBETT
Department of Zoology
University of Oxford
Oxford
UK
harriet.ibbett@bangor.ac.uk

REBECCA M. JARVIS
Institute for Applied Ecology New
Zealand
School of Science, Auckland
University of Technology
Auckland
New Zealand
Sydney Institute of Marine Science
Mosman
Australia
rjarvis@aut.ac.nz

MAHLON C. KENNICUTT II
Texas A&M University
College Station

USA
mckennicutt@gmail.com

HYEJIN KIM
German Centre for Integrative
Biodiversity Research (iDiv)
Leipzig
Germany
Institute of Biology
Martin Luther University Halle
(Saale)
Germany
hyejin.kim@idiv.de

PENI LESTARI
Wildlife Conservation Society
Indonesia Program
Bogor
Indonesia
plestari@wcs.org

LIBBY LESTER
College of Arts, Law and Education
University of Tasmania
Hobart
Australia
Elizabeth.Lester@utas.edu.au

BARBARA LIVOREIL
Fondation pour la recherche sur la
biodiversité
Paris
France
barbara.livoreil@fondationbiodi
versite.fr

BILJANA MACURA
MISTRA-EviEM
Stockholm Environment Institute
Stockholm
Sweden
biljana.macura@sei.org

PERNILLA MALMER
Swedbio

Stockholm Resilience Centre
Stockholm University
Stockholm
Sweden
pernilla.malmer@su.se

ELISABETH MARQUARD
Helmholtz-Centre for
Environmental Research – UFZ
Leipzig
Germany
lisa.marquard@ufz.de

VANESSA MASTERSON
Swedbio
Stockholm Resilience Centre
Stockholm University
Stockholm
Sweden
vanessa.masterson@su.se

MADELEINE C. MCKINNON
Bright Impact
San Francisco
USA
Madeleine@brightimpact.org

E.J. MILNER-GULLAND
Department of Zoology
University of Oxford
Oxford
UK
ej.milner-gulland@zoo.ox.ac.uk

CLIVE MITCHELL
Scottish Natural Heritage
Edinburgh
UK
clive.mitchell@snh.gov.uk

HANS COSMAS NGOTEYA
Landscape and
Conservation Mentors
Organisation
Katavi

Tanzania
hanscosmas@gmail.com

NANCY OCKENDON
Cambridge Conservaton Initiative
The David Attenborough Building
Cambridge
UK
n.ockendon@jbs.cam.ac.uk

TOBY PARK
The Behavioural Insights Team
London
UK
toby.park@bi.team

NATHALIE PETTORELLI
Institute of Zoology
Zoological Society of London
London
UK
Nathalie.Pettorelli@ioz.ac.uk

ANDREW S. PULLIN
Collaboration for Environmental
Evidence, UK Centre
Bangor University
Bangor
UK
a.s.pullin@bangor.ac.uk

STEPHEN MARK REDPATH
Institute of Biological &
Environmental Sciences
University of Aberdeen
Aberdeen
UK
s.redpath@abdn.ac.uk

ANDREW ROBINSON
Centre of Excellence for Biosecurity
Risk Analysis
University of Melbourne
Parkville

Australia
apro@unimelb.edu.au

DAVID C. ROSE
School of Agriculture, Policy and
Development
University of Reading
Reading
UK
d.c.rose@reading.ac.uk

ALEX RUSBY
Land, Environment, Economics
and Policy Institute (LEEP)
University of Exeter
Exeter
UK
Harrow School
Harrow
UK
rusbyalex@gmail.com

GABBY SALAZAR
Department of Life
Sciences
Imperial College London
Ascot
UK
gabby.r.salazar@gmail.com

KAREN SCHNEIDER
Centre of Excellence for
Biosecurity Risk Analysis
University of Melbourne
Parkville
Australia
k.schneider@unimelb.edu.au

GREG SMITH
CSIRO, Land and Water
Hobart
Australia
Greg.S.Smith@csiro.au

REBECCA K. SMITH
Department of Zoology
University of Cambridge
Cambridge
UK
r.k.smith@zoo.cam.ac.uk

ROBERT J. SMITH
Durrell Institute of Conservation and
Ecology
School of Anthropology and
Conservation
University of Kent
Canterbury
UK
R.J.Smith@kent.ac.uk

JOSEPH STARINCHAK
US Fish and Wildlife Service
Falls Church
USA
joe_starinchak@fws.gov

WILLIAM J. SUTHERLAND
Department of Zoology
University of Cambridge
Cambridge
UK
w.j.sutherland@zoo.cam.ac.uk

JESSICA J. TAYLOR
Canadian Centre for Evidence-Based
Conservation
Carleton University
Ottawa
Canada
JessicaTaylor3@cunet.carleton.ca

MARIA TENGÖ
Swedbio
Stockholm Resilience Centre
Stockholm University
Stockholm
Sweden
maria.tengo@su.se

LAURA A. THOMAS-WALTERS
Durrell Institute of Conservation and
Ecology
School of Anthropology and
Conservation
University of Kent
Canterbury
UK
lat42@kent.ac.uk

DIOGO VERÍSSIMO
Oxford Martin School
University of Oxford
Oxford
UK
Department of Zoology
University of Oxford
Oxford
UK
Institute for Conservation Research
San Diego Zoo Global
Escondido
USA
verissimodiogo@gmail.com

JULIET A. VICKERY
RSPB Centre for Conservation
Science
Royal Society for the Protection of
Birds
Sandy
UK
Conservation Science Group
Department of Zoology
University of Cambridge
Cambridge
UK
Juliet.Vickery@rspb.org.uk

JESSICA C. WALSH
School of Biological Sciences
Monash University
Clayton

Melbourne
Australia
jessica.walsh@monash.edu

PAULO WILFRED
The Open University of Tanzania
Dar Es Salaam
Tanzania
paulo.wilfred@yahoo.co.uk

BONNIE C. WINTLE
School of BioSciences
University of Melbourne
Parkville
Australia
Centre for the Study of Existential
Risk
University of Cambridge
Cambridge
UK
bonnie.wintle@unimelb.edu.au

HEIDI WITTMER
Helmholtz Centre for
Environmental Research – UFZ

Leipzig
Germany
heidi.wittmer@ufz.de

CLAIRE F.R. WORDLEY
Department of
Zoology
University of Cambridge
Cambridge
UK
cfw41@cam.ac.uk

JULIETTE YOUNG
Centre for Ecology and
Hydrology
Edinburgh
UK
AgroSup Dijon
INRA
Université Bourgogne Franche-
Comté
Dijon
France
jyo@ceh.ac.uk

# Introduction and scene setting

# CHAPTER ONE

# Making a difference in conservation: linking science and policy

WILLIAM J. SUTHERLAND

*University of Cambridge*

PETER N. M. BROTHERTON

*Natural England*

NANCY OCKENDON

*Cambridge Conservation Initiative*

NATHALIE PETTORELLI

*Zoological Society of London*

JULIET A. VICKERY

*RSPB Centre for Conservation Science*

and

ZOE G. DAVIES

*Durrell Institute of Conservation and Ecology (DICE)*

## 1.1 Introduction

Jamie Gundry's dramatic image of a white-tailed eagle (*Haliaeetus albicilla*) on the cover of this book reflects the twisting changes in fortune experienced by this species, with a revival that can be attributed to a successful interplay of science, policy and practice. White-tailed eagles were historically much more widely distributed than they are today (Yalden, 2007), once breeding across much of Europe, but by the early twentieth century the species was extinct across much of western and southern Europe. The main cause of its decline was persecution by farmers and shepherds, who considered the eagles a threat to their livestock, but, along with other raptors, white-tailed eagles were also seriously affected by DDT in the 1960s and 1970s, which had disastrous effects on the breeding success of remaining populations. However, over the past four decades the species has seen a remarkable reversal in its fortunes. Changes in public attitude and policy have resulted in several reintroductions of the species, returning breeding populations to Scotland and Ireland (Evans et al., 2009; O'Rourke, 2014), and a recent licence has been approved for a release on the Isle of Wight in southern England. White-tailed eagles also recently started nesting in the Oostvaardersplassen, part of the Netherlands that just over 50 years ago was reclaimed polder destined for industrial development, but has since become the most influential example of the concept of rewilding. The recovery of this species has required a significant shift in perception among a diverse range of stakeholders; this has resulted in positive changes in both policy and practice, with bans on the use of organophosphate pesticides and the re-setting of attitudes from those that allowed persecution, to create a context which allowed populations to be reintroduced. The spectacle of this wonderful species in locations where it was once absent is a tribute to the successful linking of science and policy, but elsewhere these links are often problematic: this book sets out to examine the range of challenges and successes.

Even before the first attempted reintroduction of a white-tailed eagle population in 1959, conservation researchers have had a long history of involvement in policy issues. One early example was Arthur Tansley, an English botanist and pioneer in the discipline of ecology. In 1913, Tansley and his colleagues established the British Ecological Society (BES), the first ever learned society in this science. By the 1940s, he was a committed conservationist, chairing the BES committee that formulated UK policies on nature reserves, and was instrumental in the formation of the Nature Conservancy, the first government agency to support ecological research. It is therefore fitting that this book has emerged from a highly successful conference entitled 'Making a Difference in Conservation: Improving the Links between Ecological Research, Policy and Practice' that was supported, in part, by the BES.

Over recent decades, conservation has evolved into a global dynamic trans-disciplinary field, which embraces the two-way relationships that occur between people and nature at many different levels (Mace, 2014). At the same time, the ways in which information is communicated have altered dramatically as a result of a progressively more complex and interconnected networks of technologies and practices. The policy landscape, both within and between nations, has also changed. The shifts in these interlinked disciplines have had a significant impact on how evidence derived from research is used in conservation decision-making. This book brings together a series of con-servation experts to share their experiences of the different aspects of, and approaches to, working constructively at the research–policy/practice interface.

The process linking science and practice is rarely linear and often complex (Owens, 2015). Policy and practice responses may be driven by a scientific discovery (such as the impact of neonicotinoids on pollinating insects), poli-tical change (such as the overhaul of land-use policies that may result from the UK's decision to leave the European Union) or even communication (such as the rapid responses of businesses, individuals and governments following the dramatic television footage of a blue whale and albatrosses consuming plastic in the BBC David Attenborough TV series *Blue Planet II*). However, dig down and each of these apparent initiation points are usually built upon other elements.

This book begins with a scene-setting chapter written by the Chief Scientific Adviser of the UK Department for Environment Food and Rural Affairs, who provides insights into how governments make decisions and the challenges of developing evidence-based policies. The remainder of the book is divided into three sections. The first covers the identification of priorities for research and approaches for collating relevant information, to ensure it is readily available for use by decision-makers. The second section examines the practicalities of engaging decision-makers and stakeholders with evidence. The final section considers how messages related to conservation can be communicated, such as by the use of social marketing or behaviour nudging, to make a tangible difference for biodiversity.

## 1.2 Identifying priorities and collating the evidence

The research–policy/practice interface may not function adequately if either there is insufficient relevant information available at the time when decisions need to be made (evidence generation failure) or information exists but is not successfully incorporated into the decision-making process (evidence use failure). If researchers are to help inform decision-making, then the emerging policy/practice issues need to be sufficiently well researched and the resulting evidence must be collated in an easily accessible form. This process may vary greatly depending on the conservation issues under scrutiny (Chapters 4–7) and can be made more effective via the considered inclusion of indigenous and local knowledge (Chapter 6), as well as meaningful engagement with a diverse array of stakeholders (Chapter 5).

One example of evidence generation failure was the sudden decision to move rapidly towards increased biofuel use announced by President George Bush in his 2006 State of the Union address, with the European Union adopting similar policies soon after. These decisions had substantial unforeseen environmental impacts. As a consequence of the policies, demand for agricultural land for biofuel crop production increased dramatically. However, uncertainties quickly emerged about the greenhouse gas benefits associated with many biofuel crops (Koh & Ghazoul, 2008). The wider problem revealed by this policy announcement was that it had not been foreseen by the environmental and conservation communities, who were therefore poorly prepared to respond, in particular lacking a relevant body of necessary evidence. A welcome development over the last decade has therefore been the growing interest in horizon scanning (Chapter 3) to identify forthcoming conservation problems.

Evidence use failure can result if the relevant evidence exists but is unavailable to decision-makers. For instance, it may be hidden behind paywalls or presented in academic papers that busy practitioners and policy-makers do not have time to find and assimilate. Alternatively, it can result from 'evidence complacency' – 'a way of working in which, despite availability, evidence is not sought or used to make decisions, and the impact of actions is not tested', by practitioners and/or policy-makers (Sutherland & Wordley, 2017). Evidence use failure occurred during the review of the Common Agricultural Policy of the European Union. The process to decide which agri-environment interventions would be supported by billions of euros in agricultural subsidies resulted in the selection of interventions that had little evidence demonstrating their effectiveness; the little evidence that did exist suggested that the chosen measures would not be effective (Dicks et al., 2014). This was despite the existence of other interventions that were both more effective and had a stronger evidence base. Tools and approaches to avoid such evidence use failure by enhancing the incorporation of evidence into policy-making at different levels are described in Chapter 8.

## 1.3 Decision-making

Incorporating evidence with other aspects of decision-making may be fraught with difficulties. This is illustrated by attempts to tackle climate change by reducing emissions of greenhouse gases. Despite overwhelming scientific consensus on the anthropogenic origins of recent changes in climactic conditions reported all over the world, many countries are still refusing to curb their reliance on fossil fuels, and little progress has been made in reducing global emissions (Tol, 2019). In contrast, the use of global research evidence successfully underpinned calls to ratify the Montreal protocol, which limited the use of CFCs that had been demonstrated to deplete the ozone layer (Mäder et al., 2010).

Pathways to influence ultimately rely on a good understanding of who to approach with evidence. The first step in the successful communication of evidence to support decision-making is a clear identification of the relevant decision-makers (Chapter 10). Decision-making among local practitioners and policy-makers involves completely different processes compared with decision-making at the global level, with the two often involving people with markedly different backgrounds and priorities (Chapter 9).

Evidence derived from research is only one of the types of evidence considered by decision-makers (Chapters 11 and 12). It is important to acknowledge that science is not, and should not be, the only factor driving decisions for society – something that can be difficult for scientists to accept (Chapter 14). In addition, evidence is never 'perfect', and ignoring the uncertainty associated with findings can lead to poor decisions (Chapter 11). However, communicating uncertainty to policy-makers and practitioners is challenging and can risk research findings being dismissed altogether. Nonetheless, innovative solutions to this problem do exist. For example, The Centre of Excellence for Biosecurity Risk Analysis (Chapter 13) has helped deliver evidence-based policy in Australia and New Zealand by establishing a formal institution through which researchers and government policy-makers take shared responsibility in the development of state-of-the-art methods (tools, guidelines, procedures) to assess and minimise environmental risks.

Differences in worldviews can result in polarised opinions and different interpretations of evidence, leading to conflict (Chapter 14). However, by engaging with the process of negotiating international conventions and agreements, scientists can contribute to making a difference (Chapter 15).

## 1.4 Communicating the message

Ultimately, most conservation issues are a consequence of human activities, meaning that a positive future for biodiversity is reliant on changing people's behaviour. Policy-makers, practitioners and researchers cannot depend on education, regulation and incentives alone, as raising awareness and delivering penalties are known to be insufficient to instigate and sustain extensive shifts in behaviour. Conservationists are therefore starting to draw on techniques and methods developed in other sectors of society, such as the business world, to alter people's behaviour through beneficial exchange mechanisms (Chapters 19 and 20). Moreover, an understanding of digital and mobile communication is becoming an increasingly powerful way to engage the public and decision-makers with conservation research. Many attempts at promoting messages through the media are ineffective (Chapter 16), but the impact of conservation communication can be enhanced by collaboration with communication scholars who are experts in media and journalism (Chapter 17). Campaigning, also described as advocacy, is a common mechanism by which non-governmental organisations try to influence decision-makers and the public, often involving media engagement. While it can be a successful approach, there are a plethora of potential pitfalls that warrant careful consideration (Chapter 18).

## 1.5 Acknowledgements

We thank the British Ecological Society and the Cambridge Conservation Initiative for their roles in organising and hosting the 'Making a Difference in Conservation: Improving the Links between Ecological Research, Policy and Practice' conference. This book is open access as a result of financial support from the British Ecological Society and Natural England. WJS is funded by Arcadia. ZGD is funded by the European Research Council (ERC) under the European Union's Horizon 2020 research and innovation programme (Consolidator Grant No. 726104).

## References

Dicks, L. V., Hodge, I., Randall, N., et al. 2014. A transparent process for 'evidence-informed' policy making. *Conservation Letters*, 7, 119–125.

Evans, R. J., Wilson, J. D., Amar, A., et al. 2009. Growth and demography of a re-introduced population of White-tailed Eagles *Haliaeetus albicilla*. *Ibis*, 151, 244–254.

Koh, L. P. & Ghazoul, J. 2008. Biofuels, biodiversity, and people: understanding the conflicts and finding opportunities. *Biological Conservation*, 14, 2450–2460.

Mace, G. M. 2014. Whose conservation? *Science*, 345(6204), 1558–1560.

Mäder, J. A., Staehelin, J., Peter, T., et al. 2010. Evidence for the effectiveness of the Montreal Protocol to protect the ozone layer. *Atmospheric Chemistry and Physics*, 10, 12,161–12,171.

O'Rourke, E. 2014. The reintroduction of the white-tailed sea eagle to Ireland: people and wildlife. *Land Use Policy*, 38, 129–137.

Owens, S. 2015. *Knowledge, Policy, and Expertise: The UK Royal Commission on Environmental Pollution 1970–2011*. Oxford: Oxford University Press.

Sutherland, W. J. & Wordley, C. F. 2017. Evidence complacency hampers conservation. *Nature Ecology & Evolution*, 1, 1215–1216.

Tol, R. S. J. 2019. *Climate Economics: Economic Analysis of Climate, Climate Change and Climate Policy*. London: Edward Elgar.

Yalden, D. W. 2007. The older history of the white-tailed eagle in Britain. *British Birds*, 100, 471–480.

# CHAPTER TWO

# Working in government: conservation research, policy and practice

IAN BOYD

*University of St Andrews*

## 2.1 Introduction

In this chapter I will provide a view of conservation research, policy and practice from within government. This has been formed as a result of my experience as Chief Scientific Adviser at the UK Department for Environment, Food and Rural Affairs. I consider how government works in relation to conservation within two broad themes: the first deals with the general political and policy context, and the second considers how the results of conservation research can be integrated into policy and practice. Some of my account, which is directed towards government officials as well as researchers, affirms the robustness of current systems and structures, but other parts challenge aspects of current thinking.

## 2.2 Governmental processes and decision-making

Government is a highly diverse, multi-layered structure. In this chapter, I refer mainly to central government, defined by the departments of state, which have ultimate responsibility for setting strategy and delivering policy outcomes. However, governmental conservation research is often most closely associated with other arms of government, including semi-independent agencies of government and those that, in Britain, are called non-departmental public bodies with their own governance structures. These bodies exist specifically to separate some aspects of governance from central government because specialised capabilities are needed to manage particular assets or public services (Anon., 2018). Even if the objectives of these organisations can be set by the parent department in central government, their operational mode and relationship with central government can be quite different and 'arm's length'.

Decision-making by government, when viewed from the perspective of problem or decision theory, is a form of multi-dimensional optimisation in which a range of variables is considered in often opaque ways. This is unattractive to people who like to deal with problems that have unequivocal solutions: this includes many researchers. Governmental decisions about the environment, however, are taken in the murky, turbulent space where the dynamism and chaos of the natural world collide with human social systems in their various cultural and structural forms. Operating in this world can be very challenging and requires special skills and resilience. It is a world where problems are wicked, in that the very act of finding a solution can make the problem worse, and where ambiguity is the norm but can, perversely, serve a useful purpose. This is because ambiguity can be used as a mechanism to sustain dialogue between groups with strong common interests – which includes most parties involved in conservation debates – but where the discourse is dominated by a narrow difference of opinion.

When viewed through a narrow scientific lens, decisions and actions in government can sometimes seem obtuse or not based on evidence. If the lens is dilated, as happens when one gets closer to the action, then other perspectives can reveal the other factors in play, and this often brings interesting insights. Governments rarely act with intentional irrationality. Apparent irrationality happens mainly because an observer is unaware of all the dimensions of the problem being addressed. Sometimes apparent irrationality only emerges post hoc, when the benefit of experience suggests that an alternative action might have been the better course to take. Government is plagued by such post-hoc analyses, unaccompanied by counterfactuals. It is easy for critics of government to assert that alternatives would have produced better outcomes based on either the benefit of hindsight or when there is no possibility of testing whether those assertions are correct. This applies as much to conservation as to any other area of policy.

Government is not a machine. It is run by people, and even if civil servants are trained to minimise value-based biases, human frailty means that the operation of government will always be imperfect. Working successfully with, or in, government requires an understanding of the social, cultural, economic, resource, structural and political stresses operating at any time. Understanding how these are integrated can be daunting; there are no fixed formulas for how to recognise and then respond to such stresses. Shifts in these stresses can result in apparent inconsistency from government, illustrated best by what happens when new political leadership appears. In the worst cases, government lurches between extremes because of the severe complexity of the problems being tackled. Sometimes these lurches are politically driven; from the perspective of research, politics can be viewed as simply one driver of stochasticity (like the climate), rather than anything that researchers can control. Thus, a degree of detachment between the researcher and the politics is important.

## 2.3 The role and typology of conservation research

Given the complexity of government decision-making, how can conservation research add value to policy and practice? Research is the supplier of knowledge, the arbiter of uncertainty and the umpire of method in governmental formulation of policy and practice. More specifically, the role of conservation research is in revealing ambiguities, helping to define objectives and then designing adaptive management practice to shift policies in the direction of achieving those objectives. 'Policy' in this context is most closely aligned with the concept of strategic solutions, while 'practice' refers to tactical or operational interventions; these differ mainly in terms of the temporal and spatial scales of delivery. In addition, practice emerges from policy. For example, the UK National Ecosystem Assessment (2011) was underpinned by a major piece of strategic research delivered by the Department of Environment Food and Rural Affairs (Defra) and partners. It supported strategic thinking about the conservation paradigm, by highlighting the utility of different policy options using cost–benefit analysis and by making trade-offs explicit. Such research can provide the broad context within which many areas of operational research, such as species conservation and habitat restoration, occur. Some of this operational research, which has followed on from the UK National Ecosystem Assessment, will have general messages, but much of it is about providing specific solutions to particular problems in particular circumstances. Generalising from these studies is a post-hoc synthesis activity, the value of which will depend greatly on circumstances.

Therefore, strategic research is arguably a more important focus for central government than operational research. There is a stronger emphasis on operational research in some of the more independent organisations at arm's length from governments that often have responsibility for delivering policy outcomes. However, at both the strategic and operational scales, research provides a systematic method for building knowledge from experience.

Although the strategic/operational typology has utility, there is perhaps a perception of greater focus on operational circumstances in conservation research, which may stem from the traditional emphasis on conservation of species rather than ecological function (Mace, 2014). This has historically led to large numbers of highly specialised studies of particular species in particular circumstances, and it is not clear whether this is the most effective approach. Conservation researchers are increasingly considering how they can develop more functionally based hypotheses, with greater emphasis on strategic solutions. While a focus on species and habitat conservation is entirely justified in many cases, conservation research could do more to lead, and question, the fundamental basis for the current policy balance between protecting species and habitats versus protecting and restoring functional ecosystems. An important outcome of strategic research should be to challenge normative thinking, allowing novel and improved policies to evolve.

Finally, the boundaries of conservation research spread far into strategic decision-making across government. For example, the effects of economic growth are at the root of many conservation problems but, as the Nobel Prize-winning economist Simon Kuznets pointed out, it is only after sufficient economic growth has occurred that a country's impact on the environment tends to decline (Kuznets, 1955, but see critique in Stern, 2004). This presents the currently unresolved conundrum: conservation relies on the products of the very processes and societal changes that create the problems that conservation is attempting to solve. It is this kind of fundamental question that more conservation research needs to address.

## 2.4 Government as a direct and indirect sponsor of research

It is important to recognise that government can be both a direct and an indirect sponsor of research. In most other contexts these two functions would be closely entwined but, at least in Britain, much government funding for research is concerned with the strategic national interest, by supporting innovation and increased productivity to achieve economic and social benefits. Government is a customer of the outputs of this research, but only in the sense that it is concerned with ensuring its investments generate wealth, generally measured in terms of growth in GDP and tax receipts. Thus, the Government benefits indirectly.

There is much less emphasis on government as the direct recipient and user of research outputs, as in the case of its sponsorship of conservation research. Therefore, where the strategic national interest is concerned, conservation research is inevitably a lower priority compared with subjects such as materials, biomedical science, computing and advanced manufacturing.

Furthermore, when central government does provide leadership by setting the agenda for strategic research priorities, it often has trouble delivering on this role. At times of budget constraint, government expenditure on strategic research for its own benefit is often reduced faster than spending on fixed costs or critical services. This is understandable, but rebalancing is needed eventually, because investment in strategic research is comparable with capital investment in skills and infrastructure (OECD, 2015). Indeed, on this basis the UK now classifies strategic research and development expenditure as part of its capital investment. This is logical, because it reframes the rationale for research investment in terms of its incremental economic and social benefits, rather than as a service to support operational needs.

Elevating conservation research within government priorities will require a much stronger business case than has been constructed to date. This needs to be based on clear examples of how its outputs lead to economic and social advantage. For example, research in environmental economics, which is broadly linked to conservation research, has helped to support the idea that nature conservation has an important indirect role in supporting economic growth and health (see Chapter 12). Emphasis also needs to be placed on the interdisciplinary nature of conservation research, requiring strengths in fields such as behavioural ecology, community ecology, taxonomy and environmental biogeochemistry. Conservation research should also be closely linked to social science because most of the problems it tackles are generated by people and the solutions also depend on people.

Much of what is classed as conservation research, such as observing and monitoring or providing a support function for environmental management, might not qualify as research at all under a strict application of the Frascati definitions used to account for research spending by governments (OECD, 2015). These definitions emphasise the process of discovery, including the investigation of systems, process and functions. It can, therefore, be difficult for government to fund 'research' activities, which cannot appear in government accounts as research when passed through the filter of international definitions.

However, in Britain, government can also be a direct customer of research, a practice established following the publication of the Rothschild Report (HMSO, 1971), which recommended that government departments should hold research budgets to directly sponsor research to deliver to their needs. Due to budget cuts, this vision has subsequently been eroded, so that government departments are now minor sponsors of research, despite a continued need for research outputs. Arguably, the idea that central government departments could be effective sponsors of research was optimistic and risky because the processes for commissioning research are highly specialised and direct sponsorship of science by a politically led organisation carries the risk of biasing the research to satisfy short-term goals and comply with politically expedient outcomes.

## 2.5 Improving the policy impact of research

The contributions of scientists, of course, involve generating new information and synthesising knowledge, but promoting the use of the emerging evidence relies on penetrating government structures and processes and building trusted relationships with decision-makers. The ambition should be to make research a highly integrated part of the policy development and delivery process (Kenny et al., 2017; see also my discussion of coproduction later in this chapter).

Seeing policies as experiments in their own right creates huge opportunity for researchers. Policy implementation can involve the components familiar to researchers: the use of controls, replicates and accurate measurement accompanied by evaluation. It can happen at a range of spatial and temporal scales from the implementation of local measures, for example to reduce eutrophication in a water body, right up to national-scale measures to improve biodiversity. If the policy outcomes differ from the prior expectation, then policies can be adjusted and the experiment repeated, in an analogy of adaptive management, even if this takes decades to play out. For example, it could be argued that the UK has been involved in a massive, long-term experiment about how to optimise the relationship between farming and environmental stewardship, which began about 60 years ago and will continue to be refined for many decades to come. In the UK, the recent drive to make publicly funded research more policy-relevant may support a shift in attitudes among both researchers and policy-makers to make more of these experimental opportunities.

Viewed from this perspective, the policy cycle does not differ greatly from the scientific process, as both, when working at their best, test options iteratively and systemically to converge on solutions. Ideally, the outputs of conservation research combined with evaluation can drive the process of policy development and implementation. Research needs to become part of the core philosophy of conservation policy, rather than a bystander to be drawn in when others think it necessary. In my view, both policy officials and researchers can do more to achieve this shift.

Two activities which could improve the policy impact of research are the technical process of synthesis and the building of relationships. In most areas of science, it is very unlikely that individuals, or even groups of individuals, with expertise in a particular field can rely on the ad-hoc accumulation of knowledge to provide advice to build robust policy. Science is mostly just too complex and the evidence base too diverse for this kind of approach to be reliable. The rise of formal synthesis has been highlighted recently as a new and important discipline within science (Donnelly et al., 2018) and this applies equally to conservation science (Sutherland & Wordley, 2018; Chapter 7, this volume). The need for synthesis to be inclusive, rigorous, transparent and accessible emphasises that it has an important social function; through building consensus it helps to build acceptance of the experience reported within the scientific literature.

Synthesis is, therefore, also a route to building trusted relationships. In general, those responsible for creating and implementing policy will prioritise the use of evidence when it is trusted and delivered through trusted intermediaries – often those people from within the scientific community who are willing to put in the effort to synthesise scientific information or are specifically employed to do this. Synthesis that integrates evidence across many different lines of research is likely to be more trusted than narrowly based opinion. Communicating ideas that originated from within the domain of science to those who operate within the domain of government needs to be worked on continuously by both parties. Patience and tolerance are needed.

Policy professionals often work to tight deadlines; however, these deadlines may appear especially difficult because of a deficit in long-term engagement and understanding between the policy professionals and conservation scientists. For example, the UK has recently published a 25-Year Plan for the environment and also plans for an independent body to hold government and others to account for the delivery of environmental outcomes, including objectives for nature conservation. This requires the consistent, cost-effective measurement of meaningful components of the environment that can work at all spatial and temporal scales and that are responsive to policy change. Early in the process of deciding these metrics, it became clear that insufficient long-term work had been applied to defining and validating these measurements for some components of the environment. While there were many reasons for this deficit in measurement capability, such a situation could arguably have been anticipated if there had been a more integrated relationship between science and policy.

## 2.6 The need for greater rigour

Conservation research is a central component of policy and practice in relevant areas of government, but the relationship between research and policy remains difficult to define. The adaptive management of policy and services calls for an intimate interaction between policy and research, recognising that the interface between ecological and social systems is complex, and that the response of both these systems is unpredictable.

As a result of this complexity, government and wider society are often guilty of applying loose definitions of what constitutes evidence. Belief-based processes, or processes that do not respect the disciplines of appropriate statistical sampling, may be used to generate evidence, which may then be used without awareness of the associated caveats. Government would be helped by the application of greater discipline in following the evidence hierarchy. This defines an ineluctable sequence, from measurement to data to information to knowledge and then finally to the generation of evidence; conservation researchers have an important role in interpreting the results they derive from scientific data so that they ultimately produce useful and relevant evidence.

While evidence is what decision-makers really seek, researchers need to take ownership and ensure that the process for generating evidence needs to be managed robustly. Data are the starting point for producing evidence, but data are not information unless one can detect structures and patterns in them, and information is not knowledge unless those patterns have been verified by statistical analysis and their implications understood. Knowledge becomes evidence when it is used to address specific questions in a given context (Donnelly et al., 2018).

The rigour of this hierarchy is under continuous challenge within government, driven by the stresses caused by the fast pace of decision-making and conflicting values. It is all too easy for researchers to acquiesce to the constraints. The considerable challenges of conservation research – lack of opportunity for replication, low statistical power and socially driven problems – mean it is especially vulnerable to loss of rigour, often because of optimism concerning the robustness of methodology at all stages of the hierarchy. For example, simply shifting the threshold of statistical significance applied in the transition from information to knowledge from 2-sigma (< 0.05) to 3-sigma (< 0.003) would render many of the conclusions from conservation research obsolete. And strong reasons exist for doing this, to help to take account of the prior probability of there being a real effect. In physics, a subject where the opportunity for controlling variables is generally much greater than in conservation research, 3-sigma is the norm. These kinds of issues are often glossed over in government, and the presentation of the significance of research results by the press and by researchers themselves often does little to promote rigour.

Research can become the servant of policy rather than its challenger. Literature reviews and evidence summaries (Donnelly et al., 2018; Sutherland & Wordley, 2018) can build pictures of what is known, but in many policy areas the outstanding knowledge gap is truly vast. Researchers are prone to dwell on the small parts of a knowledge landscape where there is information, rather than the huge areas where information is sparse. For example, there is an increasing and impressive flow of information from citizen science about the distribution of species across the country, but this remains a sparse data set; similarly, we focus on the conservation of species or habitats because they are well known and valued, such as birds, while we largely ignore others, such as keystone species in the soil microbiome. The result is that even apparently robust research can be biased and misleading in the hands of policy-makers who may not understand the difference between certainty and uncertainty (see Chapter 11).

None of this is helped by an imagined but ingrained notion that scientists can be 'independent' and therefore unbiased. The very concept of scientific independence is arguably a politically motivated doctrine promoted readily by the scientific community itself. Perhaps the most difficult task for any researcher working in a politically contentious field is to remain an honest broker and avoid becoming an advocate for one cause or another (Pielke, 2007). This is particularly important for those involved in conservation research, because of its frequently close association with applied problems and because nature conservation itself is a values-based concept. Those who work at the interface between science and policy need an acute sense of their own position in the resulting social mix, because such sensitivity can mould better outcomes.

These challenges mean there is a danger that research is conducted to reinforce, rather than to challenge, normative views and this can lead to confirmation bias. The suspicions of bias devalue the outputs of research in the eyes of policy-makers and have led researchers to attempt to present the evidence on controversial subjects in policy-neutral terms (e.g. Godfray et al., 2014).

External pressure groups often operate in very subtle ways to promote confirmation bias when in their interests. The result can be that government may take a very sceptical view of evidence generated by independent organisations, even though government itself is equally susceptible to promoting confirmation bias when it supports a favoured political point of view. However, in general, the level of external scrutiny of government activities probably reduces this effect.

Separating science from politics in conservation research is fundamentally challenging because conservation is value based. This is true at all geo-political scales. Nature conservation is potentially impacted by the current politics of globalisation and nationalism because of the global connectedness of environmental issues and because national boundaries rarely match the appropriate scales for environmental governance. Transboundary concerns make conservation a natural ally of global solutions and multi-lateral treaties and accords, such as the Convention on Biological Diversity, making conservation an increasingly political subject (Owens, 2016). Arguably, this leaches power and influence in environmental decisions from the local and national levels to bigger but much more remote institutions. Whether this has led to greater equity is a debatable point, and in some circumstances conservation can present itself as a form of cultural imperialism, promoting one set of values over another, and there may be a strong correlation between these values and wealth and power (see Chapter 14 for further discussion of this subject). These are difficult issues for scientists to address, especially when the results of their research get caught up in such highly contentious issues.

Conservation research is challenged by the need to remain objective and balanced in these circumstances, and it often fails. For example, research underpins the idea that quantifiable cost–benefit trade-offs could be a rational basis for decision-making, formalised in the concepts of ecosystem services and natural capital. These are becoming increasingly important in environmental management and conservation (Costanza et al., 1997; Chapter 12, this volume), yet can be disempowering at a local level. While proposing and supporting these solutions, conservation researchers also need to consider alternatives that might avoid further centralisation of decision control.

## 2.7  Skills and the role of specialists in government

These challenges of bias and rigour mean that the way in which government accesses scientific expertise has an important effect on how it uses knowledge in decision-making. The institutional, social and cultural source of expertise and knowledge will affect how it is interpreted and used as evidence.

Specialists can be broadly divided in to those employed by government and those external to government who mainly operate in a research market place. External expertise in the case of conservation research includes commercial companies, non-governmental organisations and academic institutions, but might also include some government employees who, in the UK, are increasingly encouraged to bid for work on a competitive basis. This covers a very broad range of research cultures, which is useful in sustaining a diversity of approaches to research-based problem-solving.

However, where there is a danger of market failure, government needs to support the existence of specialists required to deliver business-critical functions including research. For example, it is unlikely that the market could sustain all the skills in taxonomy needed to support species-based conservation or the statutory commitments of government to meet particular conservation objectives. For government, there will always be a trade-off between supporting a market solution to the supply of research and the risk of market failure in critical research capacity. To negotiate this balance, government needs expert commissioners and translators of research. These should be a cadre of generalists with skills in research specification and management, and a breadth of knowledge not normally associated with deep specialists, as well as a capacity for criticism and synthesis. These are skills that are not always taught or valued in higher education and, while this needs to change, government itself also has a role in promoting and supporting the development of these skills.

Transparency about how government uses the results of scientific research is important in building trust, and there is a role for scientific generalists embedded within government to make this happen. Promoting this trust can also be achieved by government sharing expertise with external organisations. Government needs to have a porous boundary across which the expertise needed to deliver functionality in government can flow. In effect, this means government should borrow some skills it needs from other organisations, through mechanisms such as secondments, student internships and fellowships.

## 2.8  Models of interactions between science and policy

In her book about the history of the Royal Commission on Environmental Pollution, Owens (2016) provided an analysis of the ways in which science interacts with policy in the government context. Based on her work, I describe three models, or modes, of behaviour (Table 2.1) which can operate within the

**Table 2.1** *A summary of the characteristics, strengths and weaknesses of different behavioural approaches to organising the interaction between scientific advice and government policy. The models, or modes, of behaviour are not mutually exclusive and operate effectively in different circumstances. Conflict can arise when different parties are operating to different models or where there is not a common understanding of the model which is most effective in particular circumstances. These are modified from the definitions given by Owens (2016)*

| Model name | Characteristics | Strengths | Weaknesses |
|---|---|---|---|
| Technical rational | • Provides external challenge<br>• Scientists operate largely independently<br>• Mainly unidirectional flow of advice from science to policy<br>• Scientists set the agenda | • Places science in the lead<br>• Encourages a challenge-based way of working<br>• Can highlight issues which are not visible to policy<br>• Can build in horizon scanning and strategic thinking<br>• Promotes 'independence' of scientific advice | • Can result in advice which is untargeted and poorly timed<br>• Scientists sometimes start to formulate policy themselves<br>• May be perceived by policy as scientists 'marking homework'<br>• Vulnerable to politicisation by interest groups or by default, resulting in advice being ignored because of suspicion about the motives driving those providing it<br>• Can promote the notion that no scientific advice is ever 'independent' |
| Political rational | • Policy is in the lead when deciding priorities<br>• Science is advisory and responsive | • Ensures scientific advice is targeted and relevant | • Requires policy to formulate the right questions |

**Table 2.1** (*cont.*)

| Model name | Characteristics | Strengths | Weaknesses |
|---|---|---|---|
|  | • Science is explicitly seen as one component of multi-dimensional problem-solving | • Builds confidence among policy professionals that they are being supported<br>• Encourages listening by policy to scientific advice | • Science becomes a service to policy<br>• Scientific advice can be a tool to achieve political ends<br>• Scientists end up trying to please their policy masters<br>• Scientists can disengage if they think that they are being manipulated by policy professionals for political ends |
| Coproduction | • Cooperation is central to activities<br>• Recognition that policy is neither incremental nor hierarchical and that science is about more than technical solutions<br>• Exploits the additional diversity in decision-making brought by the cognitive differences between policy professionals and scientists<br>• Ensures equal stake in the outcome | • Builds a common understanding of the problem being addressed<br>• Promotes listening on the part of scientists and policy professionals<br>• Creates constructive personal relationships between scientists and policy professionals<br>• Builds scientific advice on trust | • Scientists may be less inclined to call out problems when they arise<br>• Policy professionals may be disinclined to challenge the standard of scientific advice<br>• Scientists become a component in the policy process and could misinterpret their position as one of *coproduction* when it is really *political rational*<br>• Requires long-term building of relationships<br>• Not all scientists will be comfortable with this way of working, where trade-offs are often needed between practicality and rigour |

context of conservation research for policy and practice, although they apply equally in any area at the interface between science and policy.

In the first mode, the *technical rational* model, researchers follow their own agenda, and largely act independently of government's policy environment. In these cases, alignment with policy can be unpredictable. Researchers typically present a technical argument to government, which can then choose how to respond. It is a linear or unidirectional transfer of knowledge from those who generate knowledge to those who might use that knowledge. Typical examples include the production of evidence syntheses or technical reports, such as lists of 'ecological indicators', without close consultation with government about what would be most useful; this also includes most peer-reviewed scientific papers.

This mode is often associated with the idea of 'independent' scientific advice occasionally delivered intentionally to challenge current policy norms and to potentially displace the direction of policy. It can create a disruptive relationship between science and policy. At its most extreme, it can be seen as scientists marking the homework of policy professionals, which is just a small step from politicising science. If the motivations of those generating the research results are not transparent it can promote a 'them and us' relationship, causing distrust of researchers' motivation by policy professionals and politicians. When promoted by interest groups, such as environmental non-governmental organisations (NGOs), it can also put researchers in the invidious, and sometimes unwelcome, position of providing the rationale for challenging government on political grounds. It is particularly good at feeding press interest in reporting division rather than unity between policy and scientific advice and can result in the politicisation of research, researchers and their scientific advice.

The technical rational model can work well in certain circumstances, such as when it is the agreed way of working and the results of research are highly technical. At times it will also be good for government to be challenged by groups external to the policy process. However, in general, the technical rational mode fails to account for the complex and multi-dimensional nature of government decision-making. It can be the default position adopted by most scientists when interacting with government; it is much easier to deliver messages unidirectionally to government than to spend time understanding the complex dynamics of the problems being addressed, especially when those working within government appear to be unwilling or unable to listen. When operating in the technical rational mode, this apparent unwillingness is rarely seen as a part of the problem being addressed, which might require modification of how the scientists communicate. These kinds of problems are especially significant in conservation research when the issues being addressed can be steeped in moral and ethical dilemmas and the scientific advice is often very uncertain. In my view, the technical rational mode of operating is not well suited to solving problems in conservation policy.

In contrast, Owens' *political rational* model takes the multi-dimensionality of these kinds of policy problems in to account. This way of working sees researchers providing a service to policy. It hands the initiative about how much weight to place on the knowledge gained from research to those responsible for designing and delivering policy. However, the political rational model also runs the risk that research becomes an internalised mechanism to achieve a pre-determined political outcome. For example, a large, but almost universally unacknowledged, proportion of the rationale for government sponsoring some conservation research will have been to assuage particular pressure groups or to delay difficult decisions. The low probability of gaining clear results from many instances of conservation research means that while there may be a genuine intention to generate new knowledge, there is a low probability of this actually happening. Deflecting problems to expert advisory committees is also symptomatic of political rationality at play. Again, this can be functional and desirable in many circumstances, but there is often too little explicit acknowledgement of the context and motivations in play.

Following Owens (2016), I complement these two common, but occasionally pathological, ways of building relationships between policy and science with a *coproduction* model. I make a distinction between passive and active coproduction (Wyborn, 2015; Beier et al., 2017). The coproduction mode of working recognises policy as a messy and nonlinear process, which is neither incremental nor hierarchical. Instead, policy development is seen as a cognitive process where everybody is learning. Researchers and policy-makers create constructive relationships that help to share information within an environment in which common objectives have been agreed or have emerged. The iterative nature of problem-solving in this mode allows both researchers and policy professionals to converge towards an optimal solution, acknowledging imperfections. The open nature of the dialogue within this kind of relationship promotes common understanding and joint solutions.

Passive coproduction usually happens when the activities of researchers, often outside government, naturally align with national-level policy objectives. This may occur as a result of government's own approach to open policy-making applied over long time scales, leading to the creation of common goals between researchers and government. Much conservation research, such as the BTO breeding bird surveys and the National Biodiversity Network system of observation, has evolved in this way. Active coproduction involves the merging of different perspectives in designed deliberative situations. For example, researchers themselves may actively engage policy specialists or expert advisory groups by adopting a mode of operation focused on positive action and problem-solving rather than challenge and criticism.

Coproduction is a more sophisticated, socially derived solution than the sequential *rational* methods. However, a downside of coproduction is that it is sometimes difficult to maintain the levels of cooperation needed, because of the high transactional overheads; it is therefore easy to slip into either of the *rational* modes. This is especially likely when researchers are working with small communities where the transactional overheads are especially challenging. In these circumstances, Sutherland et al. (2017) suggest a co-assessment approach can be adopted, which integrates local knowledge with scientific evidence.

When working in the coproduction mode there is also a danger that the discipline needed to sustain the knowledge hierarchy (see above) is allowed to slip, because the researchers have to negotiate trade-offs with their policy colleagues that will be a source of tension when there are relatively strict standards to maintain. For example, in fisheries management honest interpretation of scientifically derived information, such as providing realistic confidence intervals around results, can produce outcomes where those involved in negotiating trade-offs use scientific uncertainty to gain advantage. If those making decisions tend to always allocate catch towards the top end of the plausible range, over-exploitation becomes almost guaranteed. This can result in a loss of transparency in the scientific advice, as scientists try to correct for this cognitive deficit on the part of those making decisions, by constructing their advice in ways which builds their own values in to evidence. The coproduction mode may also select for particular researchers who are more amenable to trading off standards in order to preserve the coproduction relationship with policy colleagues. We need to be sensitive to these pitfalls.

These different modes of operating are very apparent to me as a scientist embedded within central government. I see examples of them on a daily basis and, as Chief Scientific Adviser, it is a central part of my job to recognise how interactions between researchers and policy professionals are constructed and, if necessary, to try and move them towards a different mode of working. All these modes have their place, but difficulties can arise when there is misunderstanding between parties about which modes they are operating in, or when the mode being used is inappropriate to the circumstances. In my view, the coproduction mode of working is the most desirable and usually reflects a mature and strategically based relationship between scientific research and policy. Both the other technical modes tend to be associated with short-term or less-mature relationships.

Conservation research is a challenging field because it has high scientific uncertainty and it often lacks a good theoretical foundation that helps draw general conclusions from research. Problems of sample size and replication can leave research practitioners and synthesisers struggling to adhere robustly to the principles of the evidence hierarchy. Moreover, the politics surrounding controversial subjects often demand research results irrespective of whether they are truly informative. Part of the skill in applying these kinds of results within policy and practice in government is to know how to weight them appropriately. There are no formulas about how to do this; it is a skill built through experience and it is greatly enhanced when decisions are coproduced between people with complementary capabilities. Interestingly, because conservation is such a values-driven subject, it may be less important that the results from conservation research are a true reflection of natural reality than a true reflection of social reality. Put simply, the results of some conservation research may say more about us than they do about nature.

I wish to end this chapter with a more personal comment. We expend immense effort attempting to solve the many practical problems in conservation and this effort includes research. While I am sure this effort is worthwhile (because we need to make incremental improvements wherever we can), from my own position looking at the breadth of the environmental problems facing people and the planet, I am drawn reluctantly to the conclusion that it is not research in nature conservation policy and practice that will solve the problems tackled by conservation. Rather, the solution lies in truly large-scale changes in governance which will lead to incentivising people to consume fewer resources. Like our burgeoning problems with waste or air pollution, nature conservation is a consequence of this fundamental problem and we will not make significant progress until that problem is addressed with a seriousness which has yet to be witnessed within any national government or international forum.

## References

Anon. 2018. Public bodies. www.gov.uk/guidance/public-bodies-reform (accessed 21 June 2018).

Beier, P., Hansen, L. J., Helbrecht, L. & Behar, D. 2017. A how-to guide for coproduction of actionable science. *Conservation Letters*, 10, 288–296.

Costanza, R., d'Arge, R., De Groot, R., et al. 1997. The value of the world's ecosystem services and natural capital. *Nature*, 387, 253–260.

Donnelly, C. Boyd, I., Campbell, P., et al. 2018. Four principles to make evidence synthesis more useful for policy. *Nature*, 358, 361–364.

Godfray, H. C. J., Blacquiere, T., Field, L. M., et al. 2014. A restatement of the natural science evidence base concerning neonicotinoid insecticides and insect pollinators. *Proceedings of the Royal Society B*, 281(1786), 20140558.

HMSO. 1971. *A Framework for Government Research and Development*. Cmnd. 4814. London: HMSO.

Kenny, C., Rose, D. C., Hobbs, A., Tyler, C. & Blackstock, J. 2017. *The Role of Research in the UK Parliament Volume One*. London: Houses of Parliament.

Kuznets, S. 1955. Economic growth and income inequality. *American Economic Review*, 49, 1–28.

Mace, G. M. 2014. Whose conservation? *Science*, 345, 1558–1560.

OECD. 2015. *Frascati Manual 2015: Guidelines for Collecting and Reporting Data on Research and Experimental Development, The Measurement of Scientific, Technological and Innovation Activities*. Paris: OECD Publishing.

Owens, S. 2016. *Knowledge, Policy, and Expertise: the UK Royal Commission on Environmental Pollution 1970–2011*. Oxford: Oxford University Press.

Pielke, R. A. 2007. *The Honest Broker. Making Sense of Science in Policy and Politics*. Cambridge: Cambridge University Press.

Stern, D. I. 2004. The rise and fall of the environmental Kuznets curve. *World Development*, 32, 1419–1439.

Sutherland, W. J. & Wordley, C. F. R. 2018. A fresh approach to evidence synthesis. *Nature*, **355**, 364–366.

Sutherland, W. J., Shackleford, G. & Rose, D. C. 2017. Collaborating with communities: co-production or co-assessment?*Oryx*, 51, 569–570. DOI:10.1017/S0030605317001296

UK National Ecosystem Assessment. 2011. *UK National Ecosystem Assessment: Technical Report*. Cambridge: UNEP-WCMC.

Wyborn, C. A. 2015. Connecting knowledge with action through coproductive capacities: adaptive governance and connectivity conservation. *Ecology and Society*, 20, 11.

# PART I

# Identifying priorities and collating the evidence

# CHAPTER THREE

# Scanning horizons in research, policy and practice

BONNIE C. WINTLE
*University of Melbourne*
MAHLON C. KENNICUTT II
*Texas A&M University*
and
WILLIAM J. SUTHERLAND
*University of Cambridge*

## 3.1 Introduction

Conservationists have long had to deal with a number of prominent, recurring issues, such as habitat loss and fragmentation, pollution, invasive species and wildlife harvesting, to name a few. On top of these well-known challenges, others have emerged. Over the last half century, these have included the impact of halogenated pesticides and defoliants, acid rain from coal-fired electricity generation, ecological impacts of biofuel production and atmospheric releases of ozone-depleting chemicals. In more recent times, concerns have emerged around microplastics and exploitation of the Arctic, although some changes also bring opportunities for conservation, such as using mobile phones to collect data. New and emerging issues tend to make policy and practice more difficult. They add to an already challenging agenda, and often require a response when knowledge of the problem is limited.

Emerging from the relatively new field of 'futures' studies, horizon scanning is still developing as a method. By crowd sourcing information and drawing on communities of practice to sort, verify and analyse that information, horizon scanning offers an efficient way to look for early indications of poorly recognised threats and opportunities (Sutherland & Woodroof, 2009; van Rij, 2010). It aims to minimise surprises by foreseeing these threats and opportunities, enabling policy-makers and researchers to respond quickly to developing problems. Horizon scanning is an approach primarily used to retrieve, sort and organise information from different sectors that is relevant to the question at hand, in a similar process to intelligence gathering. It can also include varying degrees of analysis, interpretation and prioritisation, but deciding which issues to act on, and how to act on them, typically takes place after the horizon scanning, and is assisted by other 'futures' tools, such as visioning, causal layered analysis, scenario planning and backcasting (e.g. Glenn & Gordon, 2009; Inayatullah, 2013; Cook et al., 2014a). Recent frameworks have also been developed to link different futures tools, such as horizon scanning and scenario planning, together (Rowe et al., 2017).

Horizon scanning outputs come in a wide range of forms. Some broadly describe a single trend that cuts across different parts of society, such as the rise of big data, or the future of a general area of interest, such as 'Environmental Sustainability and Competitiveness' (Policy Horizons Canada, 2011). These outputs are usually aligned with more general foresight programmes. Other exercises look at a set of more specific potential threats, such as invasive species that may arrive in the UK and threaten biodiversity (Roy et al., 2014), and compare them in an approach similar to risk assessment. For the last 10 years, conservation scientists have run annual horizon scans to identify emerging issues with the potential to impact global conservation (e.g. Sutherland et al., 2018). A similar approach has also been used to identify important scientific questions that, if answered, would help guide conservation practice and policy (e.g. Sutherland et al., 2009).

As with any policy advisory work, there is always a risk that useful information is gathered but not followed up, as decisions are often driven by other, usually non-scientific, factors. This risk may be higher with unsolicited (grassroots scans produced by a community of practitioners, researchers or academics) rather than solicited scans (called for by policy- and decision-makers). It can be unclear where the responsibility lies for integrating outputs into policy-making, and uptake depends on the organisational culture at the time (Delaney & Osborne, 2013). Schultz (2006) pointed to a conceptual contradiction between evidence-based policy and horizon scanning, where the latter searches for issues that may not be fully supported by a definitive body of evidence. A more optimistic perspective is that horizon scanning needs to be embedded in a broader strategic foresight framework, to increase the likelihood that findings are translated into practice (e.g. van Rij, 2010; Cook et al., 2014a). As mentioned above, horizon scanning *identifies* emerging and novel threats and opportunities as a first step, but other foresight tools serve different purposes along the pathway to adopting appropriate policy. These other foresight tools are not explicitly covered in this chapter, but we provide an example, *The Antarctic Science Scan and Roadmap Challenges Exercise*, of a hybrid horizon scanning activity where an accompanying road map was also produced to outline actionable recommendations (Box 3.2).

In this chapter, we introduce both general and specific approaches to horizon scanning, outline some ways of achieving and measuring impact and explore how horizon scanning may evolve in the future.

## 3.2 Approaches to horizon scanning

'Exploratory horizon scanning' identifies novel issues by searching for the first 'signals' of change across a wide range of sources (such as an early scientific paper describing a potentially impactful new technology). 'Issue-centred scanning' monitors issues that have already been identified by searching for additional signals that confirm or deny that the issue is truly emerging (Amanatidou et al., 2012). Signals can be organised into clusters (multiple pieces of information) that can either contribute to the evidence base around pre-identified issues, or form a long list of novel issues that are potentially emerging (Figure 3.1). The long list of issues can be further analysed and prioritised into a shortlist using methods detailed below. Some horizon scanning exercises take further steps to make the output more useful for the end user, for example, by assessing the policy relevance of the issues or the feasibility of addressing them, and by identifying those that warrant ongoing monitoring (Sutherland et al., 2012).

There is a range of different ways to carry out horizon scanning; we introduce the main stages and provide some specific examples in the boxed texts and Table 3.1. Because our definition of horizon scanning concentrates largely

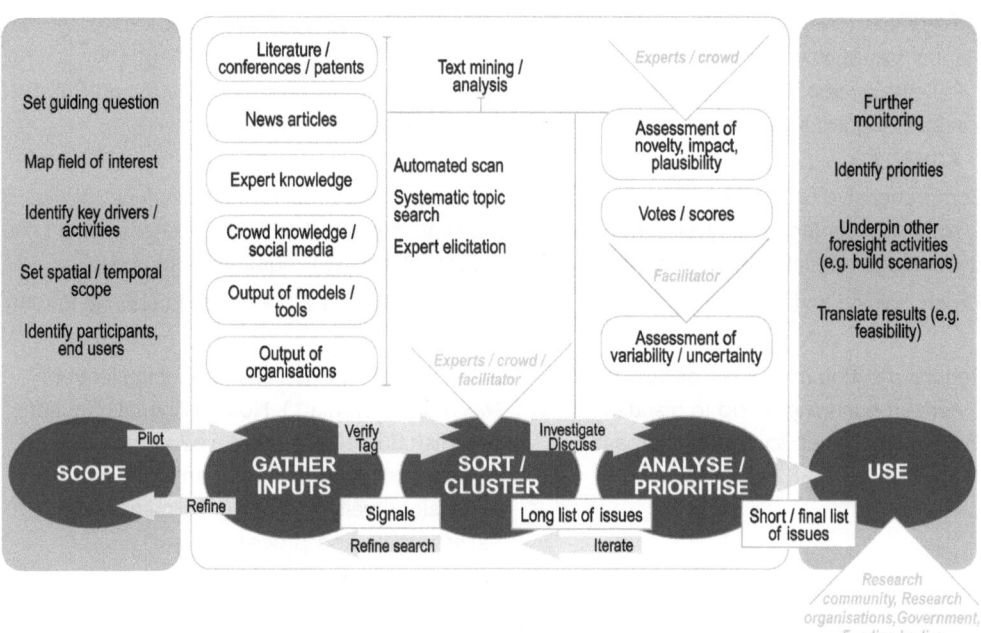

**Figure 3.1** General framework for horizon scanning, reflecting the key steps in the procedure (ovals), inputs and products (rounded rectangles), key outputs (rectangles), actors and end users (triangles), and activities and methods (floating text). Process adapted from Amanatidou et al. (2012). (A black and white version of this figure will appear in some formats. For the colour version, please refer to the plate section.)

Table 3.1 *Approaches to horizon scanning (some activities and examples overlap)*

| Approach | Examples |
|---|---|
| Manual search of an invited expert group with Delphi-style prioritisation | Global conservation (e.g. Sutherland et al., 2018), Antarctic science (e.g. Kennicutt et al., 2015), bioengineering (Wintle et al., 2017), Mediterranean conservation (Kark et al., 2016) |
| Manual search of a large crowd-sourced group (open call) with Delphi-style prioritisation (invited) | Future of the Illegal Wildlife Trade (Esmail et al., 2019) |
| Automated open-source search and manual analysis/prioritisation (usually by a community of experts) | IBIS (Grossel et al., 2017), Global Disease Detection Program (Centers for Disease Control and Prevention, www.cdc.gov/globalhealth/health protection/gdd/index.html), HealthMap (www.healthmap.org/en/), ProMed (www.promedmail.org/) |
| Advanced text analytics to identify emerging issues and research areas (e.g. sentiment analysis, machine learning) | FUSE Program (www.iarpa.gov/index.php/research-programs/fuse), Meta (https://meta.org/), X risk database (www.x-risk.net/) |
| Manual searches within an organisation (by employees, interns or volunteers), results tagged and catalogued in a database | US Forest Service (Hines et al., 2018), UK Department for Environment, Food and Rural Affairs (Garnett et al., 2016) |
| Comprehensive programme (including scanning, sentiment analysis, scenario planning; manual and automated) | Singapore's Centre for Strategic Futures (www.csf.gov.sg/), partnered with the Risk Assessment and Horizon Scanning Programme Office |
| Expert opinion (voting, survey) | Global Risks Report 2019 (World Economic Forum, 2019) |
| Regular meeting of a cross-disciplinary horizon-scanning group to discuss emerging issues and build database | Australasian Joint Agencies Scanning Network (www.ajasn.com.au/), Human Animal Infections and Risk Surveillance group (www.gov.uk/government/collections/human-animal-infections-and-risk-surveillance-group-hairs#risk-assessments-and-process) |

on information retrieval, sorting and, to some extent, analysis and prioritisation, we focus here on methods that facilitate these activities.

## 3.2.1 Scoping

Like any major project, horizon scans need to be scoped and clear guidelines developed to assist scanners. A comprehensive scoping exercise addresses the following questions.

- What is the guiding question that defines what you want to know?
- How broadly or narrowly defined is the field of interest?
- What are the key drivers of change and activities in the field? It is common to organise thinking around a STEEP (Social, Technological, Economic, Environmental and Political factors) framework.
- What is the spatial scope? For instance, are you seeking issues with global or more localised impact?
- How far into the future should scanners be projecting?
- Who should be involved?
- Who are the potential end users?

Many of these considerations will be constrained by the resources available and the needs of the end user, but tools such as stakeholder analysis (Reed et al., 2009), domain mapping (Lesley et al., 2002) and issues trees (Government Office for Science, 2017) can be useful. Scoping exercises may also involve some pilot scanning to get a feel for how well-defined the task is. For example, preliminary scanning in a US Forest Service project that aimed to identify emerging issues that could affect forests and forestry in the future revealed that 'natural resources and the environment' was too broad a topic for their exercise. Instead, it was narrowed to 'forests' (Hines et al., 2018).

Horizon scans that rely heavily on people rather than computers to do the scanning reflect the biases of those participants. A well-structured procedure for obtaining judgements from participants (e.g. Figure 3.2) will go a long way to mitigate psychological biases (Burgman, 2015b), but in order to capture a broad array of perspectives, involving a diverse group of people to identify and prior-itise candidate issues is critical. A cognitively diverse group – comprising indi-viduals who think differently – is thought to maximise collective wisdom and objectivity (Page, 2008). A good proxy for cognitive diversity is demographic diversity. Achieving demographic diversity can be challenging in practice. For example, there may be language barriers to overcome, and people with certain occupations (e.g. scholars) may be over-represented in horizon scans conducted by researchers. Inviting contributions from further afield, both geographically and from outside immediate peer circles, broadens the scope of issues consid-ered. This might be achieved by putting out an open call for issues online and advertising it through relevant websites and email lists (e.g. Esmail et al., 2019), or posting a call for ideas on social media.

### 3.2.2 Gathering inputs

Inputs to a scan can either be gathered manually (by people) or with the aid of automated software, which is then (usually) analysed by people. Manual scan-ning typically involves a group of people monitoring current research and relevant trends (e.g. technology trends, disease trends or population trends) via desktop searches, attending conferences and consulting other people in their networks. Information can be manually scanned in news articles, social media, publications, grey literature and other output of relevant organisations (such as models and projections). This is typically the first step in a 'Delphi-style' method that then goes on to analyse and prioritise candidate issues in a structured approach, usually involving one or more expert workshops (see Boxes 3.1 and 3.2 for examples and further descriptions of the procedure). Scanners could be provided with guidelines by a facilitator to direct their search, including suggestions of where to look. Manual methods have the advantage of accessing content that may not exist online (e.g. grey literature or unpublished research), or content that may be difficult to locate in the absence of known keywords to direct database and online searches. The downside of manual methods is that they are labour-intensive and may be exposed to the biases of the searcher, as they are less systematic.

### Box 3.1. A Delphi-style method for horizon scanning in conservation

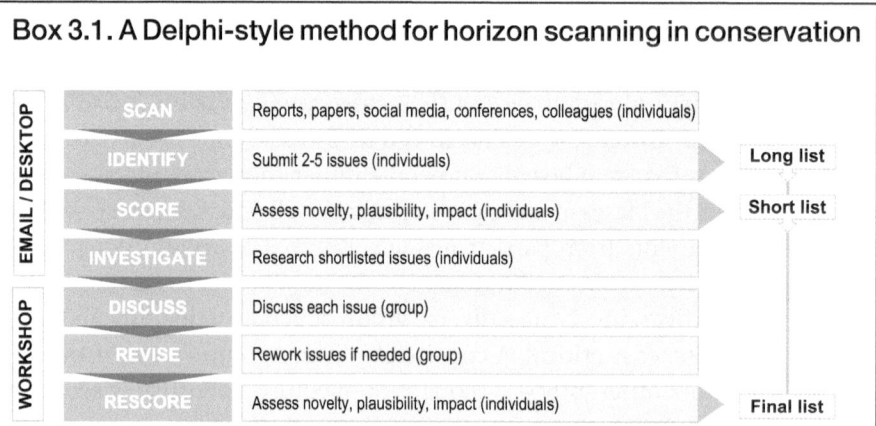

**Figure 3.2** The Delphi-style horizon-scanning approach often used in conservation (Sutherland et al., 2011). Figure reproduced from Wintle et al. (2017), published under the Creative Commons Attribution 4.0 Licence. (A black and white version of this figure will appear in some formats. For the colour version, please refer to the plate section.)

With its foundations in the Delphi Method (Linstone & Turoff, 1975; Mukherjee et al., 2015), this structured approach (Figure 3.2) was first applied in horizon scanning for conservation by Sutherland et al. (2008). There are now several variants. The key features that make this approach 'Delphi-style' are iteration (issues are submitted, scored, discussed and scored again) and anonymity of submissions and scoring. Typically, about 25 conservation experts from around the world participate in the following procedure. Over the course of several months, participants independently scan material from a variety of sources (e.g. papers, reports, websites, conferences) looking

**Box 3.1. (cont.)**

for issues (threats or opportunities) that are relatively novel, but that we should start planning for. Over email, each participant anonymously submits short summaries of two to five issues they have selected as the best 'horizon-scanning' candidates, defined as reflecting a combination of novelty, plausibility and potential future impact on global conservation. The facilitator compiles the issue summaries and circulates them back to the group, who anonymously score each issue in terms of its suitability as a 'horizon-scanning' item (using the definition above). A shortlist of the top scoring issues, containing perhaps twice the total number sought, is recirculated back to participants. Each participant is assigned approximately five issues (not their own) to investigate further, gathering further evidence to support or oppose the issues' suitability. This means each issue will be cross-examined by at least two to three people. These five issues are usually assigned to people who are *not* considered experts in that subject matter, in the hope that they will have fewer preconceptions about the issue and that the experts will add their knowledge anyway. The whole group then meets at a workshop and systematically discusses each of the shortlisted issues (e.g. to consider new perspectives, relevant research, and whether the issue is genuinely novel or just a repackaging of an old issue). The issues are kept anonymous to reduce biases and allow for an open discussion. After the discussion, participants individually score the issues a second time. The top-scoring 15 are redrafted by one of the other group members and published each year in *Trends in Ecology & Evolution* (e.g. Sutherland et al., 2018).

**Box 3.2. Antarctic science scan and Roadmap Challenges project**

The international Antarctic community came together to horizon scan the highest priority scientific questions that researchers should aspire to answer in the next two decades and beyond. The approach included online submission of questions from the science research community, followed by a subset of 75 representatives (by nomination and voting) attending a workshop. At the workshop, approximately 1000 submitted questions were winnowed down to the 80 most important through methodical debate, discussion, revision and elimination by voting. All information used, including the 1000 submitted questions, was made publicly available in a database at a horizon scan website (Kennicutt et al., 2014). The horizon scan was followed by the Antarctic Roadmap Challenges project that was designed to delineate the critical requirements for delivering the highest priority research identified. The project addressed the challenges of

**Box 3.2. (cont.)**

enabling technologies, facilitating access to the region, providing logistics and infrastructure and capitalising on international cooperation. The process uniquely brought together scientists, research funders and those that provide the logistics for field research in the Antarctic. Online surveys of the community were conducted to identify the highest priority technological needs, and to assess the feasibility (time to development) and cost of these requirements. Sixty experts were assembled at a workshop to consider a series of topic-specific summary papers submitted by a range of Antarctic communities, survey results and summaries from the horizon scan, as well as existing documents addressing future Antarctic science directions, technologies and logistics requirements (Kennicutt et al., 2015).

Computer-assisted scanning is increasingly used for automating the process of gathering a vast quantity of inputs, often crowd-sourced and usually from the internet (Palomino et al., 2012). Several such tools are now used in agriculture and health biosecurity to provide early detection of disease outbreaks (see Table 3.1 and Box 3.3 for examples) (Salathé et al., 2012; Kluberg et al., 2016; Grossel et al., 2017). Early online information, such as a tweet about a Tasmanian devil with a tumour on its face, or a YouTube video about a new device for targeting an invasive species, although unverified to begin with, may be critical for establishing the first in a series of signals that suggests a new or emerging threat (Grossel et al., 2017). Information on the internet can be retrieved in a number of ways. Keywords can be inserted into whole web search engines and/or particular websites can be targeted in more depth (e.g. Twitter can be searched using search terms, handles and hashtags). Research, news and current affairs can also be accessed via the RSS feeds of particular news and science sites, or by email and subscription to social media and blogs. Online data are often retrieved with the help of web scraping (accessing and storing particular web pages) and web crawling (accessing and storing links, and links of links from that page) (Hartley et al., 2013). With the recent increase in 'fake news', web searches require some form of quality control and vetting of sources: a process that can also be useful for *exposing* fake news. Large volumes of text scraped from the web, articles, patents, reports and other publications can be mined and filtered for potential relevance using automated software, such as machine learning algorithms.

Automated scanning is fast, systematic and comprehensive in its scope, but often relies on people – sometimes experts – to screen, review, and perhaps investigate all reports before on-posting or incorporating them (Lyon, 2010). For tools that scan across a wide range of topics, and those that use ongoing surveillance, this can be onerous and time-consuming. There are three other notable challenges to relying on online content for horizon scanning. First, material needs to already be posted on the web, and there may be a delay before an event, such as an invasive species incursion, is reported online. The second is that useful content is not always publicly available, as it can lie behind pay walls, be stored on intranets (e.g. grey literature), or secured because it is commercially, politically or personally sensitive. The third challenge is that most methods for obtaining online content rely on using the right keywords, which requires some idea of what you are looking for.

### 3.2.3 Sorting, cataloguing and clustering

Tagging and cataloguing content derived from both manual and automated scans (e.g. by relevance, credibility, source type, sectoral origin) (e.g. Garnett et al., 2016; Hines et al., 2018) occurs concurrently with input gathering by scanners. Content can be further reorganised and vetted at a later stage. During this process, new search terms to direct further scanning can be generated, or existing search terms refined. Content can be organised according to a framework that also considers the level of response required and the strength of the evidence, which can help prioritise risks and other identified issues at a later stage (Garnett et al., 2016). Clustering methods, such as network analysis (Könnölä et al., 2012; Saritas & Miles, 2012), are useful for capturing cross-cutting issues that affect a number of topics of interest.

### 3.2.4 Analysing and prioritising

At this stage, a long list of issues will have been compiled, with some more suitable to the project aims than others. This can be an opportune time to reiterate objectives. Do you seek issues that most have not heard of? Do you intend to identify broad, developing topics or very specific developments (for example, the 'increase in hydropower' versus 'fragmentation effects of hydropower in the Andean Amazon')? Are you interested in issues likely to arise soon or events that have a smaller probability of playing out in the long-term future? Does the output need to be useful to policy-makers? Many exercises, especially those with follow-up plans, aim to prioritise a select number of 'most suitable' issues, and the precise manner in which such prioritisation decisions are made makes a real difference to the quality of the output (Sutherland & Burgman, 2015). Our experience with exercises that aim to identify novel issues is that participants gravitate towards well-known although important issues. Avoiding this requires strong chairing and

## Box 3.3. Online horizon scanning: intelligence-gathering for biosecurity

The International Biosecurity Intelligence System (IBIS) is a generic web-based application that focuses on animal, plant and marine health, and provides continuing surveillance of emerging pests and diseases, including environmental ones (Grossel et al., 2017). It also detects other environmental issues, such as harmful algal blooms. It is open source, in that it gathers articles from regular feeds of trusted sources (e.g. industry news, research) and publicly available online material, like news reports, blogs, published literature and Twitter feeds. Searches can be directed by broadly relevant keywords, such as 'disease' or 'outbreak' or 'dead', in addition to specific diseases of concern (e.g. 'oyster herpes virus'). Articles can also be manually submitted by registered users to the application directly. A large expert community – the registered users, who are self-selected and approved by the administrator – then filter the articles, promoting those that they deem important and relevant to the home page, and demoting those that appear to be irrelevant or junk. Automated tools also assist with filtering (e.g. with machine learning and network cluster analysis), but as machine learning is still in its infancy, its use is limited to disease outbreaks from trusted sources. Items classified as junk by people are retained in a database to help the system's artificial intelligence (AI) algorithms learn. The broader user community (anyone who signs up online) is alerted to items that have been flagged by the registered users as important, via a daily email new digest. IBIS is also 'open-analysis', meaning that analysis of the publicly available information is performed openly by registered users. They can create or contribute to an emerging/ongoing issues dashboard that features a window for adding content, a Delphi-based forecasting section, links to related reports, share functions, comments and a map showing the location of events of interest (e.g. an outbreak). Registered users can also conduct their own searches and use integrated analytical tools to construct intelligence reports. IBIS has been effective for guiding policies and active risk management decisions for the Australian Government since 2006. The system may produce up to five Intel briefs a week on major issues affecting biosecurity and trade, allowing the government to respond to threats much faster than before. For instance, the system picked up a report of oyster herpes virus from a UK farm, which had previously purchased used aquaculture equipment from a disease-stricken oyster farm in France. Intelligence from IBIS revealed that businesses that had been closed down by the disease had been liquidating their equipment and selling to other countries. In response to this, the Australian Government changed its biosecurity policy to decontaminate all used aquaculture equipment on arrival (Burgman, 2015a).

a group that accepts the objective. To help overcome the problem, each participant can be asked whether they have heard of each issue, so that well-known topics can be excluded from the shortlist.

Within a manual Delphi-style approach (described in Boxes 3.1 and 3.2), issues are prioritised through an iterative scoring or voting process, usually facilitated online or in a workshop with a group of experts. The goal is to reduce a pool of potential horizon scanning items or ideas to a smaller subset. The number of items, or issues, covered in the final list can vary, but tends to reflect around 10–30% of the initial items put forward (e.g. Kennicutt et al., 2014; Parker et al., 2014; Kark et al., 2016; Wintle et al., 2017; Sutherland et al., 2018). As a point of comparison, the horizon scans described in Box 3.1 describe 15 issues annually, while the Antarctic hybrid horizon scan identified 80, shorter, priority scientific questions (Box 3.2). The final number may be constrained by how many the end user can realistically give their attention to (for a busy policy-maker, this may only be 15–20 half-page summaries), but is also driven by the number of (in)appropriate issues submitted. The main purpose of prioritisation is to remove issues that do not satisfy the selection criteria (novelty, plausibility, potential impact) and select those that are the most urgent or time-sensitive. Prioritisation of issues will inevitably involve trade-offs, especially where different group members have different perspectives. Because individuals' diverging opinions can be masked in aggregated scores, analysing interrater concordance (e.g. with Kendall's W) affords insights into the level of agreement between contributors. In a diverse group, we would expect a wide variety of viewpoints to be voiced, but a core of shared opinions is often discernible (e.g. Wintle et al., 2017).

Items identified in a computerised scan (e.g. articles returned from a keyword search) are also prioritised by groups of people with varying levels of content expertise. People may be employed to sort through material, such as in governmental horizon-scanning programmes like in Singapore, or they may volunteer to do so because they are interested in the output, such as a farmer or epidemiologist concerned with news of disease outbreaks. Initially, items are sorted according to their relevance to the scanning aims (often done in the initial tagging/sorting process). Irrelevant items are discarded or moved to low priority. A second form of prioritisation involves flagging issues or topics that are particularly noteworthy (Grossel et al., 2017). This can be because signals have grown stronger (more evidence is gathered to suggest an issue is becoming a threat or presenting an opportunity for action) (Cook et al., 2014b), or it might be because the potential consequences are so severe that the issue warrants immediate attention, even when evidence is limited or the probability is low ('wild cards').

### 3.2.5  Using the output

The previous step described prioritisation *within* the horizon scan to reduce a candidate set of issues. In that step, issues are ideally not judged according to importance, but rather according to less-subjective criteria, such as the likelihood of occurring or exceeding some threshold within a given timeframe. Prioritising which issues are the most important, and therefore should be acted on, is a different goal, and might be decided through follow-up, explicitly values-driven exercises involving representatives from government or relevant organisations (e.g. Sutherland et al., 2012).

Bringing together a cross-section of policy-makers in a follow-up exercise can be useful, not only to identify those issues that require further monitoring or evidence before being acted on, but also to encourage prioritisation of cross-organisational issues, knowledge sharing, and collaborative development of policy. Ideally, feasibility assessments of the options available would be included (as carried out in the extension of the recent Antarctic scan, Box 3.2).

### 3.2.6  Evaluating the process

Assessing the success of horizon scans in identifying emerging issues is challenging, and has rarely been attempted. However, a recent review by Sutherland et al. (2019) examined the first of the annual global conservation scans described in Box 3.1 (Sutherland et al., 2010) to consider how the issues identified in 2009 had developed. This was assessed using several approaches: a mini-review was carried out for each topic; the trajectory of the number of articles in the scientific literature and news media that mentioned each topic in the years before and after their identification was examined; and a Delphi-style scoring process was used to assess each topic's change in importance. This showed that five of the 15 topics, including microplastic pollution, synthetic meat and environmental applications of mobile-sensing technology, appeared to have shown increased salience and effects. The development of six topics was considered moderate, three had not emerged and the effects of one topic were considered low.

As part of the same exercise, 12 global conservation organisations were questioned in 2010 about their awareness of, and current and anticipated involvement in, each of the topics identified in 2009 (Sutherland et al., 2012). This survey was repeated in 2018 (Sutherland et al., 2019). Awareness of all topics had increased, with the largest increases associated with microplastic pollution and synthetic meat; the change in organisational involvement was highest for microplastics and mobile-sensing technology. Perhaps the most surprising result was the number that had not heard of what are now mainstream issues: 77% for microplastics, 54% for synthetic meat and 31% for the use of mobile sensing technology. A decade ago the idea of collecting environmental data using phones was cutting-edge.

Thus, efforts have begun to examine the development of previously identi-fied horizon-scan topics, but further research into the impact of horizon scans, and a consideration of issues that may have been 'missed' (not identified but subsequently emerged as important) is needed.

### 3.3  Making a difference with horizon scanning

Gauging the extent to which horizon-scanning outputs inform policy, future research directions and resource investments is not always straightforward and no-one has yet tested the effectiveness of this process. In instances where the primary decision-making organisation uses horizon scanning internally to assist with deliberations (e.g. scans to set priorities for a government agency), actions can be mapped directly against outcomes. In these cases, implement-ing the actions indicates impact. In other cases, scans can be driven by a community outside of government to set agreed future directions that can then be used to persuade external resource allocators. Even in cases where policy appears to reflect issues flagged in a horizon scan, it is difficult to trace direct influence, as inputs from multiple sources are often blended in final policy decisions without attribution. It also may take years for real-world impact to be realised. Nevertheless, there are ways in which uptake of horizon-scanning output can be encouraged.

As a starting point, horizon scanning outputs can be matched to the organisations they are most relevant to. For example, policy-makers and practitioners can come together in a follow-up workshop to assess the importance of previously identified horizon-scanning issues for their orga-nisation (Sutherland et al., 2012, 2019). Or, the end user (e.g. policy-makers and practitioners) can be engaged in the horizon scan from the outset, as in a recent scan of research priorities for protected areas (Dudley et al., 2018). Similarly, horizon-scanning networks involving representatives from a range of government agencies, such as the Australasian Joint Agencies Scanning Network, or the UK Human Animal Infections and Risk Surveillance group, provide an ongoing forum for sharing information on new and emerging issues that potentially impact different departments and organisations. Regular meetings and reports are used to deliver this informa-tion to policy-makers in a timely way (Delaney & Osborne, 2013).

In-depth follow-up analyses of horizon-scanning issues may also help policy-makers decide which to target first. A formal risk analysis of likelihood and consequences might be most appropriate for horizon-scanning outputs that compare similarly well-defined issues, for example, comparing one invasive species with another (e.g. Roy et al., 2014). It may be more challenging if some of the issues in the candidate set are more coarse-grained than others (e.g. comparing ocean warming with a specific emerging fungal disease in some snakes). Nonetheless, risk-based prioritisation at least offers a framework for comparing and forecasting issues (Brookes et al., 2014) and for formally considering the strength of evidence for each (Garnett et al., 2016).

Simply making horizon-scanning outputs known and available to policy-makers can encourage uptake. For example, issues identified in the annual global conservation scans (Box 3.1; Sutherland et al., 2018) have previously helped inform the UK's Natural Environment Research Council 'Forward Look' strategic planning, but when a decision-maker does not already have a use in mind, it may be unclear what to do with horizon-scanning information without more context and guidance. Detecting signals and potential issues is only the first step towards making a difference: further intelligence about drivers is then needed to make sense of that information. For example, incorporating available data and modelling on air traffic movements with disease surveillance data might have helped anticipate the emergence of West Nile Virus in the United States in 1999 (Garmendia et al., 2001; Brookes et al., 2014). It is the combination of horizon scanning, intelligence analysis (which provides context for the scanning output) and forecasting the chances of events unfolding that is particularly helpful in translating scanning outputs for policy-making. This can be embedded in a workflow, parts of which can be automated, such as compiling the context, narrative and structure into a digestible report on an important emerging issue (e.g. Box 3.3). When forecasting and open-analysis communities are already in place, this workflow can be delivered efficiently (Grossel et al., 2017).

Horizon scanning that occurs within organisations is evolving into a more effective tool than it was in its infancy. To facilitate the spread of best practice and reduce duplication, the UK has seen greater integration of horizon-scanning activities between different government departments, mainly in response to the Day Review (2013). The review recommended that horizon scans: (i) look beyond short-term agendas and parliamentary terms, (ii) focus on specific areas rather than broad topics in order to get more traction, (iii) are championed by those who use them in strategic decision-making, (iv) produce shorter outputs that are more likely to get the attention of senior decision-makers and (v) draw on inputs and existing analyses sourced from a 'wide range of external institutions, academia, industry specialists and foreign governments'. The extent to which all these recommendations have been implemented is unclear, but they represent a clear set of guidelines to follow.

There are a range of other useful frameworks that can be used for translating scanning outputs including roadmapping the steps towards acting on different horizon-scanning issues, for example, by assessing the feasibility and estimating how long it would take to develop technologies needed to address particular research gaps (Box 3.2; Kennicutt et al., 2015). The Antarctic science scan and roadmap has since been used to set National Antarctic Program goals, judge the effectiveness and relevance of past investments, and guide investment of other national programmes (National Academies of Sciences Engineering and Medicine, 2015; www.nsf.gov/fund ing/pgm_summ.jsp?pims_id=505320&org=OPP&from=home).

## 3.4 Future directions

We have discussed some of the pros and cons of different approaches to horizon scanning. If using a manual approach, structured methods are essential for mitigating the social and psychological biases that human horizon scanners are prone to, especially when forecasting complex and uncertain futures (Hanea et al., 2017). Although historically it has been criticised for confusing opinion with systematic prediction (Sackman, 1975), an iterative Delphi-style approach offers the advantage of drawing on the collective wisdom of a group, while affording individuals the oppor- tunity to give private, anonymous judgements and revise them in light of information and reasoning provided by others. Compared with other elici- tation approaches, such as traditional meetings, the Delphi method has also been found to improve forecasts and group judgements (Rowe & Wright, 2001). Manual approaches could be further improved by making the search for issues more systematic. Semi-automated tools and AI will increasingly enable searches uninfluenced by the biases of the manual searcher. For example, the Dutch 'Metafore' horizon-scanning approach (De Spiegeleire et al., 2016), developed in The Hague Centre for Strategic Studies, already uses some automated approaches to systematically collect, parse, visualise and analyse a large 'futures' database to complement their manual scanning.

Future horizon scanning and intelligence gathering may also see more open-analysis, 'citizen science' tools becoming adopted. While organisa- tions are increasingly scanning open-source material (including news and social media), analyses typically remain internal (Grossel et al., 2017). This means the analyses are generally not available to external users in an unfiltered form or in a timely way, which is particularly important for risks such as disease spread. Governments may opt for confidentiality for both security and political reasons. For instance, negative public percep- tions about a suspected emerging herpes virus in oysters might affect trade, which might delay the disclosure of this information by authorities, in turn delaying risk mitigation actions (Grossel et al., 2017). Intelligence tools (e.g. Box 3.3) that draw on a community of users to openly analyse news and information on potentially emerging issues offer more timely and transparent synthesis of information, which encourages more respon- sive decision-making. Examples of this can be seen in citizen science, for example where citizen volunteers have helped analyse satellite-based information in the wake of natural disasters to help emergency responders to rapidly assess the damage (Yore, 2017). In conservation science, involving a broader community of people in a participatory process like open-analysis may also increase public support for science and the environment (Dickinson & Bonney, 2012). More open-source and open-analysis scanning tools in the future will also likely be complemented with better information visualisation and GIS (e.g. including maps that indicate where a relevant incident has taken place) (Dickinson et al., 2012), not only for identifying novel issues and monitoring issues that are already emerging, but also for locating and efficiently communicating this information.

Advanced text analytics, including text mining, will also provide a more comprehensive and systematic approach to future horizon scans. Indeed, some horizon-scanning centres, such as Singapore's Risk Assessment and Horizon Scanning programme, already use sentiment analysis – a way of computationally categorising subjective opinions expressed in text (e.g. positive, negative or neutral) – to uncover themes in content retrieved by their analysts. Even more sophisticated text analytics are becoming available, for example, to explore areas of disagreement, conflict or debate in the text of scientific literature to help track developments in science and technology (Babko-Malaya et al., 2013). They can also be used to detect language expressing excitement about a new idea, and other indicators of emergence, such as the increasing use of acronyms and abbreviations indicating that the scientific community is beginning to accept a technology or idea as established (Reardon, 2014). Through automation, new computational tools have the capacity to process a massive volume of papers and patents to anticipate which developments will have the biggest impact in the future (Murdick, 2015). These advances in text analytics have recently led to the development of a particularly powerful open-source AI tool, Meta (https://meta.org/), to help biomedical scientists and funders to connect emerging research areas and potential collaborators and inform investment. Due to the complexity of emerging issues (and complex environment for machines to learn in), progress towards detecting issues effectively through AI is slow. Computers may never outperform humans at natural language understanding, but steady improvements in the technology, coupled with the speed at which text can be processed by computers – in a range of languages – will undoubtedly add value to horizon scanning in the future.

### 3.5 Acknowledgements
BCW was supported by the Centre for the Study of Existential Risk at the University of Cambridge and funded by the Templeton World Charity Foundation (TWCF). WJS is funded by Arcadia. Geoff Grossel provided useful insights into online scanning methods and applications.

# References

Amanatidou, E., Butter, M., Carabias, V., et al. 2012. On concepts and methods in horizon scanning: lessons from initiating policy dialogues on emerging issues. *Science & Public Policy*, 39, 208–221.

Babko-Malaya, O., Meyers, A., Pustejovsky, J., et al. 2013. Modeling debate within a scientific community. In *2013 International Conference on Social Intelligence and Technology* (pp. 57–63). New York, NY: IEEE.

Brookes, V. J., Hernandez-Jover, M., Black, P. F., et al. 2014. Preparedness for emerging infectious diseases: pathways from anticipation to action. *Epidemiology and Infection*, 143, 2043–2058.

Burgman, M. 2015a. Governance for effective policy-relevant scientific research: the shared governance model. *Asia & the Pacific Policy Studies*, 2, 441–451.

Burgman, M. A. 2015b. *Trusting Judgements: How to Get the Best out of Experts*. Cambridge: Cambridge University Press.

Cook, C. N., Inayatullah, S., Burgman, M. A., et al. 2014a. Strategic foresight: how planning for the unpredictable can improve environmental decision-making. *Trends in Ecology & Evolution*, 29, 531–541.

Cook, C. N., Wintle, B. C., Aldrich, S. C., et al. 2014b. Using strategic foresight to assess conservation opportunity. *Conservation Biology*, 28, 1474–1483.

Day, J. 2013. *Review of Cross-government Horizon Scanning*. London: Cabinet Office.

De Spiegeleire, S., van Duijne, F. & Chivot, E. 2016. Towards Foresight 3.0: the HCSS Metafore Approach – a multilingual approach for exploring global foresights. In Daim, T. U., Chiavetta, D., Porter, A. L. & Saritas, O., editors, *Anticipating Future Innovation Pathways Through Large Data Analysis* (pp. 99–117). Cham: Springer International Publishing.

Delaney, K. & Osborne, L. 2013. Public sector horizon scanning-stocktake of the Australasian joint agencies scanning network. *Journal of Futures Studies*, 17, 55–70.

Dickinson, J. L. & Bonney, R. 2012. *Citizen Science: Public Collaboration in Environmental Research*. Ithaca, NY: Cornell University Press.

Dickinson, J. L., Shirk, J., Bonter, D., et al. 2012. The current state of citizen science as a tool for ecological research and public engagement. *Frontiers in Ecology and the Environment*, 10, 291–297.

Dudley, N., Hockings, M., Stolton, S., et al. 2018. Priorities for protected area research. *Parks*, 24, 35–50.

Esmail, N., Wintle, B.C., Sas-Rolfes, M., et al. 2019. Emerging illegal wildlife trade issues in 2018: a global horizon scan. SocArXiv. April 25. https://doi.org/10.31235/osf.io/b5azx

Garmendia, A. E., Van Kruiningen, H. J. & French, R. A. 2001. The West Nile virus: its recent emergence in North America. *Microbes and Infection*, 3, 223–230.

Garnett, K., Lickorish, F. A., Rocks, S. A., et al. 2016. Integrating horizon scanning and strategic risk prioritisation using a weight of evidence framework to inform policy decisions. *Science of the Total Environment*, 560–561, 82–91.

Glenn, J. C. & Gordon, T. J., editors. 2009. *Futures Research Methodology – Version 3.0*. The Millennium Project.

Government Office for Science. 2017. *The Futures Toolkit, Edition 1.0*.

Grossel, G., Lyon, A. & Nunn, M. 2017. Open-source intelligence gathering and open-analysis intelligence for biosecurity. In Robinson, A. P., Burgman, M. A., Nunn, M. & Walshe, T., editors, *Invasive Species: Risk Assessment and Management* (pp. 84–92). Cambridge: Cambridge University Press.

Hanea, A. M., McBride, M., Burgman, M.A., et al. 2017. Investigate Discuss Estimate Aggregate for structured expert judgement. *International Journal of Forecasting*, 33, 267–279.

Hartley, D. M., Nelson, N. P., Arthur, R. R., et al. 2013. An overview of Internet biosurveillance. *Clinical Microbiology and Infection*, 19, 1006–1013.

Hines, A., Bengston, D. N., Dockry, M. J., et al. 2018. Setting up a horizon scanning system: a U.S. federal agency example. *World Futures Review*, 10, 136–151.

Inayatullah, S. 2013. Futures studies: theories and methods. In Gutierrez Junquera, F., editor, *There's a Future: Visions for a Better World* (pp. 36–66). Madrid: Banco Bilbao Vizcaya Argentaria Open Mind.

Kark, S., Sutherland, W. J., Shanas, U., et al. 2016. Priority questions and horizon scanning for conservation: a comparative study. *PLOS ONE*, 11, e0145978.

Kennicutt, M. C., Chown, S. J., Cassano, J. J., et al. 2014. Polar research: six priorities for Antarctic science. *Nature*, 512, 23–25.

Kennicutt, M. C., Chown, S. J., Cassano, J. J., et al. 2015. A roadmap for Antarctic and Southern Ocean science for the next two decades and beyond. *Antarctic Science*, 27, 3–18.

Kluberg, S., Mekaru, S., McIver, D., et al. 2016. Global capacity for emerging infectious disease detection: 1996–2014. *Emerging Infectious Diseases*, 22(10), e151956.

Könnölä, T., Salo, A., Cagnin, C., et al. 2012. Facing the future: scanning, synthesizing and sense-making in horizon scanning. *Science and Public Policy*, 39, 222–231.

Lesley, M., Floyd, J. & Oermann, M. 2002. Use of MindMapper software for research domain mapping. *Computers Informatics Nursing*, 20, 229–235.

Linstone, H. A. & Turoff, M. 1975. *The Delphi Method: Techniques and Applications*. Reading, MA: Addison-Wesley Pub. Co.

Lyon, A. 2010. Review of online systems for biosecurity intelligence-gathering and analysis. ACERA Project 1003.

Mukherjee, N., Huge, J., Sutherland, W. J., et al. 2015. The Delphi technique in ecology and biological conservation: applications and guidelines. *Methods in Ecology and Evolution*, **6**. 1097–1109.

Murdick, D. 2015. Foresight and Understanding from Scientific Exposition (FUSE): predicting technical emergence from scientific and patent literature. IARPA, editor. US Office of the Director of National Intelligence, www.iarpa.gov/images/files/programs/fuse/04-FUSE.pdf.

National Academies of Sciences Engineering and Medicine. 2015. *A Strategic Vision for NSF Investment in Antarctic and Southern Ocean Research*. Washington, DC: Author.

Page, S. E. 2008. *The Difference: How the Power of Diversity Creates Better Groups, Firms, Schools, and Societies*. Princeton, NJ: Princeton University Press.

Palomino, M. A., Bardsley, S., Bown, K., et al. 2012. Web-based horizon scanning: concepts and practice. Foresight, 14, 355–373.

Parker, M., Acland, A., Armstrong, H. J., et al. 2014. Identifying the science and technology dimensions of emerging public policy issues through horizon scanning. *PLoS ONE*, 9, e96480.

Policy Horizons Canada. 2011. *Leading the Pack or Lagging Behind: A Foresight Study on Environmental Sustainability and Competitiveness*. Government of Canada.

Reardon, S. 2014. Text-mining offers clues to success: US intelligence programme analyses language in patents and papers to identify next big technologies. *Nature News*, 509, 410.

Reed, M. S., Graves, A., Dandy, N., et al. 2009. Who's in and why? A typology of stakeholder analysis methods for natural resource management. *Journal of Environmental Management*, 90, 1933–1949.

Rowe, E., Wright, G. & Derbyshire, J. 2017. Enhancing horizon scanning by utilizing pre-developed scenarios: analysis of current practice and specification of a process improvement to aid the identification of important 'weak signals'. *Technological Forecasting & Social Change*, 125, 224–235.

Rowe, G. & Wright, G. 2001. Expert opinions in forecasting: role of the Delphi technique. In Armstrong, J. S., editor, *Principles of Forecasting: A Handbook for Researchers and Practitioners* (pp. 125–144). Norwell, MA: Kluwer Academic Publishers.

Roy, H. E., Peyton, J., Aldridge, D. C., et al. 2014. Horizon scanning for invasive alien species with the potential to threaten biodiversity in Great Britain. *Global Change Biology*, 20, 3859–3871.

Sackman, H. 1975. *Delphi Critique: Expert Opinion, Forecasting, and Group Process*. Lexington, MA: Lexington Books.

Salathé, M., Bengtsson, M., Bodnar, T. J., et al. 2012. Digital epidemiology. *PLoS Computational Biology*, 8, e1002616.

Saritas, O. & Miles, I. 2012. Scan-4-Light: a Searchlight function horizon scanning and trend monitoring project. *Foresight*, 14, 489–510.

Schultz, W. L. 2006. The cultural contradictions of managing change: using horizon scanning in an evidence-based policy context. *Foresight*, 8, 3–12.

Sutherland, W. J., Adams, W. M., Aronson, R. B., et al. 2009. One hundred questions of importance to the conservation of global biological diversity. *Conservation Biology*, 23, 557–567.

Sutherland, W. J., Allison, H., Aveling, R., et al. 2012. Enhancing the value of horizon scanning through collaborative review. *Oryx*, 46, 368–374.

Sutherland, W. J., Bailey, M. J., Bainbridge, I. P., et al. 2008. Future novel threats and opportunities facing UK biodiversity identified by horizon scanning. *Journal of Applied Ecology*, 45, 821–833.

Sutherland, W. J. & Burgman, M. 2015. Policy advice: use experts wisely. *Nature*, 526, 317–318.

Sutherland, W. J., Butchart, S. H. M., Connor, B., et al. 2018. A 2018 horizon scan of emerging issues for global conservation and biological diversity. *Trends in Ecology and Evolution*, 33, 47–58.

Sutherland, W. J., Clout, M., Côté, I. M., et al. 2010. A horizon scan of global conservation issues for 2010. *Trends in Ecology and Evolution*, 25, 1–7.

Sutherland, W. J., Fleishman, E., Clout, M., et al. 2019. Ten years on: a review of the first global conservation horizon scan. *Trends in Ecology and Evolution*, 34, 139–153.

Sutherland, W. J., Fleishman, E., Mascia, M. B., et al. 2011. Methods for collaboratively identifying research priorities and emerging issues in science and policy. *Methods in Ecology and Evolution*, 2, 238–247.

Sutherland, W. J. & Woodroof, H. J. 2009. The need for environmental horizon scanning. *Trends in Ecology & Evolution*, 24, 523–527.

van Rij, V. 2010. Joint horizon scanning: identifying common strategic choices and questions for knowledge. *Science and Public Policy*, 37, 7–18.

Wintle, B. C., Boehm, C. R., Rhodes, C., et al. 2017. A transatlantic perspective on 20 emerging issues in biological engineering. *Elife*, 6, e30247.

World Economic Forum. 2019. *The Global Risks Report 2019: 14th Edition*. Geneva: WEF.

Yore, R. 2017. Here's how citizen scientists assisted with the disaster response in the Caribbean. *The Conversation*, 18 October 2017.

# CHAPTER FOUR

# Generating, collating and using evidence for conservation

JOHN D. ALTRINGHAM
*University of Leeds*
ANNA BERTHINUSSEN
*Conservation First*

and

CLAIRE F. R. WORDLEY
*University of Cambridge*

## 4.1 Introduction

Does scientific evidence really matter in conservation? In this chapter we will argue that generating, collating and using scientific evidence is key to effective conservation, illustrated by a case study from our own work: how to get bats to safely cross roads. We tell the story of bat 'gantries' or bridges, and show what can go wrong in the absence of robust studies that test the effectiveness of conservation interventions. We will also discuss the importance of collating or synthesising multiple strands of evidence to identify the factors that make a conservation measure effective or ineffective, using a case study on under-passes under roads. Finally, we explore a key challenge – getting scientific evidence accepted and used routinely in conservation policy and practice.

Evidence takes a multitude of forms and can be defined in many ways, but in this chapter we will mostly use 'evidence' to refer to scientific tests of treatments or interventions, which compare the 'treatment' to a 'control' in some way and measure the effect quantitatively. We define evidence in this way as it is a broad description that can still address causality for interventions – did treatment X cause reaction Y? For example, it is not enough to know that some bats flew along bat gantries – we need to know, at a minimum, how many, and how many still flew low across the road. But more on that later.

## 4.2 Why do we need evidence-based conservation?

Modern medicine has many examples illustrating why the discipline needs a robust evidence base. However, basing medical treatments on scientific evidence was not always the norm. The use of randomised controlled trials to test medical treatments was initially considered unnecessary and unethical, and it was hotly contested. A good example comes from an early champion of evidence-based medicine, Archie Cochrane, who demonstrated that randomised controlled trials were necessary and that expert judgement alone could be flawed. In 1971, he presented preliminary results from a trial comparing home care for heart patients with care in the new Coronary Care Units (note that the findings may not be the same now). He had been criticised for risking the lives of patients allocated to the 'home care' group. What follows is in his own words:

The results at that stage showed a slight numerical advantage for those who had been treated at home. I rather wickedly compiled two reports: one reversing the number of deaths on the two sides of the trial. As we were going into the committee, in the anteroom, I showed some cardiologists the results. They were vociferous in their abuse: 'Archie', they said 'we always thought you were unethical. You must stop this trial at once'. I let them have their say for some time, then apologized and gave them the true results, challenging them to say as vehemently, that coronary care units should be stopped immediately. There was dead silence and I felt rather sick because they were, after all, my medical colleagues.

(Maynard & Chalmers, 1997)

Results such as these – where the preferred treatment of the time did not work, or actually made things worse – are used to demonstrate why scientific studies of impacts are important when treating people. A growing body of literature suggests that impact studies are also necessary for treating the health of the biosphere, although the 'gold standard' of randomised controlled trials is not always possible in this discipline (Pynegar et al., 2018). As we test more and more measures to conserve species and habitats, we find that many do not work. For example, studies have shown that widely used methods to make water voles move prior to building works were ineffective, risking accidental killing of the protected mammals (Gelling et al., 2018); that re-introduction programmes of species from macaws (Volpe et al., 2017) to tamarins (Beck et al., 1991) have resulted in high or total mortality for the released animals; and that artificial bat roosts built to replace those destroyed during building works often failed to attract any bats and, even when occupied, hosted about half the number of bats that the destroyed roost had (Stone et al., 2013). These results underline the need to test conservation solutions and not to simply assume that good intentions will lead to good outcomes.

Our case studies focus on the environmental impacts of roads. Road construction has been shown to harm animals through habitat degradation, loss and fragmentation, direct mortality and barrier effects (Laurance et al., 2009; Benítez-López et al., 2010; Rytwinski & Fahrig, 2012). Figure 4.1 summarises these cumulative impacts, which are likely to act at different rates and through a long extinction debt. Unfortunately, studies on a wide range of

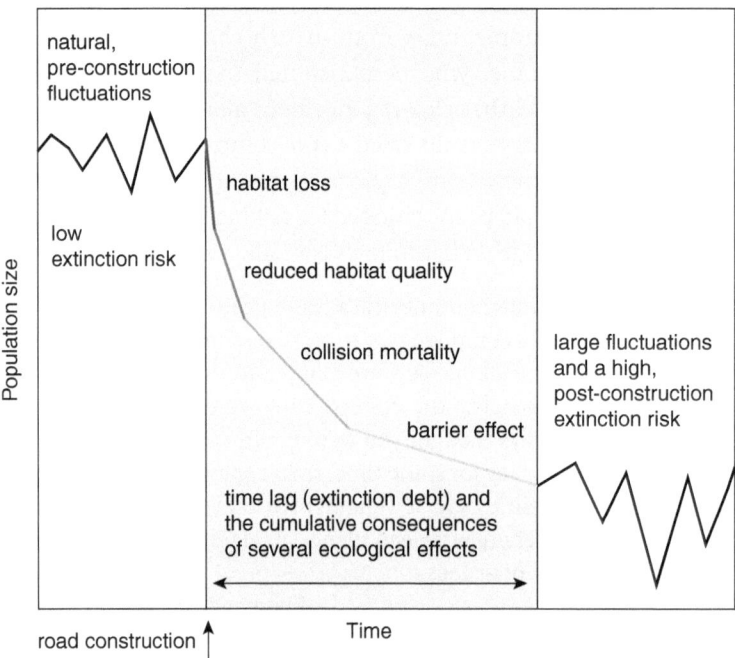

**Figure 4.1** The multiple causes of bat population reduction by road construction and the delayed response (extinction debt). Adapted from Forman et al. (2003). (A black and white version of this figure will appear in some formats. For the colour version, please refer to the plate section.)

habitats and taxa from grasslands to vertebrates show that many road mitigation options simply do not work. A growing list of papers points to not only poor design of mitigation and monitoring, but a wider context of poor target setting, weak implementation, inadequate reporting and poor or absent enforcement (e.g. Rundcrantz, 2006; Tischew et al., 2010; Beebee, 2013; Drayson & Thompson, 2013; Villarroya & Puig, 2013). We will not address all of these factors in this chapter, but they are important to consider when asking why ineffective measures have persisted for so long.

## 4.3 Case study: bats and roads

Why do we need mitigation for bats crossing roads? In the EU, all bat species have been protected under the EUROBATS agreement since 1994, in recognition of the declining populations of many species. As a consequence, whenever populations may be adversely affected by human activity, impact assessment and mitigation are a legal obligation. Over the last 10 years evidence for significant effects of roads on bats has grown and the need for effective mitigation has become increasingly evident (e.g. Altringham, 2008; Russell et al., 2009; Lesiński et al., 2010; Berthinussen & Altringham 2012b, 2015). There are clear specifications within EUROBATS to mitigate against the impacts of roads on bats.

### 4.3.1 The need to test mitigation: bat gantries

The EUROBATS commitment to mitigate against the impact of roads on bats is very positive, but are the mitigation strategies being used actually working? Early studies assessed the use by bats of underpasses and overpasses primarily built for other purposes, such as to carry minor roads, footpaths or streams. If bats were seen near to these structures it was generally assumed that they were effective mitigation tools (Highways Agency, 2001, 2006; reports reviewed by O'Connor et al., 2011). Underpasses, culverts, footbridges and bridges for vehicles, all of various sizes, were widely adopted as mitigation solutions (Figure 4.2). Many were not subsequently surveyed for use by bats, or qualitative surveys were written up in often confidential reports. Many studies reported 'use' – small numbers of bats observed in underpasses or flying over bridges of various kinds, without reference to the number still crossing the road unsafely, or not crossing at all (see Highways Agency, 2006; O'Connor et al., 2011), and many lacked convincing definitions of use. This meant that future projects could not learn from the success or failure of previously built mitigation structures.

In addition to multi-use structures, some 'bespoke' structures were built and 'bat gantries' or 'wire bridges' (Figure 4.3) were widely adopted. Bat gantries were assumed to act as navigational aids to echolocating bats, encouraging them to continue using existing 'commuting routes' from roosts to feeding areas (which often follow linear features such as hedgerows) after road construction, but lifting them above the traffic. Ideally, crossing points should be built on known bat commuting routes determined by pre-construction surveys, as bats tend to be faithful to particular routes. However, many were built away from known bat commuting routes for engineering reasons, to fit in with landscape topography, to combine bat routes with minor roads or footpaths, or simply to reduce cost. It was assumed bats would find the new crossing points (Highways Agency, 2001), and in some cases new hedge planting was designed to guide them to these structures. In many guidance documents, environmental statements and mitigation plans it was implicit, or even explicit, that the bats would respond as predicted (Highways Agency, 2001; Limpens et al., 2005).

In 2008, JDA was asked to provide evidence to a public inquiry for the effectiveness of these strategies (Altringham, 2008). No quantitative evidence was found to suggest that any of the strategies implemented were effective in protecting bats, particularly at the population level. However, neither was

**Figure 4.2** Two underpasses found to vary in effectiveness in guiding bats safely under roads. (a) An effective underpass on the A590, Cumbria, UK; (b) an ineffective underpass on the A66, Cumbria, UK. Boxplots show the number of bats crossing per survey using the underpass and crossing over the road above at safe and unsafe heights (above and below 5 m, traffic height). The variable success of underpasses underlines the need to understand the details of conservation interventions; in this example, the location of the underpasses impacted on how effective they were. From Berthinussen and Altringham (2012b). (A black and white version of this figure will appear in some formats. For the colour version, please refer to the plate section.)

there evidence to suggest that they were ineffective. This prompted us to conduct our own research to determine the effects of roads on bats and the effectiveness of mitigation (Berthinussen & Altringham, 2012a, 2012b, 2015).

In our research we emphasised the difference between qualitative assessments of the 'use' of a structure by a small number of bats and measures of

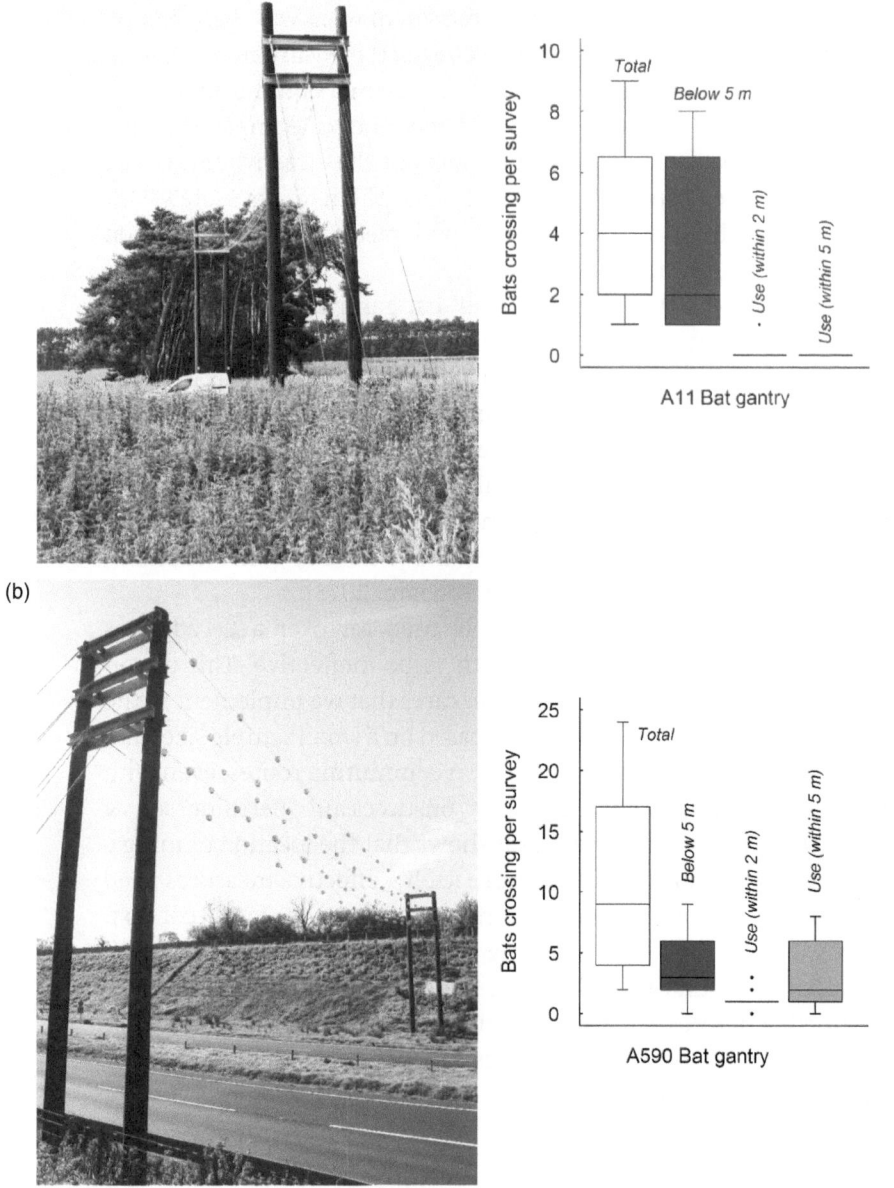

**Figure 4.3** Two bat gantry designs: (a) wire mesh design on the A11, Norfolk, UK; (b) wire and ball design on the A590, Cumbria, UK. Boxplots show the results of surveys carried out to test the effectiveness of the gantries in guiding bats safely over the road. Data were recorded for the total number of bats crossing per survey, the numbers crossing at unsafe heights (below 5 m, traffic height) and the numbers using the gantry according to two definitions of 'use' (flying within either 2 m or 5 m of the wires above traffic height). The bat gantry story neatly demonstrates the need to test conservation interventions before rolling them out on a wide scale. From Berthinussen and Altringham (2012b, 2015). (A black and white version of this figure will appear in some formats. For the colour version, please refer to the plate section.)

effective protection at the population level. We also stressed that the number of bats present pre-construction is rarely assessed, meaning that post-construction bat numbers may already be a fraction of what was there before. We proposed two broad measures of effectiveness: (1) measurements of local bat activity and of the movement of bats along severed commuting routes before and after road construction, to assess landscape-scale impact and the permeability of new roads; and (2) measurements of the effectiveness of the crossing structures – the proportions of bats that use them to cross safely. Our research was limited by logistics to the second measure – do mitigation structures guide bats safely across roads?

The headline result was that 'wire and ball' bat gantries did not alter the behaviour of bats crossing roads – they were wholly ineffective (Berthinussen & Altringham, 2012a; Figure 4.3b). This was a disturbing finding, as over the previous decade about 15 gantries had been built in the UK and continental Europe. Although our study showed that one design of bat gantry was ineffective, it was suggested that other designs would have greater success in guiding bats to fly at safe heights above roads. Our next study found that 'wire mesh' gantries (Figure 4.3a, of a different design to the 'wire and ball' structures) were equally ineffective (Berthinussen & Altringham, 2015).

In summary, a mitigation measure widely used for over a decade was essentially untested and subsequently shown to be ineffective. This underscores the need for rigorous testing of the measures that we implement in the name of conservation. We also found (albeit based on a small sample size) that building all types of crossing away from known commuting routes, even with new planting to guide bats to them, was unsuccessful (Berthinussen & Altringham, 2012a). This is important, as it shows that the location of mitigation measures is as important as the measure itself – effective measures need to be implemented with a good understanding of the local context. Furthermore, we found evidence that some underpasses were used by a high proportion of bats, and that the one green bridge tested in the UK – a large structure planted with trees, shrubs and ground cover – was used by over 90% of bats crossing the road in that area, suggesting that effective ways to allow bats to safely cross roads do exist.

## 4.4 Synthesising evidence

The bat gantry case study provides some insight into why we need to rigorously test the effectiveness of measures aiming to protect the natural world. However, this is just the first step towards implementing a truly evidence-based approach to conservation. The next step is to systematically bring together all the evidence, from many studies, on particular conservation measures. This approach is also borrowed from evidence-based medicine, where it has proven to be a lifesaver.

One of the most important developments in evidence-based medicine was the Cochrane Collaboration, an organisation set up to conduct systematic reviews of the scientific evidence on topics such as how well different treatments worked. In medicine – as in conservation – natural variation in populations means that it often takes large numbers of replicates for beneficial or detrimental effects to become apparent. Modern doctors, making potentially life-changing decisions, want to have the information on every study on a particular treatment to hand, not just the results from a single trial that may not be representative. The goal should be the same in conservation: to bring together all the evidence for an intervention to assess whether it works, whether it does harm, or whether it only works in certain situations or with certain variations of the intervention.

There are many examples of the importance of collating evidence in medicine. For example, a systematic review on cot death or sudden infant death syndrome (SIDS), using the studies already available in the 1970s, could have saved the lives of an estimated 60,000 babies. Due to a lack of evidence synthesis and an overreliance on expert opinion, medical practitioners advised parents to put children to sleep on their fronts until the 1990s, when studies and reviews led to the realisation that this sleeping position increased the risk of SIDS (Gilbert et al., 2005).

In conservation, collating or synthesising the data is as critical as it is in medicine (Sutherland et al., 2004). While the most rigorous method, systematic review, is very important (see Chapter 7, the Collaboration for Environmental Evidence and Mistra EviEM), more cost-effective methods of collating the evidence may also be desirable in this underfunded discipline, where the evidence itself can be scarce and variable in quality (Sutherland & Wordley, 2018). 'Synopses' published by Conservation Evidence (www .conservationevidence.com) follow one such method, known as subject-wide evidence synthesis (Sutherland & Wordley, 2018). Researchers draw up lists of all the interventions that could benefit a given taxa or habitat, classified according to potential threats based on IUCN criteria (Threats Classification Scheme Version 3.2); the scientific studies for the effectiveness of each intervention are then collated and summarised. For example, we produced the Bat Synopsis (Berthinussen et al., 2013, updated 2019), which provides key messages and summaries of the relevant studies, to help conservationists see which interventions for bat conservation are likely to be the most effective, and under which circumstances. The summary of this synopsis in What Works in Conservation (Berthinussen et al., 2018) takes this a step further, by using expert scoring to categorise the interventions based on levels of effectiveness, certainty in the evidence available and potential harms.

The first Bat Synopsis (Berthinussen et al., 2013) listed 78 interventions that could be implemented to conserve bats, covering areas as diverse as logging,

roost provision and wind turbine operation regimes. No evidence for effectiveness was found for 48 of the 78 interventions, many of which are used
routinely in the UK and elsewhere. This does not mean they are ineffective,
but simply that they had not been tested quantitatively when we checked the
literature. For a further 12 interventions the evidence was too limited for
assessment. This demonstrates the scarcity of experimental evidence for
many possible management actions, severely limiting the ability of conservationists, ecological consultants, developers and government agencies to
undertake evidence-based conservation or mitigation.

Of the 18 remaining interventions, 14 had some proven value as conservation tools for bats. These included using selective logging instead of conventional logging, turning off wind turbines at low wind speeds and minimising
light pollution. An update to this synopsis was published in 2019 (Sutherland
et al., 2019), expanding the list of interventions to 190 and adding new studies
that were published in the intervening years. There are many interventions
which have had valuable evidence added in this update, but we have not yet
seen a shift to a majority of interventions being tested via multiple high-
quality experiments.

### 4.4.1 Example of evidence synthesis: road underpasses

For many of the interventions addressed in the Bat Synopsis, our greatest
contribution was to demonstrate that no evidence existed for the efficacy
of these measures – hopefully spurring more research and a more critical
eye towards choosing conservation measures. But for a handful, we could
begin to tease out what made an intervention effective in some circumstances but not others – one of the many benefits of summarising multiple studies. One such intervention is the use of underpasses to get bats to
cross roads safely.

In the 2013 Bat Synopsis we found four studies, from Germany, Ireland and
the UK, which between them showed that at least nine bat species used
underpasses (none purpose-built for bats), with up to 96% of the bats crossing
through underpasses rather than the road above (although this varied greatly)
(Berthinussen et al., 2013). By summarising the key details of each study, we
can see that some species use underpasses frequently while others do not
appear to use underpasses at all, and that only a few species appear to use
small underpasses, such as drainage pipes of diameter less than 1.5 m. There
are indications that effectiveness increases with diameter and when underpasses are placed on known bat commuting routes – conclusions supported by
ongoing studies (Davies, 2019). The 2019 update of the Bat Synopsis added two
further studies, which tested much larger underpasses and still found the
largest structures to be the most effective, but also explored the differing
responses of various functional guilds of bats. These details are critical.

Further testing and refining of underpasses, followed by evidence synthesis, should help to ensure that future underpasses are as effective as they can be.

## 4.5 Getting the evidence used

We are trying to bring this work, demonstrating the importance of generating and using evidence on the effectiveness of interventions, to as large an audience as possible, to ensure that those responsible for commissioning, designing, approving and testing mitigation structures are aware of it. The bat gantry studies have been reported in national newspapers, radio and television. This was achieved through press releases, by approaching media contacts directly and by being approachable when contacted. The work has also been reported in several books and papers (Altringham, 2011; Abbott et al., 2015; Altringham & Kerth, 2016; Sutherland & Wordley, 2017). JDA and AB have run seven workshops for practitioners on road mitigation measures for bats and talked at over 10 conferences in the UK and abroad. CFRW has mentioned this study in around 50 talks to conservationists and government agencies and used it as an example in an opinion piece on evidence use in conservation (Sutherland & Wordley, 2017). The Bat Synopsis and What Works in Conservation, which contains a summary of the Bat Synopsis, have also been widely promoted.

This awareness resulted in tens of thousands of views of the paper and relevant parts of the Bat Synopsis, and this exposure has translated into further successes. The impact of early work (Berthinussen & Altringham, 2012a, 2012b) led to a Defra-funded project to develop better mitigation monitoring protocols (Berthinussen & Altringham, 2015) and a statutory conservation agency guidance note summarising the protocols. The approximately £1 M spent on bat gantries in the UK as of 2017 (Sutherland & Wordley, 2017) was brought up in the House of Lords by Lord John Krebs in January 2018, who used it to demonstrate why the UK government's 25-year environment plan needed to explicitly commit to being evidence-based. Some road-building projects have taken heed of the evidence. The A40 Penblewin to Slebech Park Improvement in Wales opted to mitigate impacts on bats using underpasses of varying sizes, many built on known commuting routes, and funded more rigorous monitoring (Davies, 2019).

However, not everyone is listening. Despite widespread reporting of the ineffectiveness of bat gantries in 2012, six gantries of a 'wire-mesh' design were built in Norfolk, England in 2014 at a reported cost of £350,000. These were probably planned before the 2012 paper was published, but plans were not modified in light of this study. In 2015 these gantries were also shown not to work (Berthinussen & Altringham, 2015). Nevertheless, seven more gantries are under construction (as of 2018) at a cost of over £1 M on the North Norfolk Distributor Road (MacDonald, 2014). In another example, the environmental statement for the proposed and controversial extension to the M4 across the

Gwent Levels in Wales (Welsh Government, 2016) draws extensively on our 2015 Defra report (Berthinussen & Altringham, 2015). However, it proposes numerous culverts for bats which, by the authors' own admission, are almost all too small to be used by the target bat species. In addition, most will not be on known commuting routes.

Furthermore, there are still inadequate mechanisms in place to assess the effectiveness of mitigation measures. A feature of many environmental statements and mitigation plans is the absence of a monitoring plan capable of assessing mitigation success or failure. There is frequently no monitoring plan at all. This appears to be due to a reluctance or inability of government agencies to enforce effective monitoring, a reluctance on the part of many developers to pay for monitoring and a lack of understanding about how to design and conduct monitoring that is fit for purpose. As a result, developers and taxpayers spend money on unproven mitigation with no prospect of improved understanding.

### 4.5.1 Why is evidence ignored?

Why are proven methods rejected, often in favour of methods that have been demonstrated not to work? Why is there an apparent reluctance to seek out, use or accept evidence, or to collect it, among some decision-makers, including some government agencies and ecological consultants? Sutherland and Wordley (2017) explored a few general psychological and structural reasons for this phenomenon, and more detail on this topic is given in Chapter 9 of this book. Here we share some of our own experiences of the failure to use evidence in road mitigation.

The real or perceived higher financial cost of effective mitigation solutions is one concern. Mitigation consumes a very small part of the total cost of a road-building project, but mitigation and monitoring are obvious targets when budgets are tight or overrun. Effective mitigation may or may not be more expensive than ineffective options, but ineffective mitigation is simply a waste of resources.

A desire to simplify the planning and implementation of mitigation is another reason why some parties are reluctant to challenge or change accepted approaches. Road building is complex, making off-the-shelf, approved mitigation solutions an attractive option. Being able to implement development projects as quickly and cheaply as possible can make mitigation a tick-box exercise – complying with regulation at minimal cost may be more important than implementing effective mitigation. Mitigation solutions such as bat gantries can be designed and built relatively cheaply and, if experts say they will work, then they fulfil all legal requirements and may be assumed to require little or no 'expensive' monitoring. To question their effectiveness can put in jeopardy budgets, work schedules, building specifications, even the project itself. A reluctance to listen to objections is understandable, but not excusable. Consultant ecologists can be placed in a difficult position. Their scientific training, personal concern about nature and professional standards all demand unbiased assessment. However, their livelihoods depend upon contracts from developers who are frequently not obliged to commit to effective mitigation and monitoring.

Finding, evaluating and applying the evidence on mitigation strategies can also be a challenge. With the existence of freely available downloadable material (such as the Bat Synopsis) in a concise, jargon-free form, decision-makers should be more aware of what works and what does not. However, ecological consultants and statutory agency staff still need time to find, read and digest the information, and require some scientific training to evaluate the evidence. The difference between quantitative evidence and anecdote is not always understood and 'professional judgement' may be relied on even when it runs counter to the evidence. However, it does not have to be painful for developers, consultants or planners to improve on current practice. Adoption of good mitigation practices early in a project can avoid the problems of making corrections during the project, and investment in effective technologies may lower the costs of solutions such as large underpasses and green bridges.

## 4.6  How can evidence use in mitigation projects be improved?

First, there should be a key requirement that mitigation structures are tested for effectiveness, not just use, and a quantitative bar set for effectiveness (Berthinussen & Altringham, 2015). Ecologists employed to assess mitigation effectiveness must be prepared to shun options proven to be ineffective. Professional bodies must fully support ecological consultants in implementing those measures shown to be effective, and sanction members who use methods known to be ineffective. Improvements may be much more evident if the enforcing statutory agencies are willing and able to deny planning permission to development projects that have poor mitigation strategies. There should be real commitment from governments to pledge to conserve species and habitats using evidence-based measures and discarding measures proven to be ineffective. This may require additional resources to assess existing and proposed legislation against evidence syntheses.

To identify effective and ineffective solutions there is a clear need for dedicated funding for rigorous tests of interventions. Monitoring interventions often requires long-term commitment which, in turn, requires adequate long-term funding. This could come from developers and government agencies, but a greater recognition by academic funding bodies of the value of applied questions would also have a huge impact. PhD projects could be encouraged to have applied components, testing interventions. Research council funding for academics to address applied questions of conservation importance and communicate them to practitioners would have a huge impact.

Greater power for statutory agencies to enforce existing laws, check up on implementation and demand replacements for ineffective solutions would dramatically improve mitigation effectiveness. A framework with greater incentives for developers to show that their mitigation has been effective would be beneficial. These could include a requirement to make the results of mitigation monitoring for effectiveness public, penalties for failures to do so and awards for new, proven effective solutions.

While many of these goals may not be realised in the near future, we can all promote approaches to conservation that are evidence-based and effective. If enough of us do it, it might just change the world.

## References

Abbott, I., Berthinussen, A., Boonman, M., et al. 2015. Bats and roads. In van der Ree, R., Grilo, C. & Smith, D., editors, *Handbook of Road Ecology* (pp. 290–299). Chichester: John Wiley & Sons. doi:10.1002/9781118568170.ch39

Altringham, J. D. 2008. *Bat Ecology and Mitigation; Proof of Evidence; Public enquiry into the A350 Westbury bypass.* Neston: White Horse Alliance.

Altringham, J. D. 2011. *Bats: From Evolution to Conservation*, 2nd edition. Oxford: Oxford University Press. doi:10.1093/acprof:osobl/9780199207114.001.0001

Altringham, J. & Kerth, G. 2016. Bats and roads. In Voigt, C. C. & Kingston,T., editors, *Bats in the Anthropocene: Conservation of Bats in a Changing World* (pp. 35–62). Cham: Springer International Publishing. https://doi.org/10.1007/978-3-319-25220-9

Beck, B., Dietz, J., Castro, L., et al. 1991.Losses and reproduction of reintroduced golden lion tamarins *Leontopithecus rosalia. Dodo*, 27, 50–61.

Beebee, T. J. C. 2013. Effects of road mortality and mitigation measures on amphibian populations. *Conservation Biology*, 27, 657–668. doi:10.1111/cobi.12063

Benítez-López, A., Alkemade, R. & Verweij, P. A. 2010. The impacts of roads and other infrastructure on mammal and bird populations: a meta-analysis. *Biological Conservation*, 143, 1307–1316. https://doi.org/10.1016/j.biocon.2010.02.009https://doi.org/10.1016/j.biocon.2010.02.009

Berthinussen, A. & Altringham, J. 2012a. Do bat gantries and underpasses help bats cross roads safely? *PLoS ONE*, 8, e38775. https://doi.org/10.1371/journal.pone.0038775

Berthinussen, A. & Altringham, J. 2012b. The effect of a major road on bat activity and diversity. *Journal of Applied Ecology*, 49, 82–89. doi:10.1111/j.1365-2664.2011.02068.x

Berthinussen, A. & Altringham, J. 2015. WC1060: Development of a cost-effective method for monitoring the effectiveness of mitigation for bats crossing linear transport infrastructure. Final report to Defra. http://sciencesearch.defra.gov.uk/Default.aspx?Module=More%26Location=None%26ProjectID=18518

Berthinussen, A., Richardson, O. & Altringham, J. 2013. *Bat Conservation: Evidence for the Effects of Interventions.* Synopses of Conservation Evidence series.Exeter: Pelagic Publishing.

Berthinussen, A., Richardson, O. C., Smith, R. K., et al. 2018. Bat conservation. In Sutherland, W. J., Dicks, L. V., Petrovan, S. O., et al., editors, *What Works in*

Conservation (pp. 67–93). Cambridge: Open Book Publishers. http://dx.doi.org/10.11647/OBP.0060

Davies, J. 2019. Monitoring the effectiveness of mitigation for horseshoe bats associated with a new road in Wales. *Conservation Evidence*, 16, 17–23.

Drayson, K. & Thompson, S. 2013. Ecological mitigation measure in English environmental impact assessment. *Journal of Environmental Management*, 119, 103–110. https://doi.org/10.1016/j.jenvman.2012.12.050

Forman, R. T. T., Sperling, D., Bissonette, J. A., et al. 2003. *Road Ecology: Science and Solutions*. Washington, DC: Island Press.

Gelling, M., Harrington, A. L., Dean, M., et al. 2018. The effect of using 'displacement' to encourage the movement of water voles *Arvicola amphibius* in lowland England. *Conservation Evidence*, 15, 20–25.

Gilbert, R., Salanti, G., Harden, M., et al. 2005. Infant sleeping position and the sudden infant death syndrome: systematic review of observational studies and historical review of recommendations from 1940 to 2002. *International Journal of Epidemiology*, 34, 874–887. doi:10.1093/ije/dyi088

Highways Agency. 2001. *Nature Conservation Advice in Relation to Bats: Design Manual for Roads and Bridges*. Volume 10, Environmental Design. Section 4, Nature Conservation. Part 3, HA80/99. Guildford: Highways Agency UK.

Highways Agency. 2006. *Best Practice in Enhancement of Highway Design for Bats*. Guildford: Highways Agency and Bat Conservation Trust.

Laurance, W. F., Goosem, M. & Laurance, S. G. 2009. Impacts of roads and linear clearings on tropical forests. *Trends in Ecology & Evolution*, 24, 659–669. https://doi.org/10.1016/j.tree.2009.06.009

Lesiński, G., Sikora, A. & Olszewski, A. 2010. Bat casualties on a road crossing a mosaic landscape. *European Journal of Wildlife*

Research, 57, 217–223. https://doi.org/10.1007/s10344-010-0414-9

Limpens, H. J. G. A., Twisk, P. & Veenbaas, G. 2005. Bats and road construction. Dutch Ministry of Transport, Public Works and Water Management Directorate-General for Public Works and Water Management, Road and Hydraulic Engineering Institute, Delft, the Netherlands and the Association for the Study and Conservation of Mammals, Arnhem, the Netherlands

MacDonald, M. 2014. The Norfolk County Council (Norwich Northern Distributor Road (A1067 to A47(T))) Order 6.2 Environmental Statement: Volume II: Chapter 2. The Scheme. PINS Reference Number: TR010015 Document Reference: 6.2 Regulation Number: 5(2)(a).

Maynard, A. & Chalmers, I., editors. 1997. *Non-random Reflections: On Health Services Research: On the 25th Anniversary of Archie Cochrane's Effectiveness and Efficiency*. London: BMJ Books. https://doi.org/10.1136/bmj.316.7143.1543

O' Connor, G., Green, R. & Wilson, S. 2011. *A Review of Bat Mitigation in Relation to Highway Severance*. Guildford: Highways Agency.

Pynegar, E. L., Gibbons, J. M., Asquith, N. M., et al. 2018. What role should Randomised Control Trials play in providing the evidence base underpinning conservation?. *PeerJ PrePrints*.

Rundcrantz, K. 2006. Environmental compensation in Swedish road planning. *European Environment*, 16, 350–367. doi:10.1002/eet.429

Russell, A. L., Butchkoski, C. M., Saidak, L., et al. 2009. Road-killed bats, highway design, and the commuting ecology of bats. *Endangered Species Research*, 8, 49–60. doi:10.3354/esr00121

Rytwinski, T. & Fahrig, L. 2012. Do species life history traits explain population responses to roads? A meta-analysis. *Biological Conservation*, 147, 87–98. http://dx.doi.org/10.1016/j.biocon.2011.11.023

Stone, E. L., Jones, G. & Harris, S. 2013. Mitigating the effect of development on bats in England with derogation licensing. *Conservation Biology*, 27, 1324–1334. doi:10.1111/cobi.12154

Sutherland, W.J., Dicks, L.V., Ockendon, N., Petrovan, S.O., and Smith, R.K. *What Works in Conservation 2019*. Cambridge, UK: Open Book Publishers, 2019

Sutherland, W. J., Pullin, A. S., Dolman, P. M., et al. 2004. The need for evidence-based conservation. *Trends in Ecology & Evolution*, 19, 305–308.

Sutherland, W. J. & Wordley, C. F. R. 2017. Evidence complacency hampers conservation. *Nature Ecology & Evolution*, 1, 1215–1216, doi:10.1038/s41559-017-0244-1

Sutherland, W. J. & Wordley, C. F. R. 2018. A fresh approach to evidence synthesis. *Nature*, 558, 364–366. doi:10.1038/d41586-018-05472-8

Tischew, S., Baasch, A., Conrad, M. K., et al. 2010. Evaluating restoration success of frequently implemented compensation measures: results and demands for control procedures. *Restoration Ecology*, 18, 467–480. doi:10.1111/j.1526-100X.2008.00462.x

Villarroya, A. & Puig, J. 2013. A proposal to improve ecological compensation practice in road and railway projects in Spain. *Environmental Impact Assessment Review*, 42, 87–94. http://dx.doi.org/10.1016/j.eiar.2012.11.002

Volpe, N. L., Di Giacomo, A. S. & Berkunsky, I. 2017. First experimental release of the red-and-green macaw *Ara chloropterus* in Corrientes, Argentina. *Conservation Evidence*, 14, 20–20.

Welsh Government. 2016. *M4 Corridor around Newport. Environmental Statement Volume 1. Chapter 10: Ecology and Nature Conservation*. http://gov.wales/topics/transport/roads/schemes/m4/corridor-around-newport/environmental-info/ecology-nature/?lang=en

# Understanding local resource users' behaviour, perspectives and priorities to underpin conservation practice

E.J. MILNER-GULLAND, HARRIET IBBETT

*University of Oxford*

PAULO WILFRED

*The Open University of Tanzania*

HANS COSMAS NGOTEYA

*Landscape and Conservation Mentors Organisation*

and

PENI LESTARI

*Wildlife Conservation Society Indonesia Program*

## 5.1 Introduction

Most of the chapters in this book focus on how best to bring science into policy, often at the national scale and mostly with a developed-world perspective. Ensuring that national policy frameworks are conducive to conservation is vital, but it is also important to improve the effectiveness of science in supporting conservation interventions on the ground. Small-scale interventions aiming to change the behaviour of local resource users in developing countries make up a large proportion of global conservation effort and funding (Brockington & Scholfield, 2010). These types of intervention are challenging to do well, and often do not produce the desired results (Larrosa et al., 2016). Typically, there is little scientific input into either the design or evaluation of these projects, and evidence of effectiveness is limited (Roe et al., 2015). Small organisations in developing countries may not have the capacity or confidence to implement scientifically informed design and monitoring, and supporting them to collate and learn from evidence may not be a major priority for researchers or donors. Increased sharing of insights and techniques to support more robust and effective interventions could transform grassroots conservation (e.g. Woodhouse et al., 2016). In this spirit, we use case studies from four locations around the world to illustrate some of the challenging steps involved in understanding conservation issues and design-ing suitable interventions. These steps are often skipped or not made explicit, but are critical to success; they ensure that interventions have a strong foun-dation in evidence, making it more likely that their desired impacts are achieved.

First, we explore how to collect robust information on the prevalence of illegal resource use, as a first step towards understanding the extent of the problem, using a case study of bird hunting in a Cambodian grassland. Next, we consider how to bring together different sources of information to under-stand both resource use and local perspectives on conservation, using a case study from Tanzania. These two case studies about evidence gathering lead on to the next stage: intervention design. We start with an example, also from Tanzania, of developing a Theory of Change for a conservation intervention, in which the process by which actions lead to a desired result is identified, assumptions are made clear and the progress of the intervention towards its desired impact can be monitored. Finally, we explore the challenges of imple-menting one particularly prevalent intervention type – alternative livelihoods projects – using an example of a shark fishery in Indonesia. Together, these case studies provide a vivid illustration of the ways in which conservation researchers and practitioners are combining efforts to ensure that interven-tions are based on robust evidence and therefore more likely to succeed.

## 5.2 Asking questions about sensitive topics

Moderating human behaviour is critical to conservation success (Gore, 2011; Milner-Gulland, 2012). However, if we are effectively to change human beha-viour, we must first ensure we understand the nature of the behaviour we want to change. Central to this is determining both the prevalence of beha-viours that are detrimental to biodiversity, and the characteristics of the people engaging in these behaviours. This is essential to ensure managers efficiently allocate resources to tackle threats, and that behavioural change interventions target the right audiences with the right incentives (St John et al., 2010, 2015). However, obtaining such information can be extremely challenging, especially if the behaviour in question is illegal (Gavin et al., 2010).

A common approach to ascertaining the true extent of illegal behaviours is asking direct questions (e.g. Gandiwa, 2011; Kiffner et al., 2015). Other studies mask the sensitivity of questions about illegal behaviours by mixing them with less-sensitive questions about other livelihood activities (e.g. Martin et al., 2012; Mgawe et al., 2012; Kiffner et al., 2015). Although direct questioning may help to cast some light on the nature of natural resource exploitation, it runs the risk of bias from untruthful responses (Nuno & St John, 2015). Respondents may be scared to answer questions honestly for fear of incriminating themselves, or the possible repercussions they might face from revealing their behaviour. They may avoid answering questions altogether, terminate interviews early or underreport activities. If respondents do answer sensitive questions, social desirability bias may lead them to moderate their responses so their actions appear more socially acceptable. This is especially true of data captured in group settings, where pressure from peers may prevent others speaking freely and truthfully about their activities. It is also important to consider the ethical implications of directly asking respondents about their illegal activity; research has an ethical responsibility to 'do no harm', yet asking such questions can cause respondents to directly implicate themselves in illegal activities, potentially leading to severe consequences.

Indirect questioning has started to become more widely used in conservation science in response to some of these challenges. The method comes from psychology, and has been used when asking questions about sensitive issues such as drug use and racial prejudice (Imai, 2011). The technique enables interviewees to respond in such a way that the interviewer cannot directly determine whether they have participated in the activity. Instead, data provide estimates of prevalence at the population level, affording both the respondent and the researcher greater levels of protection.

One form of indirect questioning increasingly applied in conservation is the Unmatched Count Technique or Item List Technique (see Gavin et al., 2010; Nuno & St John, 2015). The technique works by devising a short 'control' list of three to five innocuous items that are non-sensitive but relevant to the research topic, and a treatment list which also contains the sensitive item of interest (Figure 5.1). The sampled population is randomly shown either the control or treatment list. Respondents are asked to report only the total number of items that apply to them. Because only a number is reported, the researcher has no way of knowing which specific items apply to a given respondent. The difference in the mean number of items reported by the two groups provides an estimate of the proportion of respondents engaging in the sensitive behaviour (Thomas et al., 2015).

### 5.2.1  Case study: Bengal florican

Ibbett et al. (2019) used the Unmatched Count Technique to investigate prevalence of illegal behaviours and to identify the characteristics of resource users in central Cambodia. In the dry season, the seasonally inundated grasslands surrounding the Tonle Sap lake are home to the world's largest remaining population of Bengal florican (*Houbaropsis bengalensis*), a critically endangered bustard species (Birdlife International, 2015). Recently, agricultural abandonment, scrub advancement and the emergence of dry-season rice – a form of intensive, irrigated rice cultivation – have dramatically reduced grassland cover. The Tonle Sap florican population is estimated to

**Figure 5.1** Using the Unmatched Count technique to ask about illegal bushmeat hunting in the Ugalla Wildlife Reserve, Tanzania. Picture by Paulo Wilfred. (A black and white version of this figure will appear in some formats. For the colour version, please refer to the plate section.)

have declined by 44–66% since dry-season rice was first cultivated on the floodplain in 2004 (Packman et al., 2014). However, conservation managers lack adequate understanding of the drivers of dry-season rice expansion. There is also evidence that hunting, a historic driver of decline, may persist in local communities (Packman, 2011).

Ibbett et al. (2019) used a mixed-methods approach to investigate these issues. Because hunting is potentially a sensitive activity (hunting wildlife is illegal in protected areas), the Unmatched Count Technique was selected to identify the prevalence of bird hunting and florican egg collection. The Unmatched Count Technique was combined with direct questioning and delivered through a household questionnaire, which captured information on household demographics, livelihood activities and awareness of bird species. Due to the florican's scarcity, Unmatched Count Technique questions concerned the hunting of larger grassland birds in general, with questions phrased as 'How many of the following animals/types of egg have people in your household caught in the last 12 months?' A warm-up question about different fruits consumed in the household was asked in order to introduce respondents to the technique.

A sample of 616 households across 21 villages was secured. The warm-up question identified a significant difference between control and treatment groups, suggesting the technique was working as expected. However, no significant difference was identified between control and treatment groups for egg collecting or large bird hunting, suggesting the prevalence of these activities did not significantly differ from zero. When questioned directly, just 8.6% of households reported hunting birds in the previous 12 months, the majority of which were small, abundant game birds, such as buttonquail and ducks. Those that reported hunting birds were more likely to come from households which also collected other wildlife products, such as frogs and crickets.

### 5.2.2 Lessons learnt

While indirect questioning techniques avoid some of the pitfalls of traditional techniques, they are not without limitations. In this case, the Unmatched Count Technique failed to detect the presence of bird hunting, unlike direct questioning. This may be explained by the generally low prevalence of this activity and the probabilistic nature of the approach, which means that confidence intervals are large. Part of the issue is that the direct question was about bird hunting in general, and showed low levels of hunting of common species, while the Unmatched Count Technique question investigated targeted hunting of large bird species. Only one or two incidences of hunting large bird species were directly reported. Similar experiences of inability to estimate prevalence have been reported by others when using the Unmatched Count Technique to investigate illegal activities (e.g. Nuno et al., 2018). Therefore, the Unmatched Count Technique is unlikely to be useful when estimating the prevalence of an extremely rare activity. Indirect questioning is also not a panacea for sensitivity; if an activity is highly sensitive, particularly if it violates social norms, respondents may still not answer truthfully when the item is in a list; this can even result in negative estimates for prevalence (e.g. Fairbrass et al., 2016).

Compared to other indirect techniques, such as the Randomised Response Technique (see Nuno & St John, 2015), the Unmatched Count Technique is often preferred because it can provide higher estimates of prevalence, is simple to understand and adaptable, and thus useful in developing countries where levels of illiteracy may be high (Gavin et al., 2010; Nuno & St John, 2015). Despite this, the concept can still be difficult for respondents to grasp. Respondents may be wary, especially if they have previously had negative encounters with researchers. Taking time to thoroughly talk through the technique, using a warm-up question and explaining each list item is essential to avoid these issues. Often, conservation researchers rely on the help of translators or local research assistants. Selecting the very best help available and providing extensive training to assistants is essential in order to prevent information from getting 'lost in translation'. Local research assistants can also provide knowledge to ensure designs are appropriate. This is particularly helpful when working in illiterate communities, or when relying on pictorial prompts.

## 5.3 Triangulating different sources of evidence to build a rounded picture

Social research methods such as focus groups, interviews and household surveys are increasingly being employed to investigate illegal behaviours and profile resource users (Young et al., 2018). The current decade has seen an increase in the use of these mixed methods approaches to gain a more holistic understanding of resource use (e.g. Kahler & Gore, 2012; Harrison et al., 2015). A combination of perspectives, using both qualitative and quantitative methods, is commonly preferred.

### 5.3.1 Case study: Ugalla Game Reserve

Ugalla Game Reserve (hereafter Ugalla; 5000 km$^2$) in western Tanzania is predominantly miombo woodland. Its conservation value is high, serving as habitat for a wide range of species (UGR, 2006). It is part of the Malagarasi-Muyovosi Ramsar Site, and facilitates connectivity between protected areas in western Tanzania (Kalumanga, 2015; Riggio & Caro, 2017). The main legal activity in the reserve is trophy hunting, mostly by overseas tourists. A number of different approaches are used to conserve Ugalla, including irregular anti-poaching patrols and seasonal permission for fishing and bee-keeping activities (July–December). These also aim to attract local support for conservation and build a sense of ownership of the reserve among local people. However, recent studies suggest that this conservation approach is ineffective (Wilfred & MacColl, 2015; Wilfred et al., 2017). Unauthorised use of natural resources (including poaching, illegal logging and fishing) is common and local communities hold negative attitudes towards the reserve and its management. In an attempt to shed light on the prevalence of illegal behaviours and inform the management of Ugalla, multiple research methods were used to gather relevant information. Between 2013 and 2016, household surveys and focus groups were conducted in villages around Ugalla, along with a survey of signs of illegal activity undertaken across the Protected Area.

For the household surveys, 533 households were randomly sampled in 2016 in the vicinity of Ugalla. The Unmatched Count Technique was used to estimate the prevalence of illegal behaviours (logging, illegal hunting and honey-gathering). The survey also included questions on households' perceptions of the main threats to Ugalla and its wildlife, and what communities would do differently to improve Ugalla's management effectiveness. Six single-sex focus groups from six randomly selected villages within 20 km of the Ugalla boundary, each with 4–6 participants, were conducted to verify findings from the household survey. Free-listed threats to Ugalla were ranked in decreasing order of their importance, and each threat was then divided by the total number of threats to calculate the salience score (Papworth et al., 2013). The overall score for each threat was obtained by calculating the average salience score across the focus groups. The greater the salience, the more important the threat.

For signs of illegal activity in the reserve, 10 patrol tracks were randomly selected in 2014. Six transect starting points were placed at 3000-m intervals along each road. At each point, two 1500-m transects were walked on opposite sides and perpendicular to the road. Signs of illegal activity (e.g. tree stumps, sawpits, meat smoking racks, snares, trees felled for honey extraction, fish smoking racks, poacher camps) were noted 50 m either side of the transect line (Figure 5.2).

The Unmatched Count Technique results suggested that poaching and illegal logging were performed by 28% (SE ± 6) and 20% (SE ± 5), respectively, of surveyed households; 18% (SE ± 6) of respondents gathered honey. The top four threats to Ugalla, as identified by respondents, were poaching (40% of respondents), logging (39%), fishing (11%) and honey gathering (8%). Of the top four threats to Ugalla free-listed and ranked by focus groups, logging had the highest salience ($S = 0.5$), followed by poaching (0.45). Within the reserve, 867 illegal activity signs were encountered. Signs related to logging had the highest frequency, followed by honey gathering, poaching and fishing (Figure 5.3). These results indicate that levels of illegal activity in Ugalla are high. The different methods consistently suggest that logging and poaching are the commonest illegal activities.

**Figure 5.2** Paulo Wilfred and his research assistant recording an illegal meat smoking rack in Ugalla Wildlife Reserve. (A black and white version of this figure will appear in some formats. For the colour version, please refer to the plate section.)

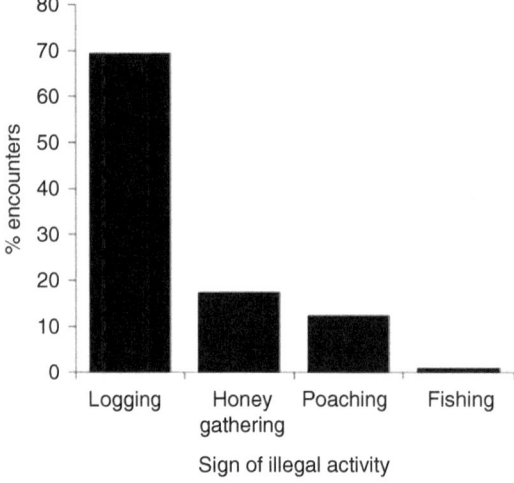

**Figure 5.3** Signs of illegal activity encountered inside Ugalla Game Reserve in 2014. Total signs = 867.

Of the activities that survey respondents and focus group participants said they would undertake if they were the Ugalla manager, the most common recommendations were to: improve the well-being of people around Ugalla (17% of respondents, $S = 0.11$); ensure that local people have adequate land for their livelihood activities (16%, $S = 0.35$); promote local participation in conservation (16%, $S = 0.13$); improve law enforcement (15%, $S = 0.14$); raise conservation awareness (15%, $S = 0.14$); and improve local people's relations with reserve managers (12%, $S = 0.54$).

### 5.3.2  Lessons learnt

Paulo Wilfred's research in Ugalla started nearly a decade ago with the overarching objective of informing conservation management. This long-term research suggests that local communities are knowledgeable about illegal activities and keen to participate in conservation efforts. For example, during household surveys, villagers from unsampled households sometimes expressed their desire to share their views and experiences about natural resources. Accordingly, researchers can facilitate liaison between reserve managers and local people.

Although Paulo's research exposes the situation on the ground, we are not yet able to connect these observations to a good understanding of the drivers of illegal behaviour or the governance context framing reserve management. To fulfil such an objective, more targeted research is required. Ideally, this should focus on individual activities, rather than trying to investigate all illegal activities at once. Different activities are conducted by different groups of people with different rationales and link to different governance issues.

The methods applied in Ugalla were resource-intensive. For example, Unmatched Count techniques typically require high sample sizes (see Nuno et al., 2013); more than 500 households were surveyed in this study, which was all that time and funding allowed. Surveying for illegal activity signs was also challenging, because it was difficult to estimate the time the signs had been present in the environment and different signs have different biases (e.g. rangers remove snares during their normal anti-poaching patrols, potentially leading to underestimates).

The main lessons learnt from Ugalla were as follows.

- Be interdisciplinary! Don't be afraid to use ecological survey methods, for example incorporating a field-based survey into the research design. This can provide a great opportunity to cross-validate findings from social research.
- Conservation researchers preferring mixed methods should not be over-ambitious. Instead, they should be realistic, choosing techniques carefully and planning activities based on the resources available, following a robust pilot study.
- While doing household surveys and focus groups, it is critical to use experienced research assistants who are neutral in the community but familiar with the study area. A survey of illegal activity signs also requires experienced field assistants, so information is collected accurately and consistently.
- Both focus group discussions and household surveys should be kept relatively short and simple to minimise participant fatigue.

## 5.4 Developing a Theory of Change for an intervention

It is vitally important to be clear about why we think that our intervention is the right thing to do, and what barriers there might be to success, before we start. This understanding needs to be set out in a logical way, so that it is understandable and appealing to project staff and donors, and so that it can later be tested. There are a number of approaches which can be used, falling under a general heading of causal chain models (Qiu et al., 2018). One such approach is Theory of Change (Center for Theory of Change, 2018), which shows how a project can reach its desired impact and goals through different pathways of change. It provides indicators that can be tested, thereby supporting evaluation of a project's success or failure. This is useful both for internal and external users, to understand what works, and to guide the allocation of project resources.

### 5.4.1 Case study: Vijana na Mazingira

In 2016, Hans Cosmas Ngoteya designed a retrospective Theory of Change for Vijana na Mazingira (VIMA), the local conservation project which he runs in the Katavi–Rukwa ecosystem of western Tanzania. The project targets youths aged 12–35, with the goal of reducing pressure on natural resources from poaching, deforestation and encroachment. The Theory of Change was designed to support an evaluation of the effects of the project on attitudes, awareness and conservation behaviours by youths aged 18–35 participating in VIMA's conservation education and alternative livelihood projects (Figure 5.4).

In order to achieve a project's desired impact, it is necessary first to understand the motivations for engaging in the behaviour that the project is aiming to modify. There are a number of frameworks available from social psychology that represent the factors that interact to influence behaviours. One of the most widely used in conservation is the Theory of Planned Behaviour (St John et al., 2013). Hans used the Theory of Planned Behaviour to identify the different factors underlying the motivations of the VIMA project's recipients (Figure 5.5). Based on Hans' local knowledge and understanding of the project, a Theory of Planned Behaviour framework was developed for four desired project impacts, each of which represented a desired behavioural change. The Theory of Planned Behaviour was then used to identify how VIMA's activities might tackle the different motivations underlying each behaviour.

A clear understanding of the motivations behind the behaviour, engendered by the Theory of Planned Behaviour exercise, can enable conservationists to map out the pathways of change the project should focus on, thereby generating a Theory of Change. The Theory of Planned Behaviour gives

**Figure 5.4** Hans Cosmas Ngoteya (second from right) setting up a beehive with local youths, as an alternative livelihood project. (A black and white version of this figure will appear in some formats. For the colour version, please refer to the plate section.)

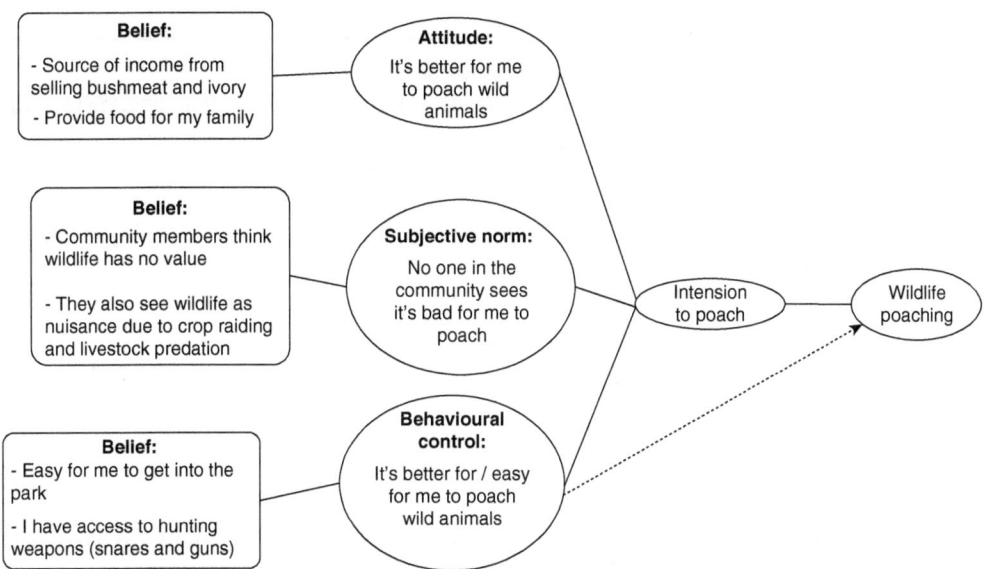

**Figure 5.5** A Theory of Planned Behaviour diagram illustrating the factors underlying the poaching behaviours of individuals targeted by the VIMA project.

a representation of what underlies an individual's behaviour, and this can be used to develop a Theory of Change for the planned intervention. In this case, the Theory of Planned Behaviour exercise highlighted that, typically, youths in Katavi–Rukwa viewed poaching as a way to feed their families and generate an income through bushmeat or ivory sales. Therefore, an intervention that developed alternative livelihood programmes could be an effective approach. This could include training youths in new income-generating activities (input), thereby providing alternative income sources (output), which will reduce their dependence on natural resources (outcome) and ultimately reduce their poaching behaviour (impact; Figure 5.6). At each step of this pathway lie assumptions; for example, that any alternative income source will replace, rather than supplement, income from hunting (Table 5.1).

Baseline surveys, focused on the elements of the Theory of Planned Behaviour (attitudes, knowledge, social norms), provide a set of indicators against which change engendered by the intervention can be measured. Progress through the Theory of Change can also be monitored, using a set of more process-based indicators. For example, an input indicator might be the percentage of VIMA's target audience engaged in the alternative income activities, an output indicator might be the income generated from the alternative livelihood, the outcome might be measured as improvements in household livelihood security and the impact might be measured using an indirect questioning technique such as the Unmatched Count Technique to quantify change in poaching prevalence.

**Table 5.1** *Assumptions underlying the Theory of Change*

| | |
|---|---|
| 1. | Participants understand the education they are given |
| 2. | If someone is educated about environmental issues it will improve their attitude towards conservation |
| 3. | Knowledge about conservation issues leads to a decrease in acceptance of environmentally harmful behaviours |
| 4. | There is dissemination of information from VIMA participants to the remainder of the community |
| 5. | If someone's attitude towards conservation improves, they will reduce their unsustainable resource-use behaviour |
| 6. | If communities are against unsustainable resource use, illegal resource exploitation will decrease |
| 7. | VIMA's alternative livelihood programmes can be put into practice and generate income |
| 8. | There is opportunity for the rest of the community to become involved in the alternative livelihood projects |
| 9. | Alternative livelihoods will be used to reduce unsustainable use of natural resources |
| 10. | Decreasing dependency on natural resources will reduce poaching, encroachment and deforestation |

### 5.4.2 Lessons learnt

The requirement for robust evaluations of the effectiveness of conservation interventions is becoming more and more apparent (Sutherland et al., 2011). Practitioners are required to ensure that their activities are based upon the best available evidence and designed for accountability and learning. However, many small NGOs (such as Hans' organisation, Landscape and Conservation Mentors' Organisation) may not feel that they have the capacity to design and implement evaluations that are both user-friendly and robust enough to be useful for adaptive management. Lacking a rigorous framework for articulating goals and assumptions, it is easy to drift through interventions without having either a strategic plan or a means of measuring success. This can lead to ineffective interventions and failure to capture the changes engendered in order to learn and adapt and demonstrate impact to funders. The development of a Theory of Change for the VIMA project enabled Hans to identify his assumptions, and develop methods to collect information which can be used to monitor future impact and test assumptions against a relevant baseline.

### 5.5 Exploring alternatives to illegal behaviour

One of the key lessons learnt in the VIMA project was the importance of having a clear understanding of the motivations behind behaviour. Unfortunately, not all conservation projects that involve communities take this approach when designing an intervention. For example, alternative

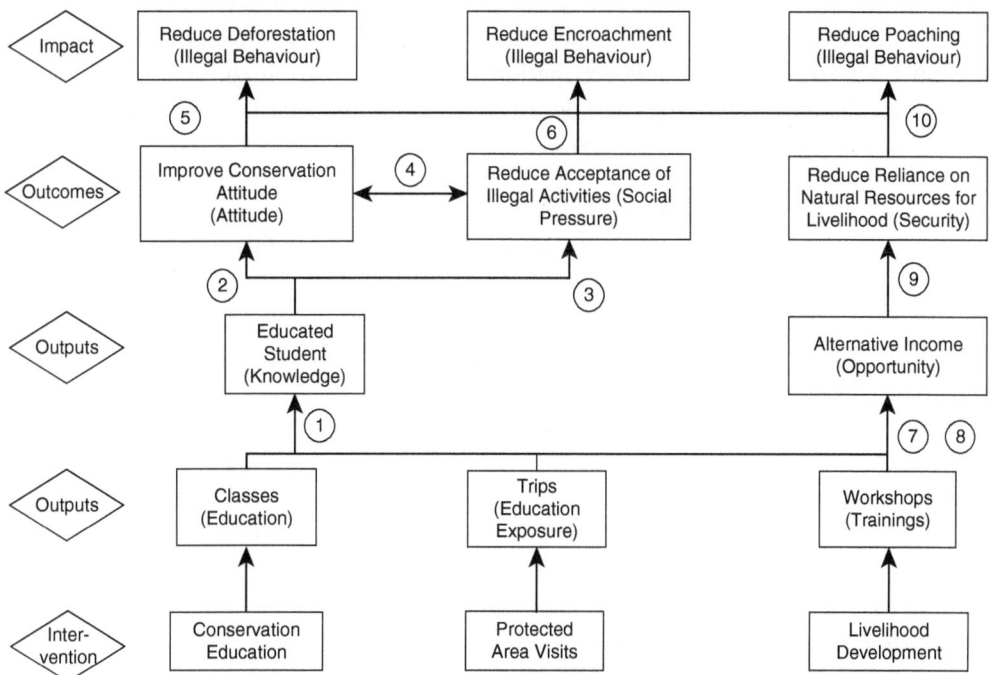

**Figure 5.6** Theory of Change for VIMA project showing interventions at the bottom and different pathways to reach the desired impacts. Numbers 1–10 are assumptions along the pathways of change (listed in Table 5.1).

livelihood projects have long been used as strategy for reducing local threats to species, habitats or resources of conservation concern. Alternative livelihood projects are designed to reduce the prevalence of behaviours that are considered environmentally damaging and unsustainable (Wright et al., 2016). However, a systematic review of alternative livelihood projects conducted by Roe et al. (2015) found insufficient evidence to understand when, where or why alternative livelihood projects work. Even though there is uncertainty regarding the effectiveness of alternative livelihood projects, they continue to be a key strategy in both terrestrial and marine conservation. However, the assumptions on which they are based are often unrealistic; for example, that the alternative livelihood projects will substitute for the undesirable behaviour, that the resource users are a homogeneous group, and that targeting interventions at individuals will scale up to population-level change in pressure on resources (Wright et al., 2016).

In marine conservation, a common response to perceived over-fishing is to provide alternative employment for existing fishers. This requires that the assumption of substitutability holds, so that fishers will willingly and happily settle into a new way of making a living (Pollnac et al., 2001; Pollnac & Poggie, 2008). Pollnac et al. (2001) added that this is based on the assumption that fishing is a hard and undesirable occupation and hence an employment of last resort, that fishers are among the poorest of the poor and that the poor care little about the type of job they have as long as they make enough to live.

### 5.5.1  Case study: shark fishers in Tanjung Luar

Fishing pressure is generally considered to be the main cause of the decline of shark populations globally (Stevens et al., 2000; Robbins et al., 2006; Dharmadi et al., 2015). Indonesia is the world's largest shark producer, with annual average production of 106,000 tons in 2000–2011, contributing 13% of global shark production (Dent & Clarke, 2015). Although the exact number is unknown, it is assumed that many Indonesian fishers are heavily dependent on shark fisheries as a source of income and food. However, shark production in Indonesia has been declining in recent years (Sub Directorate of Capture Fisheries Data and Statistics, 2016), which could be leading to a decline in income and livelihood security for fishers.

From 2014 to the present, the Wildlife Conservation Society (WCS) Indonesia Programme has carried out a study of shark fishers in Tanjung Luar, a shark-fishing community in East Lombok, in order to understand whether providing alternative livelihoods could help to reduce fishing pressure on sharks. Tanjung Luar is one of the main shark landing sites in Indonesia. It is home to a targeted shark fishery comprising approximately 50 boats employing surface and bot-tom longlines and one of the biggest fish markets on Lombok Island, with more than 5000 fishers using it to sell their catch. Fishing is the main livelihood of Tanjung Luar's population and there are at least 150 households heavily depen-dent on the shark industry, either as fishers, meat processors or traders. Shark fishers in Tanjung Luar use 4–25 gross tonnage boats, with three or four crew members, and the average fishing trip is 14 days.

Due to growing international concern regarding their conservation status, several shark and ray species have been listed on CITES Appendix II. As a CITES member, Indonesia is required to implement management mea-sures, such as quotas, size limits and export bans to ensure that international trade in these species is not detrimental to wild populations. These measures could have negative impacts on the income and livelihood security of Tanjung Luar's fishers, who are already vulnerable to market fluctuations, particularly in export markets (Jaiteh et al., 2017). WCS Indonesia Programme's study aimed to: (1) collect data on biological and operational characteristics of the fishery (Figure 5.7), (2) understand shark fishers' cur-rent socioeconomic status and aspirations, (3) explore alternative livelihood options and (4) create dialogue between fishers and the management authorities.

**Figure 5.7** WCS Indonesia team members measuring guitarfish at Tanjung Luar port. Photo provided by WCS-Indonesia. (A black and white version of this figure will appear in some formats. For the colour version, please refer to the plate section.)

Livelihoods options explored with the fishers included diversifying the target catch to more resilient species (e.g. squid, tuna and reef fish) and tourism, yet WCS Indonesia Programme's surveys showed that shark fishing offered higher revenues than other fisheries. An independent fisheries assessment by Masyarakat dan Perikanan Indonesia also showed similar results (MDPI, 2017). Tanjung Luar was known for its squid fishery in the 1980s, but the number of squid fishermen has increased rapidly, increasing competition and making the addition of new fishers unsustainable (MDPI, 2017). Some fishers in Tanjung Luar who catch tuna or skipjack mentioned that their catch is also declining, and The Indian Ocean Tuna Commission classifies yellowfin tuna as overfished (IOTC, 2017). Some shark fishers have already started to fish for groupers and snappers on the side, but the value of this catch is far less than their earnings from sharks. Tourism is promising, but the industry is still under-developed. To date, identifying feasible alternatives that provide economic incentives to shift away from shark fishing has proven challenging, as there are no legal or sustainable marine alternatives that offer similar profits.

Our research showed that shark fishers wish to remain shark fishers. Fishing is the only skill they know, and most of them said that they would continue to fish as usual even if their catch declined by 50%. Our landings survey showed that some commonly caught sharks are over-exploited. When findings were shared with fishers, although not all agreed with the results, shark fishers acknowledged that it is now harder to catch sharks and the sharks that are caught are smaller, a view also shared by shark fishers in eastern Indonesia (Jaiteh et al., 2017). The Tanjung Luar fishers' response is not surprising, as similar reactions were also reported by Pollnac and Poggie (2006), with fishermen refusing to leave their existing fishery even though their incomes were declining; it is potentially their best option in the short run if they are still making a profit.

### 5.5.2 Lessons learnt

Based on the results of this research, instead of deploying alternative livelihood projects for shark fishers in Tanjung Luar, WCS's Indonesia Programme chose to:

(1) strengthen the existing fisher institutions, which focus on tourism development, in order to help that industry to develop, become more attractive and profitable;
(2) maintain close interaction with shark fishers by regular home visits and conducting informal meetings; and
(3) facilitate formal meetings between shark fishers and the management authorities, to foster dialogue on developing management measures that ensure the sustainability of both shark and ray populations and fishers' livelihoods.

It is challenging to establish a direct connection between livelihood interventions and conservation. Rather than trying to find new livelihoods, sometimes it is more appropriate to focus on enhancing existing livelihood strategies which do not involve exploiting the natural resource of concern, targeting those most vulnerable to conservation-imposed resource access restrictions (Wright et al., 2016). It may also be possible to establish a clearer link between livelihood sustainability and conservation as a means of building good community relations, as we opted to do. It is important to have a clear pathway demonstrating how an intervention is expected to lead to the desired outcome, e.g. by using theory of change to design the intervention after gaining a thorough understanding of community dynamics.

### 5.6  Discussion: interlacing research and practice

The four case studies presented here take us from research to practice; in so doing, they illustrate how integrated the two are. By starting with a strong theoretical framework (such as the Theory of Planned Behaviour) underpinning an intervention's Theory of Change, unwarranted assumptions can be avoided, such as those which plague alternative livelihoods projects. Engaging with resource users before embarking on interventions can reveal dead ends, as illustrated in Tanjung Luar, where plans for an alternative livelihood project needed to be replaced by a more indirect process of advocacy and engagement with different parties, while building capacity for a livelihoods shift. A clear understanding of what the actual problem is, based on evidence rather than supposition, is vital; the example from Cambodia suggested that hunting was actually not a major threat to floricans, enabling conservation practitioners to focus on other threats.

Although a range of techniques is available for collecting information to underpin management, these should not be applied lightly. As the Ugalla example showed, the ideal of using mixed methods to gain a nuanced understanding takes time and resources, as well as expertise. Approaches such as the Unmatched Count Technique can look superficially appealing and easy to administer, but there are technical challenges in developing appropriate item lists, administering the questions in a way that makes respondents comfortable, and in data analysis. Even then, as the Cambodian example shows, the results may not be as informative as might be hoped. Time invested in foundational studies is well spent, but not all small NGOs can afford extensive research. Even then, however, it is possible to develop a robust Theory of Change, as a tool for exposing assumptions and supporting ongoing monitoring and evaluation, as the VIMA example showed.

Our case studies have specific lessons, but they also tell universal stories. The role of research in facilitating positive interactions between managers and local people is an interesting observation that was seen in both Ugalla and Tanjung Luar, while both the Cambodian and Ugalla case studies highlighted the importance of good local research assistants. All four case studies emphasised how research and practice need to intertwine more often and more routinely. This will enable conservationists (whether from governments or NGOs) to think through their interventions in advance, use appropriate methods to understand existing behaviour and local perspectives on ways forward, and thereby design locally appropriate, participatory interventions that support adaptive management.

## References

Birdlife International. 2015. *Houbaropsis bengalensis*. The IUCN Red List of Threatened Species 2015: e. T22692015A82252874. http://dx.doi.org /10.2305/IUCN.UK.2015-4 .RLTS.T22692015A82252874.en (accessed 19 April 2016).

Brockington, D. & Scholfield, K. 2010. Expenditure by conservation nongovernmental organizations in sub-Saharan Africa. *Conservation Letters*, 3, 106–113.

Center for Theory of Change. 2018. Theory of Change. Available from www .theoryofhcange.org.

Dent, F. & Clarke, S. 2015. State of the global market for shark products. FAO Fisheries and Aquaculture Technical Paper. Rome: FAO.

Dharmadi, Fahmi & Satria, F. 2015. Fisheries management and conservation of sharks in Indonesia. *African Journal of Marine Science*, 37, 249–258.

Fairbrass, A., Nuno, A., Bunnefeld, N., et al. 2016. Investigating determinants of compliance with wildlife protection laws: bird persecution in Portugal. *European Journal of Wildlife Research*, 62, 93–101.

Gandiwa, E. 2011. Preliminary assessment of illegal hunting by communities adjacent to the northern Gonarezhou National Park, Zimbabwe. *Tropical Conservation Science*, 4, 445–467.

Gavin, M. C., Solomon, J. N. & Blank, S. G. 2010. Measuring and monitoring illegal use of natural resources. *Conservation Biology*, 24, 89–100.

Gore, M. L. 2011. The science of conservation crime. *Conservation Biology*, 25, 659–661.

Harrison, M., Baker, J., Twinamatsiko, M., et al. 2015. Profiling unauthorized natural resource users for better targeting of conservation interventions. *Conservation Biology*, 29, 1636–1646.

Ibbett, H., Lay, C., Phlai, P., et al. 2019. Conserving a globally threatened species in a semi-natural, agrarian landscape. *Oryx*, 53, 181–191.

Imai, K. 2011. Multivariate regression analysis for the item count technique. *Journal of the American Statistical Association*, 106, 407–416.

IOTC. 2017. 2017 Status summary for species of tuna and tuna like species under the IOTC mandate. Available from www.iotc.org /sites/default/files/ Summary_of_Stock_Status.pdf

Jaiteh, V. F., Loneragan, N. R. & Warren, C. 2017. The end of shark finning? Impacts of declining catches and fin demand on coastal community livelihoods. *Marine Policy*, 82, 224–233.

Kahler, J. S. & Gore, M. L. 2012. Beyond the cooking pot and pocket book: factors influencing noncompliance with wildlife poaching rules. *International Journal of Comparative and Applied Criminal Justice*, 36, 103–120.

Kalumanga, E. 2015. How elephants utilize a miombo-wetland ecosystem in Ugalla landscape, Western Tanzania. Doctoral dissertation, Department of Physical Geography, Stockholm University.

Kiffner, C., Peters, L., Stroming, A., et al. 2015. Bushmeat consumption in the Tarangire–Manyara ecosystem, Tanzania. *Tropical Conservation Science*, 8, 318–332.

Larrosa, C., Carrasco, R. L. & Milner-Gulland, E. J. 2016. Unintended feedbacks: challenges and opportunities for improving conservation effectiveness. *Conservation Letters*, 9, 316–326.

Martin, A., Caro, T. & Mulder, M. B. 2012. Bushmeat consumption in western Tanzania: a comparative analysis from the same ecosystem. *Tropical Conservation Science*, 5, 352–364.

MDPI. 2017. Tanjung Luar Fishery Assessment – internal report for Wildlife Conservation Society Indonesia Program. Denpasar, Indonesia.

Mgawe, P., Mulder, M. B., Caro, T., et al. 2012. Factors affecting bushmeat consumption in the Katavi-Rukwa ecosystem of Tanzania. *Tropical Conservation Science*, 5, 446–462.

Milner-Gulland, E. J. 2012. Interactions between human behaviour and ecological systems. *Philosophical Transactions of the Royal Society Biological Sciences*, 367, 270–278.

Nuno, A. N. A., Bunnefeld, N., Naiman, L. C., et al. 2013. A novel approach to assessing the prevalence and drivers of illegal bushmeat hunting in the Serengeti. *Conservation Biology*, 27, 1355–1365.

Nuno, A. & St John, F. A. V. 2015. How to ask sensitive questions in conservation: a review of specialized questioning techniques. *Biological Conservation*, 189, 5–15.

Nuno, A., Blumenthal, J. M., Austin, T. J., et al. 2018. Understanding implications of consumer behavior for wildlife farming and sustainable wildlife trade. *Conservation Biology*, 32, 390–400.

Packman, C. E. 2011. Seasonal landscape use and conservation of a critically endangered bustard: Bengal florican in Cambodia. Doctorate of Philosphy thesis, University of East Anglia.

Packman, C. E., Showler, D. A., Collar, N. J., et al. 2014. Rapid decline of the largest remaining population of Bengal Florican Houbaropsis bengalensis and recommendations for its conservation. *Bird Conservation International*, 24, 429–437.

Papworth, S., Milner-Gulland, E. J. & Slocombe, K. 2013. The natural place to begin: the ethnoprimatology of the Waorani. *American Journal of Primatology*, 75, 1117–1128.

Pollnac, R. B., Pomeroy, R. S. & Harkes, I. H. T. 2001. Fishery policy and job satisfaction in three southeast asian fisheries. *Ocean and Coastal Management*, 44, 531–544.

Pollnac, R. B. & Poggie, J. J. 2006. Job satisfaction in the fishery in two Southeast Alaskan towns. *Human Organization*, 65, 329–339.

Pollnac, R. B. & Poggie, J. J. 2008. Happiness, well-being and psychocultural adaptation to the stresses associated with marine fishing. *Human Ecology Review*, 15, 194–200.

Qiu, J., Game, E. T., Tallis, H., et al. 2018. Evidence-based causal chains for linking health, development, and conservation actions. *BioScience*, 68, 182–192.

Riggio, J. & Caro, T. 2017. Structural connectivity at a national scale: wildlife corridors in Tanzania. *PLoS ONE*, 12, 1–16.

Robbins, W. D., Hisano, M., Connolly, S. R., et al. 2006. Ongoing collapse of coral-reef shark populations. *Current Biology*, 16, 2314–2319.

Roe, D., Booker, F., Day, M., et al. 2015. Are alternative livelihood projects effective at reducing local threats to specified elements of biodiversity and/or improving or maintaining the conservation status of those elements? *Environmental Evidence*, 4, 1–8.

St John, F. A. V., Edwards-Jones, G., Gibbons, J. M., et al. 2010. Testing novel methods for assessing rule breaking in conservation. *Biological Conservation*, 143, 1025–1030.

St John, F. A. V., Keane, A. M. & Milner-Gulland, E. J. 2013. Effective conservation depends upon understanding human behaviour. *Key Topics in Conservation Biology*, 2, 344–361.

St John, F.A. V., Mai, C. & Pei, K. J. C. 2015. Evaluating deterrents of illegal behaviour in conservation: carnivore killing in rural

Taiwan. *Biological Conservation*, 189, 86–94. Elsevier Ltd.

Stevens, J. D., Bonfil, R., Dulvy, N. K., et al. 2000. The effects of fishing on sharks, rays, and chimaeras (chondrichthyans), and the implications for marine ecosystems. *ICES Journal of Marine Science*, 57, 476–494.

Sub Directorate of Capture Fisheries Data and Statistics. 2016. *Capture Fisheries Statistics of Indonesia by Province, 2015.* Jakarta: Directorate General of Capture Fisheries.

Sutherland, W. J., Bardsley, S., Bennun, L., et al. 2011. Horizon scan of global conservation issues for 2011. *Trends in Ecology and Evolution*, 26, 10–16.

Thomas, A. S., Gavin, M. C. & Milfont, T. L. 2015. Estimating non-compliance among recreational fishers: insights into factors affecting the usefulness of the randomized response and item count techniques. *Biological Conservation*, 189, 24–32.

UGR. 2006. A checklist of plants, animals and birds in Ugalla Game Reserve. Unpublished report, Ugalla Game Reserve Project, Tabora, Tanzania.

Wilfred, P. & MacColl, A. D. C. 2015. Local perspectives on factors influencing the extent of wildlife poaching for bushmeat in a game reserve, Western Tanzania. *International Journal of Conservation Science*, 6, 99–110.

Wilfred, P., Milner-Gulland, E. J. & Travers, H. 2017. Attitudes to illegal behaviour and conservation in western Tanzania. *Oryx*, 1–10.

Woodhouse, E., de Lange, E. & Milner-Gulland, E. J. 2016. *Evaluating the Impacts of Conservation Interventions on Human Well-being. Guidance for Practitioners.* London: IIED.

Wright, J., Hill, Ni., Roe, D., et al. 2016. Reframing the concept of 'alternative livelihoods'. *Conservation Biology*, 30, 7–13.

Young, J. C., Rose, D. C., Mumby, H. S., et al. 2018. A methodological guide to using and reporting on interviews in conservation science research. *Methods in Ecology and Evolution*, 9, 10–19.

# CHAPTER SIX

# Mobilisation of indigenous and local knowledge as a source of useable evidence for conservation partnerships

PERNILLA MALMER, VANESSA MASTERSON
*Swedbio, Stockholm Resilience Centre*

BEAU AUSTIN
*Charles Darwin University*

and

MARIA TENGÖ
*Swedbio, Stockholm Resilience Centre*

## 6.1 Introduction

Rapid and interlinked changes in the biosphere, including degradation of the biodiversity and ecosystems that underpin human well-being, are reported with increasing regularity. As such, there is an urgent need for conservation initiatives that are capable of countering the speed and veracity of change, while meeting the needs of human societies on a crowded planet. While significant advancements in scientific knowledge in the fields of sustainability and conservation continue to be achieved, the forecasted rate of rapid ecological and social change requires the production of innovative mechanisms for management and policy.

   One way of contributing to new solutions in a timely manner is to more effectively mobilise multiple knowledges, values and governance systems that can complement Western approaches to science. Together these can extend the collective knowledge base and contribute to collaboratively designing ways forward for looking after people and the biosphere. Compared with Western-based approaches, indigenous and local knowledge systems represent alternative ways of learning from and with the environment, through close and continuous observation framed by distinct worldviews with particular strengths and limitations (like all knowledge systems). Knowledge is embodied by the actors and in their practices, tools, and technologies, as well as in the institutions that organise the production, transfer and use of knowledge (Cornell et al., 2013). There has recently been more attention focused on the urgent need for science and policy to recognise and mobilise the knowledge of indigenous people and local communities who steward substantial biodiversity across the globe (Brondizio & Le Tourneau, 2016; Mistry & Berardi, 2016). Collaborative ways for mobilising knowledge and learning across diverse knowledge systems can contribute complementary knowledge, innovations and new solutions. Involvement of multiple actors and knowledges can strengthen usefulness and legitimacy in decision-making and implementation (Sterling et al., 2017a; Gavin et al., 2018).

In this chapter, we draw attention to the potential for mobilising local and indigenous knowledge systems, institutions and actors in ways that allow meaningful use of their knowledge about landscapes and their functions as evidence for conservation. By doing this, we propose that innovative and collaborative mechanisms can be designed and implemented that will create opportunities for long-term sustainable governance and conservation of biodiversity.

We introduce the Multiple Evidence Base (MEB) approach to guide the design and implementation of conservation partnerships that enable engagements with indigenous and local knowledge as evidence as an entry point to promote sustainable governance of interrelated ecosystems and human well-being (Tengö et al., 2014, 2017). The approach was developed to guide inclusive processes for collaborations across knowledge systems, based on equity and usefulness for all actors involved. It emphasises that indigenous, local and scientific knowledge systems are complementary, equally valid and useful for informing sustainable governance of biodiversity and ecosystems. The MEB focuses on the theoretical and practical potential for collaborative knowledge-weaving processes to mobilise indigenous and local actors, institutions and practices to achieve long-term conservation and sustainability targets. We argue that collaborative approaches to conservation must be equitable and fair to be effective in the long term (Brondizio & Le Tourneau, 2016; Sterling et al., 2017a; Gavin et al., 2018).

The utility and value of the MEB approach will be discussed in light of its aim to support more informed and efficient local, national and international policy processes and governance decisions for the integrated benefits of conservation, sustainable use and human well-being. We describe the current and potential role that a MEB approach may have in enhancing the efficacy of conservation science and policy by clarifying and strengthening synergies with indigenous knowledges and practices. To achieve this, we first review the peer-reviewed and grey literature to reflect on the extent of uptake of the MEB and how it has been applied in both science and policy-practice processes. Second, to illustrate the approach and reflect on successes and practical challenges, we take a deeper look at three case studies of piloting a MEB approach. The cases demonstrate the potential for the MEB approach to be used as both a framing tool for collaborative partnerships and a practical guide to weaving multiple knowledge systems. Lastly, we discuss ways forward to nurture conservation and mobilise partnerships that build on knowledge collaborations. We find that a MEB approach has potential to support the inclusion of a wider range of evidence in conservation practice, strengthen active participation of local actors and improve conservation partnerships through the recognition and revitalisation of local knowledge systems and governance.

## 6.2 The need for new approaches to collaborative conservation

There is a long history of attempts to reconcile conservation objectives with local livelihoods in integrated development and conservation processes, which have often been framed as 'win–win' opportunities with social–ecological benefits (Adams et al., 2004). In the conservation literature, the importance of involving local people is well established, with mounting evidence that processes that meaningfully engage local people are more likely to succeed in protecting biodiversity (Waylen et al., 2010; Sterling et al., 2017a) and that failure to do so can lead to lack of trust and commitment, project failure, and in the worst case, lingering conflicts (Oldekop et al., 2015). While many indigenous peoples and local communities continue to be evicted from their ancestral lands and experience colonisation in the name of conservation, there is now a move towards recognising their connections to land and endogenous obligations to care for it as synergetic with biodiversity conservation outcomes (Knox, 2017). This provides a foundation for enabling local people and conservation organisations to be strategic allies. Furthermore, there is increasing evidence that involving local actors in monitoring enhances management responses at local spatial scales, and increases the speed of decision-making to tackle environmental challenges at operational levels of resource management (Danielsen et al., 2010; Sterling et al., 2017a).

Despite these generally acknowledged realities about the usefulness of engaging with indigenous peoples and local communities, they are often included as stakeholders in conservation, without recognition of their knowledge and expertise (Danielsen et al., 2010). In the literature much attention is given to the uniqueness and utility of indigenous and local knowledge systems, which is often holistic, providing an understanding of integrated social–ecological systems, biocultural values and belief systems (Sheil et al., 2015; Sterling et al., 2017a). However, in practice, there often exists scepticism about the contemporary existence and/or effectiveness of indigenous and local knowledge as useful evidence in conservation. Similarly, holders of indigenous and local knowledge can be sceptical of the claims generated through western scientific approaches due both to the unfamiliarity of the epistemic practices employed and recent or ongoing experiences of colonisation and disempowerment

(Nadasdy, 1999; Johnson et al., 2015; Kealiikanakaoleohaililani & Giardina, 2016; Mistry & Berardi, 2016).

## 6.3 The multiple evidence base approach: connecting knowledge systems for the benefit of conservation and human well-being

The need to engage with diverse sources of knowledge for conservation has been recognised in high-level science–policy processes, such as the Convention on Biological Diversity, and the Intergovernmental Science Policy Platform on Biodiversity and Ecosystem Services (IPBES). From the outset, IPBES had the ambition to *recognise and respect the contribution of indigenous and local knowledge to the conservation and sustainable use of biodiversity and ecosystems* (Díaz et al., 2015). This was used as a window of opportunity to start an open dialogue to explore current divides between indigenous, local and scientific knowledge systems, and to elicit methods for collaborations based on equity, reciprocity and useful- ness for all involved (see Tengö et al., 2014). A science–policy–practice dialogue process brought together knowledge-holders and experts from diverse knowl- edge systems, convened by SwedBio at Stockholm Resilience Centre in colla- boration with key partners representing indigenous peoples and local communities, such as the International Indigenous Forum on Biodiversity and the African Biodiversity Network. The active engagement from these networks, representing a diversity of knowledge systems and linking practices on the ground with global policy and science, created legitimacy and recognition of outcomes from the dialogues. The starting point was the pivotal dialogue meet- ing prior to the establishment of IPBES in the indigenous territory of Guna Yala, Panama, where essential principles for exchange across knowledge systems were identified: *trust, respect, reciprocity, equity, transparency* and *free prior and informed consent* (Tengö & Malmer, 2012). Since then, the MEB approach has developed in parallel to the IPBES, while carefully paying attention to other interests and needs of the partners.

The MEB can be understood as a deep approach to collaborative knowledge- sharing that explicitly acknowledges that challenges are fundamentally due to different perspectives and practices concerning human–nature relationships, approaches to knowledge validation, knowledge governance and who quali- fies as an 'expert'. Also, it recognises that scientists have tended to dominate the design and implementation of collaborations across knowledge systems both historically and contemporarily (Nadasdy, 1999; Mistry & Berardi, 2016). Another key component of an MEB approach is its emphasis on the need for mobilisation and validation of knowledge within knowledge systems them- selves. That is, if scientific methods that often are specific and partial are applied to local knowledge that is practical, multidimensional and holistic, there is a risk of omission, misinterpretation and rejection of critical and useable knowledge.

The MEB approach views different knowledge systems as complementary and emphasises that joint analysis assists in working both with convergence and divergence (e.g. Molnár et al., 2016a; Hohenthal et al., 2018). For example, Molnár et al. (2016a) highlight that when discussing approaches to conservation in the Hungarian steppe, local herders focus on primarily utilitarian purposes, such as how they can manage the behaviour of their grazing animals in order to promote the health and diversity of grass assemblages for production. In comparison, conservationists working in the same landscapes focus almost solely on the protection of the plants themselves, with little regard to the impact on grazing animals. If this difference is ignored, or framed as a problem, it has the potential to create tension when attempting to collaboratively design and implement conservation initiatives in the region. Conversely, these different perspectives can be worked together to provide an enriched picture of exactly what is necessary for maintaining and enhancing biodiversity and social–ecological system function in the steppe.

In order to build evidence – whether new knowledge or existing – that is legitimate and useful for all actors in such collaborations for conservation, there is a need to engage with local knowledge systems and knowledge-holders from the outset, co-defining a common problem and facilitating equitable engagement through all activities, including mobilising and assessing knowledge. This process is outlined in the three phases of the MEB approach (Figure 6.1a). Collaboratively analysing and interpreting the complementary evidence from diverse sources is a way to triangulate information, strengthen legitimacy and relevance of existing knowledge and build a base for further learning.

As guidance for how to implement an MEB approach, five tasks were identified as critical (Figure 6.1b; Tengö et al., 2017). First, to *mobilise* knowledge – to ensure that the knowledge is articulated, validated internally and free to be shared with others. Second, to *translate* knowledge, reciprocally, so that all actors can comprehend each others' knowledge and where it is derived from. Third, to *negotiate*, to jointly address convergence and divergence between knowledge systems, and the extent to which the latter can be resolved, for example by understanding differences in underlying assumptions and values (Gagnon & Berteaux, 2009; Molnár et al., 2016a). Fourth, to *synthesise*. Here we emphasise synthesis based on a joint process that does not require that all knowledge is validated by one knowledge system (e.g. empirical validation by science). Lastly, to *apply* – and this is where we iterate the need to recognise the different needs and interests by different actors. Knowledge collaborations need to be designed in a way that is is perceived as useful and leads to constructive outcomes for all involved. The bridging of knowledge systems therefore requires the creation of settings for exchange of multiple

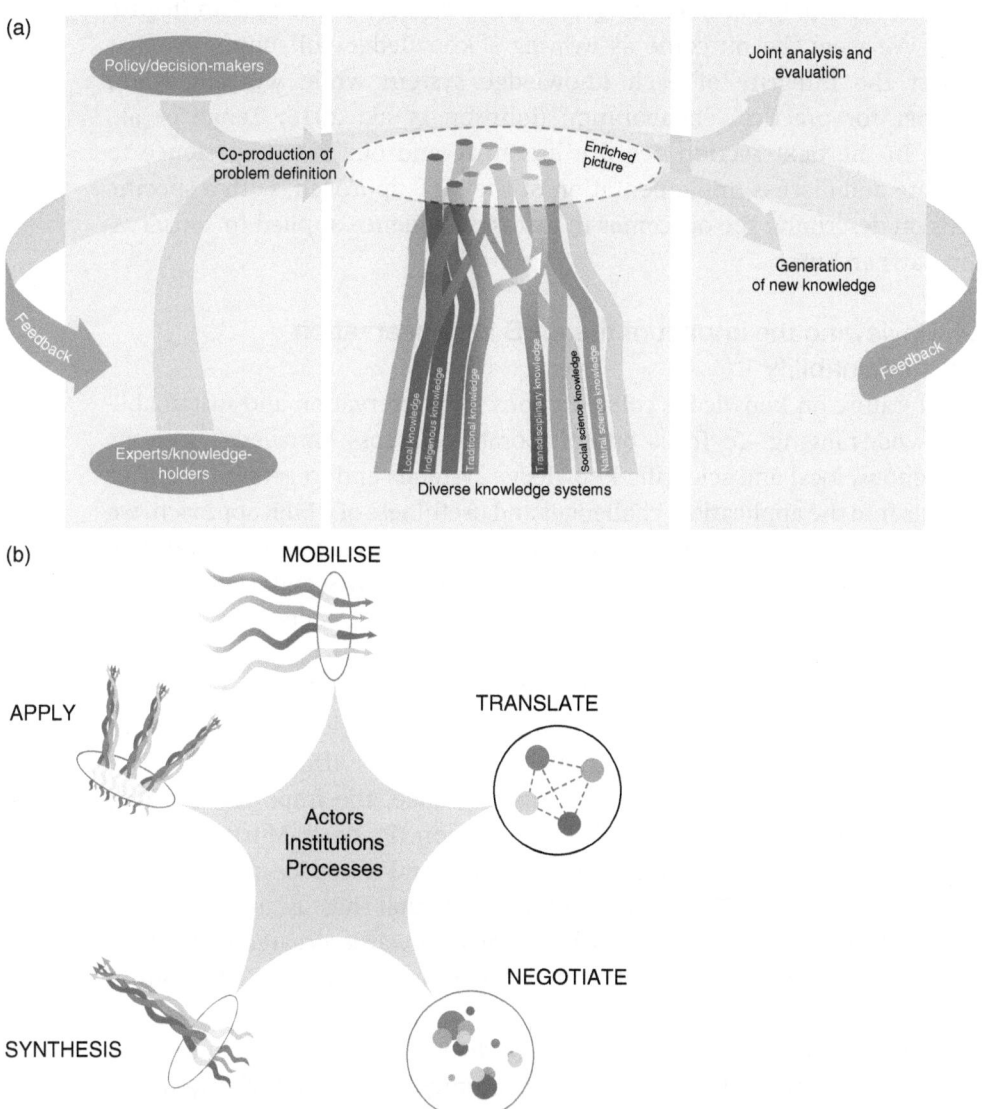

**Figure 6.1** The Multiple Evidence Base approach in action. (a) The three phases of a MEB approach: joint problem formulation, generating an enriched picture with contribution from multiple sources of evidence and joint analysis and evaluation of knowledge (Tengö et al., 2014). (b) Actors, institutions and processes are at the core of the five tasks required for successful collaboration across diverse knowledge systems. The different colours of the lines and dots in parts (a) and (b) represent different knowledge systems, or streams of knowledge within knowledge systems (Tengö et al., 2017). (A black and white version of this figure will appear in some formats. For the colour version, please refer to the plate section.)

forms of knowledge and learning across key aspects of the system (Figure 6.1b). We view the outcome as *weaving* – knowledge collaborations that respect the integrity of each knowledge system while working them together for practical collaboration (Johnson et al., 2016; Tengö et al., 2017). In the next section, we use literature and our own experience to evaluate and discuss implementation of the MEB approach, with a specific focus on describing the outcomes in terms of evidence applied in conservation partnerships.

## 6.4 Reviewing the impact of the MEB in conservation and sustainability

The literature on knowledge collaborations for conservation and sustainability is wide-ranging. To focus on collaborations across knowledge systems (indigenous, local and scientific knowledge systems) and to generate further insights into the application, challenges and usefulness of a MEB approach, we reviewed articles that cite Tengö et al. (2014) or that mention 'Multiple Evidence Base' in the academic literature, represented by Scopus (123 articles), and the grey literature (219 results), represented by Google Scholar (as of 2018-02-01).

The results of this review demonstrate that the MEB approach has contributed to a general move towards broader participation of knowledge-holders in multi-level ecosystem assessments (Díaz et al., 2015; Nesshöver et al., 2016), as well as citizen science, the importance of the plurality of knowledge systems in conservation (Prado & Murrieta, 2015) and knowledge application in public policy and resource management (Bruckmeier, 2016). This is part of a 'shift that has occurred in the science–policy–society interface with a move towards greater inclusivity, and efforts to transcend traditional reductionist approaches' (Jabbour & Flachsland, 2017, p. 196).

The MEB approach is finding traction in diverse discussions including citizen science (Buytaert et al., 2014) and community-based monitoring (Johnson et al., 2015; Lyver et al., 2017), collaborative management and decision-making (Mathevet et al., 2016), community-based conservation (Nkambule et al., 2016; Sterling et al., 2017a), measuring resilience (Quinlan et al., 2016; Sterling et al., 2017b), approaches to modelling global change processes (Verburg et al., 2016), indigenous autonomy and cultural revitalisation (Gonzales, 2015), value pluralism in ecological economics (Martín-López & Montes, 2015; Kenter, 2016; Pascual et al., 2017), biocultural values and diversity (Gavin et al., 2015; Sterling et al., 2017b) and political ecology, law and environmental justice (Gambon & Rist, 2018; Hohenthal et al., 2018).

The majority of articles reviewed (51 percent) engage with the MEB approach in a relatively superficial manner to illustrate that combining multiple knowledge systems is a sustainability challenge. The literature is awash with programmatic articles with calls to include, combine and integrate knowledges to find solutions to sustainability problems (e.g. Balvanera et al., 2017; Vasseur et al., 2017). However, still very little attention is paid to exactly how this will be done. Additionally, 20 percent of articles reviewed represent collaborative processes in practice but do not apply a MEB approach. Many articles view actors as stakeholders and talk about 'open participation and open consultation' (e.g. Livoreil et al., 2016) rather than addressing their role as knowledge-holders and experts and the need for equitable platforms for engagement, mobilisation and translation of indigenous and local knowledge.

The MEB approach has also received significant attention in the grey literature and science–policy–practice community. For example, it is called for as a way of ensuring equitable participation for indigenous, local and scientific knowledge in monitoring of the Convention on Biological Diversity. For example the Convention's Aichi target 18 on traditional knowledge, innovation and practices, along with the Community-Based Monitoring and Information Systems, is a bottom-up approach developed by indigenous peoples and local communities to contribute their experiences and observations through monitoring (CBD, 2014; Farhan Ferrari et al., 2015). Further, a MEB approach has been encouraged in traditional knowledge inventories, as well as in the development of safeguards for biodiversity financial mechanisms and Reducing Emissions from Deforestation and Degradation (REDD+) under the United Nations Framework Convention on Climate Change.

To illustrate the implementation of the MEB approach in the literature, we have selected a small set of pertinent case studies. Table 6.1 presents an analysis using key features of the MEB approach – joint problem formulation, validation within knowledge system and the five tasks illustrated in Figure 6.1b.

The cases illustrate that, in different contexts, specific phases of the MEB approach presented by Tengö et al. (2014, 2017) are more or less useful, and are operationalised in different ways. The process of co-defining the problem and questions together with all knowledge-holders appears to be a challenge not taken up in all cases, often with scientists or project proponents defining a problem, and then approaching indigenous and local knowledge-holders and local communities through consultation sessions to join and support the collaboration (e.g. Strangway et al., 2016; Lyver et al., 2017; Smith et al., 2017). However, other papers do emphasise the critical role of joint problem formulation for the success of conservation interventions (Brondizio et al., 2016; Galvin et al., 2016).

Maintaining the integrity of diverse knowledge systems throughout collaborative knowledge processes also appears to be a particular challenge in science-driven processes. Actively thinking about what validation of knowledge within knowledge systems means (rather than using science to validate local knowledge) and how it may be embedded in practice is absent from most papers. There are notable exceptions that explicitly reflect upon this challenge (e.g. Austin et al., 2017) and suggest new approaches, such as peer-to-peer validation by farmers (Smith et al., 2017; Table 6.1). Other papers do not address this explicitly, but still engage with how local knowledge systems evaluate knowledge (e.g. through interactions with internally acknowledged experts and their local institutions) (Molnár et al., 2016; Nguyen et al., 2017; see Table 6.1). Additionally, joint discussion and analysis of data across knowledge systems has sometimes been incorporated through formal consultation structures or committees (e.g. Strangway et al., 2016; Austin et al., 2017; Reed & Abernethy, 2018). The articles also illustrate the progress in development of methods to facilitate the phases and activities defined in Tengö et al. (2014, 2017) to combine and relate multiple data through e.g. participatory scenario planning, focus groups (Danielsen et al., 2014), fuzzy cognitive maps and community monitoring with digital devices (Brammer et al., 2016). The use of art (Rathwell & Armitage, 2016; Polfus et al., 2017), participatory maps (Robinson et al., 2016) or film (Molnár et al., 2016) to mobilise, translate and present knowledge on an equitable platform has facilitated joint analysis and negotiation. Articles also illustrate practical ways of maintaining equity, such as creating research agreements or protocols concerning intellectual property; free, prior and informed consent; the roles and responsibilities of each member of the project team (Robinson et al., 2016); and recognising indigenous and local knowledge-holders as authors on scientific articles (Molnár et al., 2016a; Smith et al., 2017; Table 6.1).

The citations suggest that the mobilisation and translation activities suggested by Tengö et al. (2017) have had particular resonance in the conservation and sustainability literature. There has been consistent progress towards the explicit mobilisation and translation of indigenous knowledge and worldviews (Gonzales, 2015; Vogt et al., 2016; Horstkotte et al., 2017; Timoti et al., 2017). In this way, the mobilisation of multiple knowledge systems contributes to a movement towards environmental justice and pluralism in decision-making (Hohenthal et al., 2018), as well as recognising indigenous peoples' autonomous actions towards dealing with climate change (Gonzales, 2015).

In the next section, we use three in-depth case studies to further explore the value of a MEB approach to contribute to conservation partnership based on diverse sources of knowledge.

**Table 6.1** *Articles applying a multiple evidence base in literature*

| Article citation | Issue investigated including location | Multiple evidence base | Evidence of joint problem formulation and usefulness for all (Tengö et al., 2014) | Evidence of validation within knowledge systems (Tengö et al., 2014) | Evidence of application of the five tasks for successful collaboration across diverse knowledge systems (Tengö et al., 2017). 1 = mobilise, 2 = translate, 3 = negotiate, 4 = synthesise, 5 = apply |
|---|---|---|---|---|---|
| Austin et al., 2017 | MEB approach to enable enriched picture of progress of an Indigenous Land and Sea Management programme run by the Wunambal Gaambera people in the Kimberley, Australia | Informing the evidence base of the Wunambal Gaambera Healthy Country Plan using western scientific and local indigenous knowledge. Parallel integration of western and indigenous monitoring data/information to support co-production of enriched picture of country and management activities by the Uunguu Monitoring and Evaluation (M&E) Committee | Collaborative and multiple evidence-based M&E committee designed the approach to conducting evaluation of progress and assessment of key targets | Yes, each stream of knowledge was internally validated and cross-checked through collaborative self-assessment by Uunguu M&E Committee | 1, 2, 3, 4, 5 Local indigenous knowledge was mobilised through the planning process and the M&E committee; translation of information from various sources via monitoring methods; further negotiation and translation occurred within M&E committee; all knowledge streams synthesised through M&E committee meetings and reporting processes; and applied through adaptive management of the Healthy Country Plan |

**Table 6.1** (*cont.*)

| Article citation | Issue investigated including location | Multiple evidence base | Evidence of joint problem formulation and usefulness for all (Tengö et al., 2014) | Evidence of validation within knowledge systems (Tengö et al., 2014) | Evidence of application of the five tasks for successful collaboration across diverse knowledge systems (Tengö et al., 2017). 1 = mobilise, 2 = translate, 3 = negotiate, 4 = synthesise, 5 = apply |
|---|---|---|---|---|---|
| Nguyen et al., 2017 | Sustainable management of eroding mangrove-dominated muddy coasts in Vam Ray, Hon Dat district, Kien Giang Province, Vietnam | Partnership between government agencies, scientists and local communities. Methods included literature review, semi-structured interviews, participatory community meetings, participatory diagramming and thematic analysis. The introduction and analysis of different knowledge systems are undertaken in participatory community meetings, semi-structured interviews, field visits, photovoice and debriefings | All parties agreed to co-formulate the problems and use local and scientific knowledge to generate and pilot new knowledge for solving them. It was agreed to build local capacity and to utilise as many local resources as possible for developing the fence and nursery construction, to solve serious erosion problems affecting the community | Local knowledge of, e.g. Melaleuca fence construction was validated by local experts based on their experience. However, facing new challenges in the community created interest in other knowledge such as scientific knowledge. The collaboration led to new knowledge about fencing for controlling coastal erosion | 1, 2, 5 Local knowledge held by individuals regarding traditional Melaleuca fences, and local contexts were systematically collected and brought together with the relevant scientific knowledge in relation to sedimentation and coastal dynamics in Kien Giang, Vietnam into ecologically based, cost-effective strategies for successfully controlling coastal erosion |

| Robinson et al., 2016 | Water management in territories of aboriginal people connected to the Girringun Indigenous Corporation (Girringun) in northern Australia | Participatory maps created in workshops of Girringun support staff, Aboriginal rangers, some Girringun elders who are also artists and some of the authors to determine the values, knowledge and management aspirations of participants for their 'fresh water country'. Second workshop to discuss the values that the participants had for native plants and trees and to identify risks to those values and the attributes of partnerships that support these values | Co-research approach, in which Girringun representatives worked with the researchers to select participants and design the participatory mapping workshops, advising on an appropriate focus, location and design for each workshop | Yes, the integrity of each indigenous knowledge system was maintained throughout process | 1, 2, 3, 4, 5 Girringun representatives and scientists created individual maps to mobilise and translate the knowledge needed for Girringun and its associated tribal groups to assess two distinct issues of concern. Collective watershed maps were also used to negotiate knowledge and although there was some variety in the information shared by different participants, the integrity of each indigenous knowledge system was maintained throughout the process. Synthesis of themes occurred through creation of targeted research 'products', including a one-page summary, that could be used by the research team and the Girringun Indigenous Community to translate the results of the project in a way that was useful to the participants, Girringun and the wider natural resource management community |
| --- | --- | --- | --- | --- | --- |

**Table 6.1** (cont.)

| Article citation | Issue investigated including location | Multiple evidence base | Evidence of joint problem formulation and usefulness for all (Tengö et al., 2014) | Evidence of validation within knowledge systems (Tengö et al., 2014) | Evidence of application of the five tasks for successful collaboration across diverse knowledge systems (Tengö et al., 2017). 1 = mobilise, 2 = translate, 3 = negotiate, 4 = synthesise, 5 = apply |
|---|---|---|---|---|---|
| Smith et al., 2017 | Developing conservation strategies for pollinators in the context of pollinator decline in Orissa, India | Peer-to-peer validation of trends and statements distilled from focus groups including 50 smallholder subsistence farmers, including tribal people, who have personal and procedural knowledge of crop production, and rural advisors; anecdotal network. This was in preparation for integration with scientific knowledge from other regions | The problem (a potential pollinator crisis) was defined by scientists, who recognised the dearth of information on diversity of crops and pollinators and together with farmers and rural advisors collated traditional and local knowledge on the same | Yes. Peer-to-peer validation of indigenous knowledge of trends and statements distilled from focus groups including 50 farmers and rural advisors | 1, 2, 4 Traditional knowledge of crop diversity and pollinators was elicited and internally validated, providing a consensus on knowledge which was collated for integration with scientific knowledge |

| Molnár et al., 2016a, 2016b | Mitigation of conflicts between cattle herding and conservation management of salt-steppe and wood pastures in Hungary | An inventory of objectives and practices of herders (representing traditional knowledge) and conservationists and ethnobotanists (scientific knowledge) were collected by participatory knowledge co-production in teamwork with the co-authors. Possible resolutions to potential conflicts were suggested. Methods include: (1) participatory observation, (2) semi-structured interviews with herders and conservationists, (3) co-author herders and conservation managers completed and clarified the contents of the tables in two rounds | Herders and ethno-ecologists jointly formulated the problem, and potential solutions were suggested by all parties | Yes. Herder interviews colleagues that explain their observations and experiences and jointly validate the relevance for the issues. Less focus on conservation manager's validation. However, importance of integrity, equity and reciprocity in their knowledge-based interactions highlighted | 1, 2, 3, 4 Herders' and conservationists' knowledge of practices were elicited in interviews. Herders' perspectives were mobilised and translated through film. Data were negotiated among diverse author group and synthesised for joint publication |

**Table 6.1** (cont.)

| Article citation | Issue investigated including location | Multiple evidence base | Evidence of joint problem formulation and usefulness for all (Tengö et al., 2014) | Evidence of validation within knowledge systems (Tengö et al., 2014) | Evidence of application of the five tasks for successful collaboration across diverse knowledge systems (Tengö et al., 2017). 1 = mobilise, 2 = translate, 3 = negotiate, 4 = synthesise, 5 = apply |
|---|---|---|---|---|---|
| Strangway et al., 2016 | A registry monitoring an aboriginal subsistence fishery in the Cree community of Waskaganish (Waskaganish Voluntary Anadromous Cisco Catch Registry) within the Environmental Impact Assessment (EIA) Follow-up Phase after diversion of the Rupert River for hydroelectric plant at Nútimesánán in Northern Quebec, Canada | Collaboration between Hydro-Quebec and the impacted community. Bringing together different monitoring reports of cisco including the Voluntary Catch Registry of the Crees of Waskaganish First Nation, biological monitoring and complementary studies | Unclear. State-owned utility, Hydro-Quebec, proposed a monitoring programme after authorisation of river diversion which would include studies on cisco spawning success, as cisco harvesting is key for Waskaganish First Nation community's cultural identity and subsistence economy. Four community members hired as monitors to collect data on cisco catch | Yes. In the voluntary cisco catch registry programme, compiled data were presented to land users at the end of the fishing season for interpretation and validation. Shared observations regarding the fishing season, fishing success and any other comments, as well as how the data will be presented outside of the community, were discussed and included in the final Registry reports | 1, 2, 4, 5  Fishers propose mitigation measures to increase fishing success under new flow rates, and stakeholders assess their potential, with the results of the various cisco monitoring programmes, including the Voluntary Registry, featuring prominently in the decision. Once measures are implemented, the Registry programme is used to evaluate their effectiveness by collecting catch data |

*Notes:* Examples assessing the experience of applying a MEB approach, showing the issue investigated, the location and the multiple evidence base, in literature that either quoted Tengö et al. (2014) or referred to MEB. The review examined evidence of joint problem formulation and usefulness for all stakeholders (defined by Tengö et al., 2014), evidence of internal validation within knowledge systems (defined by Tengö et al., 2014) and evidence of application of each of the five tasks for successful collaboration across diverse knowledge systems (as defined by Tengö et al., 2017).

**Table 6.2** *Summary of MEB tasks to guide knowledge collaborations (Tengö et al., 2017) as applied in the three case studies*

| MEB phases | Multiple evidence-base case examples | | |
| --- | --- | --- | --- |
| | 4.1. Piloting the MEB: Tharaka's river is running dry | 4.2. Mobilising indigenous knowledge systems for saltwater country across the Kimberley, Australia | 4.3. Justice and conservation: Global Dialogue on Human Rights and Biodiversity Conservation |
| 1. Mobilise Develop knowledge-based products through a process of innovation and/or engaging with past knowledge and experience | The preparatory process for the ecocultural mapping, where the elders of the clans start to engage and document their experiences. The process of making ecocultural maps and calendars, which mobilised and synthesised knowledge on the landscape and how it has changed over time | Project objectives and research activities identified by an intercultural collaborative Working Group (WG) to ensure focus on local priorities. At the individual community workshop level, each of the indigenous ranger groups designed the specific activities, venue and participants. Focus group discussions and knowledge-holder interviews were selected as appropriate methods for indigenous people to use their knowledge to inform the process. Ranger groups were all equally resourced to facilitate and participate in research activities | Participation occurred before the actual dialogue, through interactions over internet. Preparation of indigenous community representatives to present and mobilise knowledge about ecology as well as human rights among the participants. The contributions from indigenous communities were planned to be presented during walking workshops in the Oigek Community. However, an outbreak of Marburg virus meant the dialogue was moved to Eldoret. The stories were told by community representatives attending the workshop |

**Table 6.2** (cont.)

| MEB phases | Multiple evidence-base case examples | | |
|---|---|---|---|
| | 4.1. Piloting the MEB: Tharaka's river is running dry | 4.2. Mobilising indigenous knowledge systems for saltwater country across the Kimberley, Australia | 4.3. Justice and conservation: Global Dialogue on Human Rights and Biodiversity Conservation |
| 2. Translate<br><br>Adapt knowledge products or outcomes into forms appropriate to enable mutual comprehension in the face of differences between actors | Occurred together with mobilisation in the ecocultural mapping event, where representatives from local authorities, regional authorities and national institutions were present<br><br>Also later in the process, in the documentation of customary laws that were considered along with modern law, and in the gazetting of sacred sites led by the National Museums of Kenya | All research results generated by indigenous workshop participants and knowledge-holders were collated and provided in short, simple reports to relevant indigenous communities for validation. A period of one month was provided to give feedback, make amendments, add anything that was missing or embargo content<br><br>Once the reports were validated they were presented to the WG who discussed how to analyse and represent results from the perspective of both indigenous people and their non-indigenous partners in collaborative management of saltwater country | The core focus for the dialogue and concerns articulated was what indigenous knowledge, practice and belief systems mean for indigenous peoples, in relation to how it is perceived by scientists and government representatives. But also, what human rights mean if applied to biodiversity conservation decisions. Dialogue was designed to encompass the very diverse ways of expression, experiences and perspectives among the participants |

| | | | |
|---|---|---|---|
| 3. Negotiate<br>Interact among different knowledge systems to develop mutually respectful and useful representations of knowledge | All actors accepted evidence of the critical situation for the river brought up by the ecocultural mapping, complemented with technical data from research and government institutions provided by regional authorities<br>Negotiation also happened in the development of action plans following the mapping process, where the community and local authorites agreed upon actions to improve the condition of the river | WG and research team had regular contact to ensure a collaborative research approach and facilitate discussions on saltwater research and monitoring at a regional scale<br>WG provided an important conduit between indigenous communities, their staff and the research community<br>WG held a final workshop attended by indigenous people, indigenous rangers, indigenous representative bodies, scientists and federal and state governments to raise awareness and seek final feedback on project outputs | Negotiations included how to interpret biodiversity data from different approaches for management and governance of ecosystems, along with knowledge about human rights principles and legislation, and cultural and socio-economic use of biodiversity |
| 4. Synthesise<br>Shape broadly accepted common knowledge bases for a particular purpose | Synthesis occurred in compiling customary laws and conventional laws together with authorities, when gazetting the sacred sites and in the development of action plans for protecting the river | Indigenous participants engaged in regular synthesis of results, from scoping, defining research questions and conducting fieldwork, to analysing results and communicating outcomes. The use of the MEB approach was a result of the WG's capacity to consider a range of possible tools and processes and choose the ones that work best for the project<br>Indigenous people had the opportunity to continuously monitor to ensure that project frameworks and tools | Agreement in place, based on evidence, that synergies are possible between conservation and human rights. In policy and practice, more efforts are needed to synthesise 'how' this can happen. The dialogue did not aim for a synthesis, that is for a later stage, with policy decisions leading to application. When presenting a summary of evidence, it was considered important to recognise convergence, but also identify and |

**Table 6.2** (cont.)

| MEB phases | Multiple evidence-base case examples | | |
| --- | --- | --- | --- |
| | 4.1. Piloting the MEB: Tharaka's river is running dry | 4.2. Mobilising indigenous knowledge systems for saltwater country across the Kimberley, Australia | 4.3. Justice and conservation: Global Dialogue on Human Rights and Biodiversity Conservation |
| 5. Apply Use common knowledge bases to make decisions and/or take actions and to reinforce and feed back into the knowledge systems | The process led to applications to improve river conditions at multiple levels: Revitalisation of rituals and enforcement of customary law at sacred sites Government recognition of the custodians as protecting the sites Enforcement of regulations of water extraction and riparian zone protection by regional authorities | fitted into the holistic, contextual and current situation in the Kimberly Saltwater Country The primary outcome: a regional network of indigenous people who have negotiated as regional knowledge brokers with their elders and knowledge-holders Short-term funding secured for the WG to support implementation, modification and compliance of the best-practice approaches developed Tools developed: – Regional saltwater monitoring framework based on indigenous knowledge identified social, cultural, economic and environmental values – Digital research protocol and application systems – Set of guidelines to describe simple processes for knowledge collaborations | recognise where there were still disagreements The evaluation showed that knowledge for conservationists about Human Rights law and implementation representing a strand of research was considered useful. There was potential for application of insights around equal benefit of conservation and human rights in all the cases brought up in the dialogue |

## 6.5 Exemplifying MEB cases and reflecting on lessons learned

Here we present three case studies that have explicitly implemented a MEB approach (Table 6.2). The first two set out processes to address local conservation and development issues. The third is an international dialogue meeting where the aim was to create a platform to discuss a fundamental crux in conservation globally – how to realise synergies between human rights and biodiversity conservation, and support local people and conservation organisations in becoming strategic allies.

### 6.5.1 Piloting the MEB approach: Tharaka's river is running dry

#### 6.5.1.1 Context

Drought is a recurring challenge to the livelihoods of the people in Tharaka, Kenya. Kathita River is the main water source and of paramount importance, economically, culturally and spiritually. Fourteen sacred natural sites along the river are protected by the communities for their cultural and spiritual values. In recent years, the government's policy guidelines and regulations for protecting the river have not been upheld and traditional ecological law has not been enforced either. This has led to excessive and often illegal abstraction along the river's course, degradation of the riverine vegetation and destruction of the catchment area. The local people, led by clan-based custodians of the sacred sites, decided to come together to find ways of protecting the river using their indigenous and local knowledge and practices and customary laws. A nongovernmental organisation in the area offered to facilitate an eco-cultural mapping process to enhance the eroded local capacity to govern the river.

A preparatory process brought together custodians of the sacred sites along the river. Local community organisations, county leaders and government institutions, including the National Museums of Kenya, were successively engaged in the process. In August 2014, community members jointly developed eco-cultural maps and calendars of the past and present, which illustrated changes in the integrity of their social–ecological system. Based on these, maps of the future envisioning different scenarios were drafted, creating a collective understanding and describing alternative pathways for the future. The maps and insights were shared and discussed with different actors beyond the local community.

A couple of years after the initial process, several of the problems identified with river governance have been addressed: strategies have been formulated for local authorities to reach out to land owners to safeguard riparian reserves. Tree seedlings are raised and distributed to land owners for planting in order to protect the riparian zone. The National Museums of Kenya have, together with the communities, gazetted the sacred sites along Kathita River, which has given them a government-recognised status. Rituals are again carried out at the sacred sites and the customary rules are enforced (Mburu, 2016).

### 6.5.1.2 *Role of the MEB*

The local non-governmental organisation convening the eco-cultural mapping in Tharaka is a member of a bridging organisation who were engaged in the initial dialogue across knowledge systems and volunteered to pilot a MEB approach. Eco-cultural mapping emerged as a culturally appropriate tool for knowledge mobilisation to enhance ecosystem governance for the society at large, beyond the community benefits expected by the clans that initiated the process. This also led to a greater understanding of the roles that different actors play in the local community and who to approach, how to formulate proposals and the utility of referring to established facts from community-based monitoring of the river.

The process contributed to unifying actors towards an enriched picture of understanding that could be shared and discussed with decision-makers outside the community. The eco-cultural mapping activity focused on how knowledge can be *translated* and *negotiated* to benefit an official process of conservation of sacred sites, and better ecosystem management of the Kathita river at large, through collaboration to protect the landscape (see Table 6.2). For this step, it was important to engage with actors with the authority to act in the customary governance system. Thus, the clans that were managing the sacred sites had a critical role in mobilising other community members.

### 6.5.1.3 *Challenges and opportunities*

The power imbalance between farmers with resources to extract and use water, and the majority of the community who did not have such resources, but were still exceptionally dependent on Kathita River as a water source, proved a challenge. Community research groups have been formed to solve specific emerging problems defined by the community.

The initiative for the eco-cultural mapping process came from the communities and the local non-governmental organisation, who contacted government and later also the Natural Museums of Kenya in order to catalyse change and ensure impact. The local actors as initiator created a solid base for trustful collaborations across knowledge systems.

## 6.5.2 Mobilising indigenous knowledge systems for saltwater country across the Kimberley region, Australia

### 6.5.2.1 *Context*

The Kimberley region in tropical north-western Australia is globally significant for its biodiversity, relatively intact ecosystems and its aesthetic and recreational values. Indigenous peoples comprise almost half of the region's population and have ownership or management rights over most of the land and sea. They are caretakers of a diverse cultural landscape dating back at least 60,000 years. The Australian public places high value on the cultural and natural assets of the Kimberley. The Western Australian Government concluded in 2011 that to ensure the best possible outcomes of conservation efforts in the Kimberley, a combination of indigenous knowledge and scientific knowledge was needed.

The Kimberley Indigenous Saltwater Science Project (KISSP) was established by a group of indigenous peoples and their organisations, research institutes, corporations and government organisations to investigate ways of co-producing collaborative monitoring, management and research regionally. A working group was established in 2014 with representatives from seven indigenous groups (Balangarra, Bardi Jawi, Dambimangari, Karajarri, Nyul Nyul, Wunambal Gaambera and Yawuru peoples) and key staff from local indigenous organisations. The working group recruited a team of researchers to assist the project. In total, there were 103 indigenous participants in five Traditional Owner workshops and one Knowledge-Holder interview.

### 6.5.2.2 Role of the MEB
Although not intentionally applied at the commencement of the project, the MEB process was followed intuitively by the experienced practitioners involved. Midway through the project, the MEB approach was formally introduced to participants, who immediately recognised its value in describing their practice. The working group agreed to adopt the MEB as an overarching framework for the KISSP project and to design regional frameworks for collaborative knowledge production, monitoring, research and management of Kimberley Saltwater Country (Table 6.2).

### 6.5.2.3 Challenges and opportunities
The biggest challenge faced by the KISSP was to establish engagement with indigenous peoples in the Kimberley. Prior to the formation of the working group, the project struggled for many years to create dialogue with indigenous peoples. Finally, a workshop was held to identify collaborative pathways towards project goals. The intervention of the indigenous-led working group demonstrated the potential for MEB approaches to ensure useful outcomes through intercultural and interdisciplinary projects.

Initially, lack of investment in the capacity of indigenous peoples and their organisations to engage in the research process limited progress. This should not be understood as a lack of knowledge or capacity to care for saltwater country, but rather as a need for support to mobilise their knowledges and practices to contribute to the KISSP as a collaborative, intercultural project. There was a prior assumption that indigenous peoples and their knowledge and practice could easily fit into a regional project that comprised indigenous and scientific knowledge systems side by side. There was no insight of the need for recognition and equity, and for explicit usefulness of the research products for all involved in collaborative practices. For example, there was consistently a subconscious assumption that flows of knowledge produced throughout the project would be channelled in a unilateral direction to scientists in the regional capital in the form of 'data' to be analysed so as to suitably inform the policy and decision-making processes of the state. The communication of this new information back to indigenous peoples in the Kimberley was more of an afterthought and, presumably, seen more as a bureaucratic demand than a practical mechanism for improving collaborative management of Saltwater Country. This assumption ignored the practical, and fairly reasonable, requirement of local indigenous peoples that any knowledge shared or co-produced through collaborative research and monitoring be made available for informing their own local decision-making and practice for looking after Saltwater Country. The indigenous-led KISSP Working Group made this point patiently and constructively and, thus, ensured that the project could produce several locally useful outputs and outcomes for indigenous peoples in Kimberley Saltwater Country.

### 6.5.3 Justice and conservation: Global Dialogue on Human Rights and Biodiversity Conservation

#### 6.5.3.1 *Context*

The Global Dialogue on Human Rights and Biodiversity Conservation was an international meeting initiated to address the conflicts that have often emerged across the globe between conservation agencies and indigenous peoples with longstanding relationships with their ancestral territories, co-organised by SwedBio at Stockholm Resilience Centre, Forest Peoples Programme, Natural Justice and the Chepkitale Indigenous Peoples Development Project as the local host in Kenya. The organisers represented actors engaged from different scales and perspectives, which created confidence and legitimacy for the dialogue. The dialogue started from the conviction that local people and conservation organisations could be strategic allies. It was attended by conservation agencies, social justice and human rights advocates, biodiversity conservation and sustainable use experts, legal and human rights professionals, members of community-based organisations, government officials, UN organisations and academics. It was designed in a global policy-setting context, while also aiming to contribute to local ways forward. The venue for the dialogue, Eldoret, Kenya, is situated between two biodiversity-rich areas conserved by indigenous peoples as their ancestral lands. The Ogiek people are an indigenous hunter-gatherer community on Mt Elgon, at the border of Uganda, while the Sengwer people are traditionally living with and taking care of the Embobut Forests. Both Ogiek and Sengwer have been faced with repeated attempts of eviction over decades in the name of conservation. In 2011, through a conservation-related mediation method called the Whakatane mechanism, the Ogiek communities in Mt Elgon reached an initial agreement to live in and govern parts of their ancestral lands. However, the Sengwer have rather experienced increased tensions in later years.

### 6.5.3.2 Role of the MEB

Globally there is an increased recognition that human rights protection can, and should, be complementary to safeguarding biodiversity and ecosystems (Knox, 2017), but there is a need to mainstream *how*, through good case examples and methods in policy and practice. A MEB approach was introduced in the preparatory process before the dialogue as part of the multi-actor dialogue method. The design process started with informal discussions between conservation agencies, indigenous peoples, human rights professionals and the organisers a year before the dialogue took place. The long preparatory process helped mobilise knowledge and confidence as a base for common understanding of the overarching ecological, legal, institutional and political challenges among participating actors. Through the dialogue process, the MEB approach provided guidance to ensure equity, reciprocity and usefulness for all actors. In the evaluation, the community representatives stressed they had never before had experience of being recognised and presenting their stories as evidence on an equal footing with science and governments.

### 6.5.3.3 Challenges and opportunities

Establishing a collaboration among different actors at national level in Kenya representing government, indigenous peoples and conservation agencies that generally do not meet was the greatest challenge. Thanks to the global context of the meeting, the presence of international actors with diverse experiences contributed to a constructive dialogue. Interactions among indigenous peoples and scientists were successful because a common understanding of the MEB approach had been established during the preparation. Persistent barriers between indigenous peoples and governments still exist in local cases, in particular the Sengwer people, and should be resolved through policy and legal processes. However, establishing MEB processes whenever governance of ecosystems and biodiversity can be enhanced through collaborative processes across multiple knowledge systems can be useful for all involved in the meantime.

## 6.6 Sharing lessons from the three cases

In the first case from Tharaka River, the importance of mobilising indigenous and local knowledge as a solid base for translation and negotiation phases was very clear. This then helped people speak about their knowledge, and also catalysed the revitalisation of eroded institutions and rules that previously served to protect the river, including the recognition and protection of the sacred sites. As the problem formulation was owned by the community, this enabled articulation of the importance of the sacred sites for understanding previous river governance, and motivated local people to restore the river. Later, they contacted the Natural Museums of Kenya, to provide support in gazetting their biodiversity-rich sacred sites for formal national recognition. This illustrates the important role that values and beliefs in diverse knowledge systems can play for conservation, how they may be identified, and how knowledge and governance capacity is embedded in the belief systems.

In the case from the Kimberley, the use of the MEB and the role of indigenous and local knowledge in collaborative management, created space and enthusiasm for experimenting with new ways of combining knowledge systems for management and governance of Saltwater Country. The KISSP demonstrated that working with multiple knowledge systems and disciplines in the context of unequal power relations requires design, support and monitoring of mechanisms that can maintain constant dialogue (e.g. the KISSP working group). Thinking of the collaboration as 'intercultural' was useful for understanding what capacity development was required for all actors. No single party had capacity deficits, but the collective needed to build joint capacity for weaving knowledge systems in ethical and equitable ways.

In the Global Dialogue on Human Rights and Conservation, reaching synergistic solutions between conservation and human rights once again was about overcoming power imbalances. The dialogue was an opportunity for key actors with different knowledges, experiences, worldviews and power to meet in a neutral context. Diversity of experiences (positive and negative) across scales and a careful mix of actors helped to overcome these imbalances during the dialogue. Mobilisation of indigenous and local knowledge and strengthening confidence among participating community representatives, but also knowledge about human rights and other legal aspects before the dialogue, was critical for deliberations. The recognition of indigenous rights and the value of their knowledge and practices for conservation expressed by researchers contributed to trust followed by constructive proposals. Positive experiences from successful collaborations in conservation of indigenous lands contributed to exploring ways forward in cases where conflicts persist. The learning across different sectors and scales, such as ecologists learning about human rights aspects, was appreciated in the evaluations. It also became clear that the deepest conflicts may not relate to conflicting evidence from different knowledge systems regarding ecology, but to controversial policy, such as the eviction of people from conservation areas.

In all three cases, the main challenge of the collaborative process was to overcome power imbalances and build trust and confidence. The focus on recognising, mobilising and discussing evidence from diverse knowledge systems was an entry point that contributed to the development of strong collaborative partnerships. Designing a process that was considered useful for all involved was critical to securing successful and sustainable outcomes, new and useful ways to combine and apply knowledge from diverse knowledge systems, and sometimes the generation of new knowledge. In all cases, the aim of creating synergies across knowledge systems for providing evidence on sustainable governance could be realised when all holders of knowledge gained from collaborations. A MEB approach, on whatever level it is conducted, emphasises the importance of collaborative processes that value multiple knowledges and practices needed to sustain the social–ecological landscape to the double benefit of sustainable livelihoods and conservation over the long term. Further, the collaborative relationships of trust developed provide new opportunities to align multiple modes of governance of ecosystems, to ensure decisions and policy are based on all available knowledge.

## 6.7 Discussion

In this chapter, we review the use of one recent and important approach to combining the knowledge of indigenous peoples, local communities and scientists for sustainability and conservation partnerships. We have focused on the MEB and its potential for building more inclusive understanding of multiple sources of evidence, how it is generated and how it is transmitted among diverse conservation actors. We argue that such an approach is important for better understanding of interlinked social–ecological systems, strengthening conservation partnerships and identifying new evidence-based pathways towards sustainability. Our review and the three case studies show examples of different ways to move forward that recognise the complementarity and integrity of knowledge systems in addressing specific problems (Molnár et al., 2016; Smith et al., 2017), create conditions (and methodologies) for full and open dialogue on how to move ahead, overcome power inequalities and navigate cultural differences (Robinson et al., 2016; Reed & Abernethy, 2018). We demonstrate reciprocal synergies between indigenous and local knowledge and conservation science and rich cases of how cross-fertilisation leads to stronger partnerships and better outcomes. The three case studies also show that the MEB requires partnerships that are underpinned by recognition, respect and understanding of diverse knowledge systems, and that the process for producing and applying common knowledge to problems cannot be viewed separately from the outcomes of partnerships. That is to say, much like the concept of adaptive management in conservation, the diversity and dynamism of knowledge systems dictate that the process of collaboration be taken as seriously as the achievement of conservation outcomes themselves (Gavin et al., 2018).

More work is needed to further elaborate how to implement a MEB approach in different processes and contexts. The IPBES process has struggled with the tension between open collaboration and the demands for structure set by the scientific knowledge governance. There is yet some way to go to better acknowledge and solve epistemic challenges, such as diverse modes of validation across knowledge systems (Löfmarck & Lidskog, 2017; Obermeister, 2017). There is also a need to continue developing tools and approaches for bridging knowledge systems that are connected to local, cultural, social and ecological conditions. Our review illustrates that indigenous peoples, local communities and scientists have begun to tackle this challenge (Molnár et al., 2016; Robinson et al., 2016; Smith et al., 2017), but further dialogue is required, both horizontally across local scales and vertically through local to global institutions.

We have shown that a MEB approach has been particularly effective in dialogues where there are power imbalances among actors and historical bias concerning the validity or usability of knowledge systems other than western approaches to science (see also Klenk & Meehan, 2015). Building trust and respect is especially pertinent in the context of ongoing and historical injustices and abuse of indigenous rights, and requires the recognition of indigenous peoples as rights-holders and defenders of biodiversity, who maintain management and governance systems of vast ecosystems (Brondizio & Le Tourneau, 2016; Mistry & Berardi, 2016).

Tengo et al. (2017) suggest five tasks that can guide processes that build trust and agency (see Figure 6.1b), while at the same time building a stronger evidence base for action. We find in our review that the mobilisation task is often neglected, or that documentation of indigenous and local knowledge is not fully recognised. More research is needed, but mobilisation of knowledge and empowerment of knowledge-holders may be critical steps for successful knowledge collaborations that also contribute to strengthening collaborative governance capacity. We also find that explicit joint problem formulation and analysis across knowledge systems is absent from many processes and is clearly a challenge in regional and global assessments with rigid scientific formats (Livoreil et al., 2016; Nesshöver et al., 2016; Oubenal et al., 2017). Our case examples clearly show the importance of creating the right conditions for joint problem formulation.

It should be acknowledged that the implementation of a MEB approach is demanding, in terms of time and other resources, and requires strong commitment from all parties. However, we reiterate that there is mounting evidence of the potential positive outcomes in terms of novel indicators, more efficient responses to and implementation of findings, as well as for synergies between conservation and human well-being, including human rights (Danielsen et al., 2010; Johnson et al., 2015; Sterling et al., 2017b, 2017c). As found by the participants in our third case example on reconciling conservation and human rights, conservation initiatives can play a positive role by engaging with communities and increasing their recognition as actors and partners who hold important and useful knowledge.

Our experiences derive mainly from dialogues and collaborations with indigenous and local knowledge-holders who have deep connections, obligations to care for and a duty to fight for their rights to actively govern their ancestral territories. We are aware that in many other contexts, local knowledge-holders may be less empowered and traditional governance systems and cultural connections may be displaced and eroded. However, we believe that insights about dialogue and partnership between indigenous peoples, local communities and scientists can also be applied in western, urban and developing settings, where local knowledge and experience may be less evident but remains critical for nurturing effective stewardship of biodiversity and ecosystems. Ultimately, the MEB approach contributes to a much-needed conceptual mind shift to mobilise all knowledge that is useful for maintaining the life-supporting ecosystems in our world.

## References

Adams, W. M., Aveling, R., Brockington, D., et al. 2004. Biodiversity conservation and the eradication of poverty. *Science (New York, N.Y.)*, 306(5699), 1146–1149. doi:10.1126/science.1097920

Austin, B. J., Vigilante, T., Cowell, S., et al. 2017. The Uunguu Monitoring and Evaluation Committee: intercultural governance of a land and sea management programme in the Kimberley, Australia. *Ecological Management and Restoration*, 18(2), 124–133. doi:10.1111/emr.12257

Balvanera, P., Calderón-Contreras, R., Castro, A. J., et al. 2017. Interconnected place-based social–ecological research can inform global sustainability. *Current Opinion in Environmental Sustainability*, 29, 1–7. doi:10.1016/j.cosust.2017.09.005

Brammer, J. R., Brunet, N. D., Burton, A. C., et al. 2016. The role of digital data entry in participatory environmental monitoring. *Conservation Biology*, 30(6), 1277–1287. doi:10.1111/cobi.12727

Brondizio, E. S., Foufoula-Georgiou, E., Szabo, S., et al. 2016. Catalyzing action towards the sustainability of deltas. *Current Opinion in Environmental Sustainability*, 19, 182–194. doi:10.1016/j.cosust.2016.05.001

Brondizio, E. S. & Le Tourneau, F.-M. 2016. Environmental governance for all. *Science (New York, N.Y.)*, 352(6291), 1272–1273. doi:10.1126/science.aaf5122

Bruckmeier, K. 2016. *Social–Ecological Transformation Reconnecting Society and Nature*. London: Palgrave McMillan.

Buytaert, W., Zulkafli, Z., Grainger, S., et al. 2014. Citizen science in hydrology and water resources: opportunities for knowledge generation, ecosystem service management, and sustainable development. *Frontiers in Earth Science*, 2 (October), 1–21. doi:10.3389/feart.2014.00026

Convention on Biological Diversity (CBD). 2014. Decision XII/12. Article 8(j) and related provisions.

Cornell, S., Berkhout, F., Tuinstra, W., et al. 2013. Opening up knowledge systems for better responses to global environmental change. *Environmental Science & Policy*, 28, 60–70. doi:10.1016/j.envsci.2012.11.008

Danielsen, F., Burgess, N. D., Jensen, P. M., et al. 2010. Environmental monitoring: the scale and speed of implementation varies according to the degree of peoples involvement. *Journal of Applied Ecology*, 47, 1166–1168. doi:10.1111/j.1365-2664.2010.01874.x

Danielsen, F., Jensen, P. M., Burgess, N. D., et al. 2014. Testing focus groups as a tool for connecting indigenous and local knowledge on abundance of natural resources with science-based land management systems. *Conservation Letters*, 7, 380–389. doi:10.1111/conl.12100

Díaz, S., Demissew, S., Carabias, J., et al. 2015. The IPBES Conceptual Framework – connecting nature and people. *Current Opinion in Environmental Sustainability*, 14, 1–16. doi:10.1016/j.cosust.2014.11.002

Díaz, S., Pascual, U., Stenseke, M., et al. 2018. Assessing nature's contributions to people. *Science*, 359 (6373), 270–272. doi:10.1126/science.aap8826

Farhan Ferrari, M., de Jong, C. & Belohrad, V. S. 2015. Community-based monitoring and information systems (CBMIS) in the context of the Convention on Biological Diversity (CBD). *Biodiversity*, 16(2–3), 57–67. doi:10.1080/14888386.2015.1074111

Gagnon, C. A. & Berteaux, D. 2009. Integrating traditional ecological knowledge and ecological science: a question of scale. *Ecology and Society*, 14(2), 19.

Galvin, K. A., Reid, R. S., Fernández-Giménez, M. E., et al. 2016. Co-design of transformative research for rangeland sustainability. *Current Opinion in Environmental Sustainability*, 20, 8–14. doi:10.1016/j.cosust.2016.03.003

Gambon, H. & Rist, S. 2018. Moving territories: strategic selection of boundary concepts by indigenous people in the Bolivian Amazon – an element of constitutionality? *Human Ecology*, 46, 27–40. doi:10.1007/s10745-017-9960-z

Gavin, M. C., McCarter, J., Mead, A., et al. 2015. Defining biocultural approaches to conservation. *Trends in Ecology & Evolution*, 30, 140–145. doi:10.1016/j.tree.2014.12.005

Gavin, M., McCarter, J., Berkes, F., et al. 2018. Effective biodiversity conservation requires dynamic, pluralistic, partnership-based approaches. *Sustainability*, 10, 1846. doi:10.3390/su10061846

Gonzales, T. 2015. An indigenous autonomous community-based model for knowledge production in the Peruvian Andes. *Latin American and Caribbean Ethnic Studies*, 10, 107–133. doi:10.1080/17442222.2015.1034433

Hohenthal, J., Räsänen, M. & Minoia, P. 2018. Political ecology of asymmetric ecological knowledges: diverging views on the eucalyptus–water nexus in the Taita Hills, Kenya. *Journal of Political Ecology*, 25, 1–19.

Horstkotte, T., Utsi, T. A., Larsson-Blind, Å., et al. 2017. Human–animal agency in reindeer management: Sámi herders' perspectives on vegetation dynamics under climate change. *Ecosphere*, 8(9), e01931. doi:10.1002/ecs2.1931

Jabbour, J. & Flachsland, C. 2017. 40 years of global environmental assessments: A retrospective analysis. *Environmental Science & Policy*, 77, 193–202. doi:10.1016/j.envsci.2017.05.001

Johnson, J. T., Howitt, R., Cajete, G., et al. 2016. Weaving Indigenous and sustainability sciences to diversify our methods. *Sustainability Science*, 11, 1–11. doi:10.1007/s11625-015-0349-x

Johnson, N., Alessa, L., Behe, C., et al. 2015. The contributions of community-based monitoring and traditional knowledge to Arctic observing networks: reflections on

the state of the field. *Arctic*, 68(5),
1-13. doi:10.14430/arctic4447

Kealiikanakaoleohaililani, K. &
Giardina, C. P. 2016. Embracing the
sacred: an indigenous framework for
tomorrow's sustainability science.
*Sustainability Science*, 11(1), 57-67.
doi:10.1007/s11625-015-0343-3

Kenter, J. O. 2016. Editorial: Shared, plural
and cultural values. *Ecosystem Services*,
21(October), 175-183. doi:10.1016/j.
ecoser.2016.10.010

Klenk, N. & Meehan, K. 2015. Climate
change and transdisciplinary science:
problematizing the integration
imperative. *Environmental Science &
Policy*, 54, 160-167. doi:10.1016/j.
envsci.2015.05.017

Knox, J. 2017. Report of the Special
Rapporteur on the issue of
human rights obligations relating
to the enjoyment of a safe,
clean, healthy and sustainable
environment. UN OHCHR A/HRC/
34/49.

Livoreil, B., Geijzendorffer, I., Pullin, A. S.,
et al. 2016. Biodiversity knowledge
synthesis at the European scale: actors
and steps. *Biodiversity and Conservation*,
25, 1269-1284. doi:10.1007/s10531-
016-1143-5

Löfmarck, E. & Lidskog, R. 2017. Bumping
against the boundary: IPBES and the
knowledge divide. *Environmental
Science & Policy*, 69, 22-28. doi:10.1016/
j.envsci.2016.12.008

Lyver, P. O. B., Timoti, P., Jones, C. J., et al.
2017. An indigenous
community-based monitoring system
for assessing forest health in New
Zealand. *Biodiversity and Conservation*,
26, 3183-3212. doi:10.1007/s10531-
016-1142-6

Martín-López, B. & Montes, C. 2015.
Restoring the human capacity for
conserving biodiversity: a social-
ecological approach. *Sustainability

Science*, 10, 699-706. doi:10.1007/
s11625-014-0283-3

Mathevet, R., Thompson, J. D., Folke, C.,
et al. 2016. Protected areas and their
surrounding territory: social-
ecological systems in the context of
ecological solidarity. *Ecological
Applications*, 26, 5-16. doi:10.1890/14-
0421

Mburu, G. 2016. *Reviving Indigenous and
Local Knowledge for Restoration of
Degraded Ecosystems in Kenya
A. Contribution to the Piloting of the
Multiple Evidence Base Approach*. Report.
Stockhom: SwedBio, Stockholm
Resilience Centre.

Mistry, B. J. & Berardi, A. 2016. Bridging
indigenous and scientific knowledge.
*Science*, 352, 1274-5.

Molnár, Z., Kis, J., Vadász, C., et al. 2016a.
Common and conflicting objectives
and practices of herders and
conservation managers: the need for
a conservation herder. *Ecosystem
Health and Sustainability*, 2(4), e01215.
doi:10.1002/ehs2.1215

Molnár, Z., Sáfián, L., Máté, J., et al.
2016b. "It does matter who leans
on the stick": Hungarian herders'
perspectives on biodiversity,
ecosystem services and their
drivers. In Roué, M. & Molnár, Z.,
editors, *Indigenous and Local
Knowledge of Biodiversity and
Ecosystem Services in Europe and
Central Asia* (pp. 42-56). Paris:
UNESCO.

Nadasdy, P. 1999. The politics of TEK:
power and the 'integration' of
knowledge. *Arctic Anthropology*, 36
(1-2), 1-18. doi:10.2307/40316502

Nesshöver, C., Vandewalle, M.,
Wittmer, H., et al. 2016. The Network
of Knowledge approach: improving
the science and society dialogue on
biodiversity and ecosystem services in
Europe. *Biodiversity and Conservation*,

25(7), 1215–1233. doi:10.1007/s10531-016-1127-5

Nguyen, T. P., Luom, T. T. & Parnell, K. E. 2017. Developing a framework for integrating local and scientific knowledge in internationally funded environment management projects: case studies from Kien Giang Province, Vietnam. *Local Environment*, 22, 1298–1310. doi:10.1080/13549839.2017.1342617

Nkambule, S. S., Buthelezi, H. Z. & Munien, S. 2016. Oppportunities and constraints for community-based conservation: the case of the KwaZulu-Natal Sandstone Sourveld grassland, South Africa. *Bothalia*, 46(2), 1–8. doi:10.4102/abc.v46i2.2120

Obermeister, N. 2017. From dichotomy to duality: addressing interdisciplinary epistemological barriers to inclusive knowledge governance in global environmental assessments. *Environmental Science & Policy*, 68, 80–86. doi:10.1016/j.envsci.2016.11.010

Oldekop, J. A., Holmes, G., Harris, W. E., et al. 2015. A global assessment of the social and conservation outcomes of protected areas. *Conservation Biology*, 30, 133–141. doi:10.1111/cobi.12568

Oubenal, M., Hrabanski, M. & Pesche, D. 2017. IPBES, an inclusive institution? Challenging the integration of stakeholders in a science–policy interface. *Ecology and Society*, 22(1), art11. doi:10.5751/ES-08961-220111

Pascual, U., Balvanera, P., Diaz, S., et al. 2017. Valuing nature's contributions to people: the IPBES approach. *Current Opinion in Environmental Sustainability*, 26–27, 7–16. doi:10.1016/j.cosust.2016.12.006

Polfus, J. L., Simmons, D., Neyelle, M., et al. 2017. Creative convergence: exploring biocultural diversity through art. *Ecology and Society*, 22(2), Art4. doi:10.5751/ES-08711-220204

Prado, H. M. & Murrieta, R. S. S. 2015. Ethnoecology in perspective: the origins, interfaces and current trends of a growing field. *Ambiente & Sociedade*, 18(4), 133–154. doi:10.1590/1809-4422ASOC986V1842015

Quinlan, A. E., Berbes-Blazquez, M., Haider, L. J., et al. 2016. Measuring and assessing resilience: broadening understanding through multiple disciplinary perspectives. *Journal of Applied Ecology*, 53, 677–687. doi:10.1111/1365-2664.12550

Rathwell, K. J. & Armitage, D. 2016. Art and artistic processes bridge knowledge systems about social–ecological change: an empirical examination with Inuit artists from Nunavut, Canada. *Ecology and Society*, 21(2), art21. doi:10.5751/ES-08369-210221

Reed, M. G. & Abernethy, P. 2018. Facilitating co-production of transdisciplinary knowledge for sustainability: working with Canadian Biosphere Reserve practitioners. *Society and Natural Resources*, 31, 39–56. doi:10.1080/08941920.2017.1383545

Robinson, C. J., Maclean, K., Hill, R., et al. 2016. Participatory mapping to negotiate indigenous knowledge used to assess environmental risk. *Sustainability Science*, 11, 115–126. doi:10.1007/s11625-015-0292-x

Sheil, D., Boissière, M. & Beaudoin, G. 2015. Unseen sentinels: local monitoring and control in conservation's blind spots. *Ecology and Society*, 20(2), art39. doi:10.5751/ES-07625-200239

Smith, B. M., Basu, P. C., Chatterjee, A., et al. 2017. Collating and validating indigenous and local knowledge to apply multiple knowledge systems to an environmental challenge: a case-study of pollinators in India. *Biological Conservation*, 211(Part A), 20–28. doi:10.1016/j.biocon.2017.04.032

Sterling, E. J., Betley, E., Sigouin, A., et al. 2017a. Assessing the evidence for stakeholder engagement in biodiversity conservation. *Biological Conservation*, 209, 159–171. doi:10.1016/j.biocon.2017.02.008

Sterling, E. J., Filardi, C., Toomey, A., et al. 2017c. Biocultural approaches to well-being and sustainability indicators across scales. *Nature Ecology and Evolution*, 1, 1798–1806. doi:10.1038/s41559-017-0349-6

Sterling, E., Ticktin, T., Kipa Kepa Morgan, T., et al. 2017b. Culturally grounded indicators of resilience in social–ecological systems. *Environment and Society*, 8, 63–95. doi:10.3167/ares.2017.080104

Strangway, R. E., Dunn, M. & Erless, R. 2016. Monitoring Nûtimesânân following the diversion of our river: a community-led registry in Eeyou Istchee, Northern Québec. *Journal of Environmental Assessment Policy and Management*, 18, 1–21. doi:10.1142/S1464333216500010

Tengö, M., Brondizio, E. S., Elmqvist, T., et al. 2014. Connecting diverse knowledge systems for enhanced ecosystem governance: the multiple evidence base approach. *Ambio*, 43, 579–591. doi:10.1007/s13280-014-0501-3

Tengö, M., Hill, R., Malmer, P., et al. 2017. Weaving knowledge systems in IPBES, CBD and beyond – lessons learned for sustainability. *Current Opinion in Environmental Sustainability*, 26–27, 17–25. doi:10.1016/j.cosust.2016.12.005

Tengö, M. & Malmer, P. 2012. Dialogue workshop on 'Knowledge for the 21st Century: Indigenous knowledge, traditional knowledge, science and connecting diverse knowledge systems'. Usdub, Guna Yala, Panama, 10–13 April,

2012. Workshop Report. SwedBio at Stockholm Resilience Centre.

Timoti, P., Lyver, P. O., Matamua, R., et al. 2017. A representation of a Tuawhenua worldview guides environmental conservation. *Ecology and Society*, 22(4), 20. doi:10.5751/ES-09768-220420

Vasseur, L., Horning, D., Thornbush, M., et al. 2017. Complex problems and unchallenged solutions: bringing ecosystem governance to the forefront of the UN sustainable development goals. *Ambio*, 46, 731–742. doi:10.1007/s13280-017-0918-6

Verburg, P. H., Dearing, J. A., Dyke, J. G., et al. 2016. Methods and approaches to modelling the Anthropocene. *Global Environmental Change*, 39, 328–340. doi:10.1016/j.gloenvcha.2015.08.007

Vogt, N., Pinedo-Vasquez, M., Brondízio, E. S., et al. 2016. Local ecological knowledge and incremental adaptation to changing flood patterns in the Amazon delta. *Sustainability Science*, 11, 611–623. doi:10.1007/s11625-015-0352-2

Waylen, K. A., Fischer, A., McGowan, P. J. K., et al. 2010. Effect of local cultural context on the success of community-based conservation interventions. *Conservation Biology*, 24, 1119–1129. doi:10.1111/j.1523-1739.2010.01446.x

# CHAPTER SEVEN

# Informing conservation decisions through evidence synthesis and communication

ANDREW S. PULLIN
*Collaboration for Environmental Evidence*
SAMANTHA H. CHENG
*Center for Biodiversity and Conservation*
STEVEN J. COOKE
*Canadian Centre for Evidence-Based Conservation*
NEAL R. HADDAWAY, BILJANA MACURA
*Stockholm Environment Institute*
MADELEINE C. MCKINNON
*Bright Impact*

and

JESSICA J. TAYLOR
*Canadian Centre for Evidence-Based Conservation*

## 7.1 Introduction

The volume of evidence from scientific research and wider observation is greater than ever. Approximately 2.5 million articles are published annually (Plume & van Weijen, 2014) and this rate is increasing at around 3–3.5% per year (Ware & Mabe, 2015). Conservation is no exception to this trend and the result is a rapidly expanding body of potentially useful information for decision-makers (Li & Zhao, 2015). While the expansion of research represents an important increase in knowledge generation, much of this information is scattered in fragments over increasingly diverse sources. This, along with the sheer volume, makes it harder for decision-makers to find, access and digest all of the relevant information on a particular topic, resolve seemingly contradictory results or simply identify a lack of evidence. Evidence synthesis is the process of searching for, and summarising, a body of research on a specific topic in order to inform decisions. The extent of relevant research may range from nothing, or one or two primary studies, to many hundreds. Despite the obvious potential value of synthesising findings from multiple studies (where two studies may be all that is needed to add value through synthesis), methods of rigorous evidence synthesis have been largely neglected until recently. We argue that it is time to place evidence synthesis as a central pillar of evidence-informed decision-making in conservation and environmental management.

As an enterprise, evidence synthesis is very broad and includes many and diverse methodologies, some more rigorous than others. For example, syntheses labelled as 'literature reviews' often lack standardised methodology, fail to report their methods and therefore lack transparency or the potential for repeatability (O'Leary et al., 2016). Additionally, these literature reviews do not deal with the risk of bias in either the primary research (e.g. poor-quality experimental design and conclusions that may not be supported by a given study) or the synthesis process (e.g. selective use of information). Meta-analysis approaches have become popular where significant amounts of quantitative data are available, but they are often biased in the way they select and include studies in their analysis (Koricheva & Gurevitch, 2014). In response to these problems, more rigorous methodologies, such as systematic reviews, have been developed. These were first used in the health sector through the work of the Cochrane Collaboration (Higgins & Green, 2011), and have subsequently been applied to conservation and environmental management by the Collaboration for Environmental Evidence (Pullin & Knight, 2009; Collaboration for Environmental Evidence, 2018).

In this chapter we make a case for rigorous evidence synthesis: we explain why these methods are appropriate, how they can benefit wider society and how evidence can be synthesised, shared and used as a public good. Although evidence synthesis can inform a broad range of decision-making contexts, we focus here on two major aspects of conservation where evidence might be useful. First, in measuring the direct and indirect impacts of human activity on the natural world, and second, the effectiveness of conservation efforts to mitigate those impacts.

## 7.2 The central role of evidence synthesis in informing decisions in conservation policy and practice

Many factors can contribute to making a decision. In contexts where social and political stakes are high, as is common for conservation policy, scientific evidence will likely only inform decisions, rather than act as the primary driving force behind them. Although evidence is sometimes crucial, it may equally be ignored or overruled by other factors, such as political context, infrastructure and capacity. Ideally, evidence synthesis should play a central role in providing reliable evidence and enabling the wider society to understand or challenge decisions that might affect them. Making decisions without considering all available evidence might perpetuate biases, increase the likelihood of taking a wrong or costly action, or lead to missed opportunities to achieve faster or more cost-efficient outcomes. In a democratic society, comprehensive and rigorous evidence synthesis and open communication makes 'sidelining' (i.e. deliberately ignoring evidence) and/or biased (i.e. selective) use of evidence by authorities more difficult without challenge and transparent justification.

Unfortunately, evidence synthesis is itself often 'bypassed' completely or manipulated to get the answer required (i.e. policy-based evidence) (Dicks et al., 2014). There may be significant resistance to the use of transparent evidence synthesis in the face of vested interests, and this may partly explain why organised and independent evidence synthesis receives so little attention or funding. Rigorous scientific evidence could also be seen as a threat to those with entrenched beliefs. Beyond outright opposition, complacency or inaccessibility of evidence might inhibit adoption of synthesis findings even when good intentions towards informed decision-making exist.

Fortunately, most decision-makers in conservation want practical advice that is grounded in the best available evidence (Cook et al., 2013). Leveraging syntheses and integrating their findings into decision-making processes requires an understanding of how and when evidence is necessary, and what level of confidence is needed to inform a decision. Such considerations will determine the choice of synthesis method(s), which should reflect practical needs to guide management decisions or future research. Syntheses can be used either to generate a new theory, conceptual framework or hypothesis (e.g. applying existing theory to a different context) or to test an existing hypothesis (e.g. evaluating the effectiveness of an intervention). In the context of effectiveness of interventions, evidence syntheses are relevant to decisions at several critical stage points in the life cycle of a programme or initiative: (1) initial scoping of a new topic early on in strategic planning (e.g. informing a new strategy on land use for a philanthropic foundation (Snilstviet et al., 2016)); (2) identification or validation of specific intervention designs (e.g. understanding how gender composition affects outcomes of resource management groups (Leisher et al., 2016)); (3) benchmarking of institutional outcomes against other programmes (e.g. investments in community forest management by the Global Environment Facility (Bowler et al., 2010)); (4) evaluation of overall effectiveness of an intervention across multiple contexts or applications (e.g. effects of property regimes in different biomes (Ojanen et al., 2017)). Understanding the purpose of the syntheses for informing the different stages of decision-making will ensure selection of a suitable method, appropriate engagement of stakeholders and relevant communication of findings.

Some evidence synthesis methods, such as systematic review, have been described as following the 'information deficit model' (Owens, 2000); that is to say, they follow the assumption that the simple production and push delivery of evidence that fills a gap will be sufficient to achieve uptake. However, this perception misrepresents the full process behind the methodology.

Systematic reviews can be socially inclusive, with extensive stakeholder contribution to formulating a question and approach, including setting the scope of the topic. This engagement attempts to ensure the findings of a review will fill a real and important synthesis gap (a knowledge need where sufficient primary research exists to allow synthesis) and respond to stakeholder demand. When engaging with stakeholders, a balance needs to be struck between involving them in the design of the review and independence from undue vested interest (Haddaway & Crowe, 2018). In the field of conservation, this balance is very much dependent on the nature of the question and the extent of vested interests (Kløcker Larsen & Nilsson, 2017). Many aspects of evidence synthesis are collective, with stakeholders having shared motivation to benefit from the findings. In other cases, evidence synthesis is conducted in contested areas, with stakeholders that hold opposing views and may be hostile to the process and its findings. In the latter case, it is important to have a process that allows consultation when appropriate but also provides independence when necessary. For example, for some key steps, such as initial formulation of the question, engagement with stakeholders is usually essential (Land et al., 2017), while other steps may need to be conducted free of such vested interests. To date, systematic reviews have engaged with a spectrum of stakeholders at different levels. Some reviews, for example those that are more academic or have specific commissioners (e.g. private goods reviews (Oliver & Dickson, 2016)), have only passively engaged stakeholders by informing or consulting them (typically only at the beginning of the review process), while others have employed more in-depth engagement, extending to co-design of review methods and scope (Land et al., 2017).

Alongside the purpose of syntheses, the level of confidence required to make a decision determines their method and scope. In some instances, where evidence of effectiveness is key, uncertainty in the evidence base hampers decision-making. In such instances one might ask 'How much evidence is enough?' or 'How much uncertainty is acceptable?' (Salafsky & Redford, 2013). The need for evidence synthesis in the conservation sector may also vary depending on aspects of spatial scale, complexity and controversy. For example, decisions regarding inexpensive and low-risk local-scale interventions (e.g. applied to improve biodiversity or habitat conditions in nature reserves) may benefit most from locally generated, rigorous evidence, or more commonly from primary research studies conducted in similar contexts. This evidence could be provided by a single, self-generated study (as in adaptive management), be internally generated by the relevant organisation, or come from collating evidence from similar case studies. In contrast, decisions regarding expensive, often large-scale, high-risk programmes (e.g. to eradicate poaching and illegal trade in wildlife), where stakeholders are likely to be global and might hold conflicting views, may benefit from an independent global-scale, multi-context evidence synthesis. This might require a rigorous analysis of what works, where and when and for whom, involving analysis of heterogeneity in outcome and identification of effect modifiers. Often within conservation, a broader set of evidence types (e.g. controlled trials, case studies, quantitative and qualitative research) is needed to fully capture the complexity of conservation contexts.

## 7.3 Key aspects of rigorous evidence synthesis and why they are needed

To be reliable, evidence syntheses should consider all available evidence and attempt to provide the most accurate and precise estimation of the truth. A suite of methodologies has been developed that maximises transparency and repeatability while minimising subjectivity, susceptibility to bias or influence of vested interest. The most widespread of these, systematic reviews and systematic maps, are well-documented secondary research methods that follow detailed guidance (e.g. Collaboration for Environmental Evidence, 2018) and use step-wise processes set out in an a-priori protocol to comprehensively identify and collate all available evidence (Table 7.1).

Systematic reviews in conservation and environmental management have most commonly aimed to answer specific cause-and-effect type questions, for example relating to the effect of a management intervention or exposure on a subject of concern. (e.g. 'What is the impact of a specific factor $x$ on a subject $z$?'). In contrast, systematic maps collate and catalogue available evidence on a relatively broad subject, describing the nature of the evidence base and highlighting evidence clusters and gaps, along with methodological patterns in primary research (Collaboration for Environmental Evidence, 2018). Systematic maps can be used as an initial step of an *evidence synthesis pathway* to identify subtopics suitable for a systematic review and subtopics where there is insufficient evidence to make synthesis of primary data worthwhile. In such latter cases, which are common in conservation, the map may identify individual primary studies that provide useful evidence (for an example of a systematic review question generated from a map, see www.eviem.se/en/projects/SR15-Prescribed-forest-burning/).

Systematic reviews were originally developed in response to an absence of easily accessible and rigorous synthesis of available evidence. However, recent assessments have shown that non-systematic reviews that aim to inform environmental policy and practice are still prevalent, but have low methodological reliability, suffering from lack of transparency and methodological rigour, and are consequently highly susceptible to bias (Woodcock et al., 2014, 2017; O'Leary et al., 2016). Moreover, the term 'systematic review' is often used by authors (and not challenged by editors or peer reviewers) when the reviews are in no way systematic. The production of substandard and

**Table 7.1** *Overview of systematic evidence synthesis stages and the issues they address. For an explanation of bias see Collaboration for Environmental Evidence (2018) or Bayliss and Beyer (2015)*

| Systematic review stage | Description | Defining features | Type of issue addressed |
|---|---|---|---|
| Review question identification and formulation (with stakeholder engagement) | Question is carefully identified and formulated with help of stakeholders | Social acceptance, relevance and legitimacy of the review process | |
| Protocol | Protocol outlines the intended method in detail. Protocol is peer-reviewed and published on an open-access platform | Public acceptance, peer review | Review bias, question creep |
| Searching for relevant literature | Comprehensive searches for grey and commercially published literature from a variety of sources | Comprehensiveness, repeatability (through transparency) | Publication bias |
| Eligibility screening | Careful screening of all identified articles according to pre-determined inclusion criteria | Consistency | Selection bias, review bias |
| Critical appraisal of study validity (optional for systematic maps) | A detailed assessment of the susceptibility to bias and generalisability of each study | Account for variability in internal validity and power of individual studies | Susceptibility to bias in individual studies and in study weighting by reviewers |
| Data coding and extraction | Transparent coding and, in case of systematic reviews, extraction of study finding | Consistency, repeatability (through transparency), minimising subjectivity | Selection bias |

**Table 7.1** (*cont.*)

| Systematic review stage | Description | Defining features | Type of issue addressed |
|---|---|---|---|
| Qualitative and/or qualitative data synthesis (not required for systematic maps) | Well-documented and comprehensive synthesis of qualitative and/or quantitative study findings | Comprehensiveness, repeatability (through transparency) | Selection bias, vote-counting, publication bias |
| Reporting and communication of review findings | Transparent reporting of the review results with extensive supplementary information | Repeatability (through transparency), avoiding overreach | Discussion bias |

'fake' systematic reviews is increasing in all fields, from public health to environmental management and education (Haddaway et al., 2016; Ioannidis, 2016; Haddaway, 2017; Pussegoda et al., 2017); they are 'fake' in the sense that they lack necessary comprehensiveness, transparency and reliability (Haddaway, 2017). This further confuses the issue for potential readers, with only a handful of environmental journals requiring authors to follow accepted standards of conduct and reporting (see Collaboration for Environmental Evidence, 2018). A potential evidence user can use keywords like 'systematic review' in their search and have it return documents that claim to be such, when in fact they are not. The misuse of the term 'systematic review' can undermine efforts towards effective decision-making and is a key reason for establishing independent standards.

Stakeholders, including scientists, rarely have the time or training to differentiate between a 'true' systematic review and one that misses critical components of the method (resulting in increased risk of bias and lack of transparency) especially when published in an outlet such as a peer-reviewed journal. To enhance the uptake of more rigorous and reliable synthesis methodologies and maximise the potential of evidence to inform decisions, independent coordinating bodies have been founded in different sectors of society to provide guidelines and standards for evidence synthesis. In the field of medicine this process began in the 1990s with the establishment of the Cochrane Collaboration, which aimed to conduct systematic reviews in order to provide healthcare professionals with the best available evidence on the effectiveness of clinical interventions (Higgins & Green, 2011). The methods were transferred to the field of conservation and environmental management in the early 2000s (Pullin & Stewart, 2006) and are now under the coordination of the Collaboration for Environmental Evidence. These independent coordinating bodies provide guidelines for and training in the conduct of systematic reviews and systematic maps, as well as registering, endorsing and publishing such evidence syntheses. Syntheses registered through the coordinating bodies are scrutinised by methodology experts, guaranteeing a level of reliability and rigour (Collaboration for Environmental Evidence, 2018).

In circumstances where vested interests might potentially influence the outcome of an evidence synthesis, these independent organisations provide a framework and platform to assist the review team to achieve and demonstrate independence of the synthesis process. The framework allows for full engagement of commissioners and other stakeholders in formulation of the review question and planning of the review protocol, followed by independent peer review and publication of the protocol prior to the conduct of the review. In cases where conflict or the risk of undue influence from particular stakeholders is high, the review process should be conducted by an independent review team and the report submitted for independent peer review. Following this process, the review findings may be endorsed by the independent organisation.

## 7.4 New developments that address barriers to evidence synthesis and communication

There are persistent barriers to the conduct of environmental evidence syntheses and communication of their findings. First, the high resource costs required have been a major disincentive to producing high-quality syntheses, despite their critical value for effective conservation. Second, efficient and effective means of communicating results and facilitating their use for real-life decision-making scenarios are haphazardly applied. These barriers limit the ability of evidence synthesis to dynamically and adaptively respond to conservation challenges. However, new developments in big data and deep learning approaches are offering exciting opportunities to harness evidence syntheses and promote them to broader audiences.

Conducting rigorous evidence syntheses, such as systematic reviews, can carry both significant monetary and human resource costs (Dicks et al., 2014). These costs are particularly prohibitive for organisations with critical needs for evidence, but who have limited time and resources to engage in such synthesis efforts or even to glean needed information from lengthy synthesis reports (Elliott et al., 2014). Moreover, high costs make updating syntheses to create a dynamic evidence base with the most up-to-date knowledge effectively impossible using current technology (Garritty et al., 2010). Additionally, the window of opportunity for decision-making may be shorter than the time in which a credible synthesis can be completed. Thus, to be useful to conservation, evidence syntheses must be optimised to efficiently find, collate and communicate existing evidence (Boyack & Klavans, 2014).

In a policy space where decision-making timelines are short and demands for rigorous, reliable evidence are high, methods assisted by advances in computing can support rapid evidence collation as well as increase cost efficiency (Shemilt et al., 2016). Computer-assisted approaches range from tools that manage data and streamline the synthesis process to tools powered by machine learning algorithms that allow rapid screening and extraction of evidence with reduced human intervention (Kohl et al., 2018). Promising computer-assisted approaches, including automatic term recognition, document clustering, automatic document classification and document summarisation (Frantzi et al., 2000; O'Mara-Eves et al., 2015) have been trialled in medical and health topics (Ananiadou et al., 2009) and are beginning to be tested in ecological topics (Westgate et al., 2015; Grubert & Siders, 2016; Roll et al., 2018).

These developments are encouraging for increased efficiency of the synthesis processes and potentially enabling dynamic syntheses that continuously update with new evidence as it becomes available. However, there are certain caveats and limitations that must be considered prior to widespread employment of computer-assisted tools. First, unlike medicine and fields such as economics, the semantics of conservation are highly heterogeneous and non-standardised (Westgate & Lindenmayer, 2017), posing difficulties for both efficient and comprehensive searching, and reliable application of machine learning algorithms to sort and mine text for desired patterns. Second, thus far, the performance of these approaches remains largely untested empirically, particular for conservation and environmental topics. As the value of evidence synthesis methods is in their transparency and credibility, reliable data on the efficacy of different computer-assisted approaches are important for uptake and expansion. Third, many existing computer-assisted platforms are fee-based or require programming skills, limiting their utility to a broader field of users. To improve global ability to address pervasive environmental threats, we need to democratise access to the tools that can help decision-making worldwide, not solely in countries or among researchers with means.

## 7.5 Mainstreaming evidence synthesis for decision support

Efforts to engage in open science and collaborative practice between conservation and technology fields will require forming collaborative partnerships and fostering conversation between evidence producers, evidence users and data scientists, to build a cohesive and engaged community of practice to open channels of communication to all users (Joppa, 2015). This will allow the broader community to use existing efforts as a starting point and avoid reinventing the wheel and wasting already limited resources (Lowndes et al., 2017). Furthermore, collaborative partnerships and creative funding can foster the long-term sustainability of tools that can live on to serve users. Too often, tools and platforms are created in good faith but require maintenance and updating and lack the ongoing funding and personnel to do so. This is particularly important as tools are most useful when they can dynamically respond to user needs and emerging technologies. This is a critical stepping stone for breaking down barriers to understanding and using evidence synthesis methodologies, as without a dynamic toolbox, synthesis methods will reman aloof from the needs of a diversifying and widening audience.

Evidence synthesis conducted to Collaboration for Environmental Evidence standards generates systematic reviews and systematic maps that are theoretically accessible to all. Yet, simply because something is available does not mean that the potential user is aware of it, knows where to find it, or even how to make sense of it. This is particularly the case for those new to the concept of evidence synthesis. Indeed, many practitioners and policy-makers rely on past experience or consult colleagues, rather than make use of the full suite of evidence (Pullin et al., 2004; Young et al., 2016). These issues create a number of inherent challenges for those decision-makers seeking to be evidence-informed and also broader potential audiences, such as stakeholders and wider society.

One of the mantras of science communication is 'know your audience' (Wilson et al., 2016; Cooke et al., 2017) and to have impact, the findings of an evidence synthesis need to be effectively tailored and communicated to different groups of people in different ways and through different media. Communication efforts should, for example, be sensitive to the fact that different groups vary in their 'trust' of the science they encounter from different sources (e.g. academic journals, colleagues, social media) (Wilson et al., 2016; Cooke et al., 2017).

A study that surveyed the willingness of practitioners to use a synopsis of relevant literature on bird conservation found that participants were more likely to use the evidence to inform decisions if it was easily accessible and in a clearly summarised format (Walsh et al., 2014). Similar summaries are needed to complement evidence syntheses. These summaries may then need to be further refined and transformed into policy briefs. Policy briefs are often written through the cultural lens of a given organisation and a given issue, meaning that these are unlikely to be useful if prepared in a generic format. Sundin et al. (2018) recently proposed the use of storytelling as a tool to effectively communicate the results of evidence syntheses. This method could give meaning to the evidence and can be communicated through videos (e.g. see https://youtu.be/4uPowxn2skg), presentations or public forums (e.g. newspapers, magazines). Nevertheless, uptake of these methods in science communication is generally slow and also could still rely on poorly conducted syntheses (McKinnon et al., 2018).

There has also been a rise in various knowledge management platforms and data-visualisation tools to explore underlying data that support evidence synthesis (e.g. www.3ieimpact.org/en/evaluation/evidence-gap-maps/, or www.cedar.iph.cam.ac.uk/resources/evidence/). These platforms present data from synthesis projects using interactive features and intuitive visualisations. For example, the Evidence for Nature and People Data Portal (www.natureandpeopleevidence.org) allows users to filter data according to desired parameters – such as diving into a data set to examine a specific intervention or outcome or geographic region, and visualising resultant trends. Syntheses, and in particular systematic maps, can be multi-layered and complex, precipitating a need for an interface that is graphical and intuitive, allowing a broader audience to use it (Figure 7.1).

| CONSERVATION INTERVENTION | | | | | | | | | | HUMAN WELL-BEING |
| Area protection | Land/water management | Resource management | Species management | Education & awareness | Law & policy | Livelihood incentives | Ext. capacity building | Sustainable use | Other | |
|---|---|---|---|---|---|---|---|---|---|---|
| 247 | 213 | 278 | 91 | 34 | 149 | 248 | 80 | 22 | 9 | Economic living standards |
| 158 | 151 | 185 | 52 | 20 | 99 | 119 | 28 | 16 | 5 | Material living standards |
| 22 | 16 | 24 | 6 | 6 | 12 | 17 | 4 | 0 | 3 | Health |
| 49 | 43 | 68 | 23 | 41 | 30 | 56 | 17 | 5 | 5 | Education |
| 102 | 105 | 140 | 45 | 18 | 75 | 89 | 47 | 13 | 2 | Social relations |
| 45 | 29 | 33 | 21 | 5 | 19 | 16 | 13 | 3 | 1 | Security & safety |
| 133 | 162 | 202 | 58 | 31 | 134 | 109 | 54 | 16 | 6 | Governance & empowerment |
| 36 | 37 | 23 | 19 | 15 | 24 | 47 | 25 | 10 | 3 | Subjective well-being |
| 21 | 17 | 10 | 11 | 5 | 6 | 21 | 8 | 2 | 3 | Culture/spirituality |
| 1 | 3 | 3 | 1 | 1 | 2 | 2 | 1 | 0 | 0 | Freedom of choice/action |
| 0 | 0 | 4 | 2 | 0 | 2 | 3 | 0 | 1 | 0 | Other |

NO. OF STUDIES

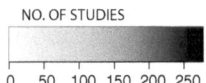

0   50   100   150   200   250

**Figure 7.1**  An example of an evidence 'heat map' linking conservation interventions with human well-being outcomes. The map allows the user to assess the evidence base for gaps and gluts as well as clicking on each box to further examine the relevant studies (after McKinnon et al., 2016). (A black and white version of this figure will appear in some formats. For the colour version, please refer to the plate section.)

If reported responsibly, these platforms and visualisations can play an important role in how stakeholders access evidence. A challenge for these approaches is to communicate that evidence syntheses are only estimates of the truth, which depend on the reliability of the evidence with which they were made. There is potential for evidence to be misinterpreted if the relative weight or reliability of a given element is misconstrued when visualised. Regardless of the output, it is important that authors of evidence syntheses communicate any uncertainty in the evidence and the risks associated with relying on studies that have high risk of bias.

Although it is laudable to communicate the findings of a topical evidence synthesis, additional efforts are also needed to communicate to practitioners the value of systematic reviews or maps, how they differ from other evidence synthesis methods and how they can be integrated with existing science advice and decision-making processes within different regions or institutions. Writing academic papers and delivering presentations at scientific conferences is unlikely to reach the typical practitioner, so creative approaches to outreach are needed to access and inform them.

Without use of rigorous evidence synthesis, policies and practice claiming to be 'evidence-informed' can be meaningless. For conservation and the environmental sector in general, the value of evidence synthesis has yet to be fully realised and we have the feeling that its time is yet to come. However, the recent methodological developments, awareness-raising and capacity development, together with new technologies for faster and more efficient conduct, suggest this time is not far away. Conservation is an interdisciplinary field and cannot remain for long in a state of relative evidence synthesis deficit in comparison with other sectors with which it seeks to be relevant. Although still marginalised, the methodology and infrastructure to build conservation's evidence base through rigorous synthesis now exist at a global level. A commitment to evidence-informed decision-making that recognises the central role of rigorous evidence synthesis is required by key actors in the sector if these potential benefits are to be achieved.

## References

Ananiadou, S. S., Rea, B. B., Okazaki, N. N., et al. 2009. Supporting systematic reviews using text mining. *Social Science Computer Review*, 27, 509–523. https://doi.org/10.1177/0894439309332293

Bayliss, H. R. & Beyer, F. R. 2015. Information retrieval for ecological syntheses. *Research Synthesis Methods*, 6, 136–148.

Bowler, D., Buyung-Ali, L., Healey, J. R., et al. 2010. *The Evidence Base for Community Forest Management as a Mechanism for Supplying Global Environmental Benefits and Improving Local Welfare: A STAP Advisory Document.* Washington, DC: Scientific and Technical Advisory Panel. Global Environment Facility.

Boyack, K. W. & Klavans, R. 2014. Creation of a highly detailed, dynamic, global model and map of science. *Journal of The Association for Information Science and Technology*, 65, 670–685. https://doi.org/10.1002/asi.22990

Collaboration for Environmental Evidence. 2018. *Guidelines and Standards for Evidence Synthesis in Environmental Management. Version 5.0.* Available from www.environmentalevidence.org/information-for-authors (accessed 8 March 2018).

Cook, C. N., Mascia, M. B., Schwartz, M. W., et al. 2013. Achieving conservation science that bridges the knowledge–action boundary. *Conservation Biology*, 27, 669–678. https://doi.org/10.1111/cobi.12050

Cooke, S. J., Gallagher, A. J., Sopinka, N. M., et al. 2017. Considerations for effective science communication. *FACETS*, 2, 233–248. https://doi.org/10.1139/facets-2016-0055

Dicks, L. V., Walsh, J. C. & Sutherland, W. J. 2014. Organising evidence for environmental management decisions: a '4S' hierarchy. *Trends in Ecology and Evolution*, 29, 607–613. https://doi.org/10.1016/j.tree.2014.09.004

Elliott, J. H., Turner, T., Clavisi, O., et al. 2014. Living systematic reviews: an emerging opportunity to narrow the evidence-practice gap. *PLoS Medicine*, 11, 1–6. https://doi.org/10.1371/journal.pmed.1001603

Frantzi, K., Ananiadou, S. & Mima, H. 2000. Automatic recognition of multi-word terms: the C-value/NC-value method. *International Journal on Digital Libraries*, 3, 115–130. https://doi.org/10.1007/s007999900023

Garritty, C., Tsertsvadze, A., Tricco, A. C., et al. 2010. Updating systematic reviews: an international survey. *PLoS ONE*, 5, e9914. https://doi.org/10.1371/journal.pone.0009914

Grubert, E. & Siders, A. 2016. Benefits and applications of interdisciplinary digital tools for environmental meta-reviews and analyses. *Environmental Research Letters*, 11, 093001. https://doi.org/10.1088/1748-9326/11/9/093001

Haddaway, N. R., Land, M. & Macura, B. 2016. 'A little learning is a dangerous thing': a call for better understanding of the term 'systematic review'. *Environment International*, 99, 356–360. https://doi.org/10.1016/j.envint.2016.12.020

Haddaway, N. R. 2017. Response to 'Collating science-based evidence to inform public opinion on the environmental effects of marine drilling platforms in the Mediterranean Sea'. *Journal of Environmental Management*, 203, 612–614. https://doi.org/10.1016/j.jenvman.2017.03.043

Haddaway, N. R. & Crowe, S. 2018. Experiences and lessons in stakeholder engagement in environmental evidence synthesis: a truly special series. *Environmental Evidence*, 7, art11.

Higgins, J. & Green, S. 2011. *Cochrane Handbook for Systematic Reviews of Interventions. Version 5.1.* Cochrane Collaboration. Available from http://handbook-5-1.cochrane.org (accessed 8 March 2018).

Ioannidis, J. P. A. 2016. The mass production of redundant, misleading, and conflicted systematic reviews and meta-analyses. *The Milbank Quarterly*, 94, 485–514. https://doi.org/10.1111/1468-0009.12210

Joppa, L. N. 2015. Technology for nature conservation: an industry perspective. *Ambio*, 44, 522–526. https://doi.org/10.1007/s13280-015-0702-4

Kløcker Larsen, R. & Nilsson, A.E. 2017. Knowledge production and environmental conflict: managing systematic reviews and maps for constructive outcome. *Environmental Evidence*, 6. https://doi.org/10.1186/s13750-017-0095-x

Kohl, C., McIntosh, E. J., Unger, S., et al. 2018. Online tools supporting the conduct and reporting of systematic reviews and systematic maps: a case study on CADIMA and review of existing tools. *Environmental Evidence*, 7(8). https://doi.org/10.1186/s13750-018-0115-5

Koricheva, J. & Gurevitch, J. 2014. Uses and misuses of meta-analysis in plant ecology. *Journal of Ecology*, 102, 828–844. https://doi.org/10.1111/1365-2745.12224

Land, M., Macura, B., Bernes, C., et al. 2017. A five-step approach for stakeholder engagement in prioritisation and planning of environmental evidence syntheses. *Environmental Evidence*, 6(25). https://doi.org/10.1186/s13750-017-0104-0

Leisher, C., Temsah, G., Booker F., et al. 2016. Does the gender composition of forest and fishery management groups affect resource governance and conservation outcomes? A systematic map. *Environmental Evidence*, 5(6). https://doi.org/10.1186/s13750-016-0057-8

Li, W. & Zhao, Y. 2015. Bibliometric analysis of global environmental assessment research in a 20-year period. *Environmental Impact Assessment Review*, 50, 158–166. https://doi.org/10.1016/j.eiar.2014.09.012

Lowndes, J. S. S., Best, B. D., Scarborough, C., et al. 2017. Our path to better science in less time using open data science tools. *Nature Ecology and Evolution*, 1, 160. https://doi.org/10.1038/s41559-017-0160

McKinnon, M. C., Cheng, S. H., Dupre, S., et al. 2016. What are the effects of nature conservation on human well-being? A systematic map of empirical evidence from developing countries. *Environmental Evidence*, 5(8).

McKinnon, M. C., Cheng, S., Dicks, L., et al. 2018 Seek higher standards to honestly assess conservation effectiveness. *Mongabay*. Available online on 17 April 2018.

O'Leary, B. C., Kvist, K., Bayliss, H. R., et al. 2016. The reliability of evidence reviews in environmental science and conservation. *Environmental Science Policy*, 64, 75–82. https://doi.org/10.1016/j.envsci.2016.06.012

O'Mara-Eves, A., Thomas, J., McNaught, J., et al. 2015. Using text mining for study identification in systematic reviews: a systematic review of current approaches. *Systematic Reviews*, 4(5). https://doi.org/10.1186/2046-4053-4-5

Ojanen, M., Zhou, W., Miller, D. C., et al. 2017. What are the environmental impacts of property rights regimes in forests, fisheries and rangelands? *Environmental Evidence*, 6(12). https://doi.org/10.1186/s13750-017-0090-2

Oliver, S. & Dickson, K. 2016. Policy-relevant systematic reviews to strengthen health systems: models and mechanisms to support their production. *Evidence and Policy*, 12(2), 235–259. https://doi.org/10.1332/174426415X14399963605641

Owens, S. 2000. 'Engaging the public': information and deliberation in environmental policy. *Environment and Planning A*, 32, 1141–1148. https://doi.org/10.1068/a3330

Plume, A. & van Weijen, D. 2014. Publish or perish? The rise of the fractional author . . . . *Research Trends*, 38. www.researchtrends.com/issue-38-september-2014/publish-or-perish-the-rise-of-the-fractional-author/ (accessed 12 December 2019).

Pullin, A. S., Knight, T. M., Stone, D. A., et al. 2004. Do conservation managers use scientific evidence to support their decision-making? *Biological Conservation*, 119, 245–252. https://doi.org/10.1016/j.biocon.2003.11.007

Pullin, A. S. & Stewart, G. B. 2006. Guidelines for systematic review in conservation and environmental management. *Conservation Biology*, 20, 1647–1656. https://doi.org/10.1111/j.1523-1739.2006.00485.x

Pullin, A. S. & Knight, T. M. 2009. Doing more good than harm: building an evidence-base for conservation and environmental management. *Biological Conservation*, 142, 931–934. https://doi.org/10.1016/j.biocon.2009.01.010

Pussegoda, K., Turner, L., Garritty, C., et al. 2017. Systematic review adherence to methodological or reporting quality.

*Systematic Reviews*, 6(131). https://doi.org/10 .1186/s13643-017–0527-2

Roll, U., Correia, R. A. & Berger-Tal, O. 2018. Using machine learning to disentangle homonyms in large text corpora. *Conservation Biology*, 32, 716–724. https://doi .org/10.1111/cobi.13044

Salafsky, N. & Redford, K. 2013. Defining the burden of proof in conservation. *Biological Conservation*, 166, 247–253. https://doi.org /10.1016/j.biocon.2013.07.002

Shemilt, I., Khan, N., Park, S., et al. 2016. Use of cost-effectiveness analysis to compare the efficiency of study identification methods in systematic reviews. *Systematic Reviews*, 5, 140. https://doi.org/10.1186 /s13643-016–0315-4

Snilstviet, B., Stevenson, J., Villar, P. F., et al. 2016. Land-use change and forestry programmes: evidence on the effects on greenhouse gas emissions and food security. www.3ieimpact.org/media/filer_ public/2016/11/17/egm3-landuse-forest.pdf (accessed 8 March 2018).

Sundin, A., Andersson, K. & Watt, R. 2018. Rethinking communication: integrating storytelling for increased stakeholder engagement in environmental evidence synthesis. *Environmental Evidence*, 7(6). https://doi.org/10.1186/s13750-018–0116-4

Walsh, J. C., Dicks, L. V. & Sutherland, W. J. 2014. The effect of scientific evidence on conservation practitioners' management decisions. *Conservation Biology*, 29, 88–98. https://doi.org/10.1111/cobi.12370

Ware, M. & Mabe, M. 2015. An overview of scientific and scholarly journal publishing. International Association of Scientific,

Technical and Medical Publishers, www .stm-assoc.org /2015_02_20_STM_Report_2015.pdf (accessed 8 March 2018).

Westgate, M. J., Barton, P. S, Pierson, J. C., et al. 2015. Text analysis tools for identification of emerging topics and research gaps in conservation science. *Conservation Biology*, 29, 1606–1614. https://doi.org/10.1111/cobi .12605

Westgate, M. J. & Lindenmayer, D. B. 2017. The difficulties of systematic reviews. *Conservation Biology*, 31, 1002–1007. https:// doi.org/10.1111/cobi.12890

Wilson, M. J., Ramey, T. L., Donaldson, M. R., et al. 2016. Communicating science: sending the right message to the right audience. *FACETS*, 1(1), 127–137. https://doi .org/10.1139/facets-2016–0015

Woodcock, P., Pullin, A. S. & Kaiser, M. J. 2014. Evaluating and improving the reliability of evidence in conservation and environmental science: a methodology. *Biological Conservation*, 176, 54–62. https:// doi.org/10.1016/j.biocon.2014.04.020

Woodcock, P., O'Leary, B. C., Kaiser, M. J., et al. 2017. Your evidence or mine? Systematic evaluation of reviews of marine protected area effectiveness. *Fish and Fisheries*, 18, 668–81. https://doi.org/10.1111/faf.12196

Young, N., Corriveau, M., Nguyen, V. M., et al. 2016. How do potential knowledge users evaluate new claims about a contested resource? Problems of power and politics in knowledge exchange and mobilization. *Journal of Environmental Management*, 184, 380–388. https://doi.org/10.1016/j .jenvman.2016.10.006

# Aligning evidence for use in decisions: mechanisms to link collated evidence to the needs of policy-makers and practitioners

LYNN V. DICKS
*University of Cambridge*
BARBARA LIVOREIL
*Fondation pour la recherche sur la biodiversité*
REBECCA K. SMITH
*University of Cambridge*
HEIDI WITTMER
*Helmholtz Centre for Environmental Research – UFZ*

and

JULIETTE YOUNG
*Centre for Ecology and Hydrology and AgroSup Dijon*

## 8.1 Introduction

We should not be surprised by the scale of the challenge when trying to link a body of scientific knowledge to the complex, shifting and see-mingly unpredictable world of policy, or to the massively decentralised, globally distributed world of conservation practice (Young et al., 2014). One side of the challenge is developing a consensual understanding of the science itself. By nature, scientific knowledge is continually progres-sing, with theories, empirical data and new interpretations emerging all the time. Even within a single discipline, it can be hard to convey what is known at a particular point in time, and this often involves presenting different scientific viewpoints. For instance, there is substantial variation around the world in public health advice regarding alcohol consump-tion, with 'safe' limits in the UK being 50% of those in the USA (Wood et al., 2018). In conservation, the challenge is even greater, as relevant research cuts across the natural, physical and social sciences.

The other side of the challenge is working out how, and when, to offer relevant scientific knowledge to decision-makers, in order to have the greatest impact on the decisions being made. This is the focus of our chapter. We argue that it is a question of correct alignment: of selecting the right knowledge to address the needs of decision-makers, ensuring that knowledge is accessible to them, and articulating it within their decision-making processes.

First, we consider how well current efforts to synthesise evidence in conservation align with the needs of decision-makers. Then we describe three mechanisms that might be used to enhance the alignment of available knowledge with decision-making, starting at small local scales and moving to the global scale: decision support tools, active knowledge exchange and large-scale scientific assessments. For each mechanism, we provide examples and draw out general guidelines regarding the circumstances in which it is likely to be most effective.

## 8.2  How well do current evidence synthesis activities align with policy and practice needs?

When scientific evidence is needed for decision-making, the process of obtaining and analysing the evidence is often demand-led. An organisation faced with a difficult management or policy decision will undertake or commission a review to answer a specific question. For example, the UK Government Department of Environment, Food and Rural Affairs (Defra) commissioned a review of evidence on the status of pollinators (Vanbergen et al., 2014) before designing the National Pollinator Strategy for England (Defra, 2014). When this happens, the evidence synthesis is well-aligned with the policy and practice needs, summarising relevant material that can be found in the time available. However, it also puts immense time pressure on the evidence synthesis process, because decision-making can only happen once the evidence has been reviewed. This tends to lead to the selection of evidence synthesis methods such as rapid evidence assessments, traditional non-systematic literature reviews and expert consultations, which are not the most rigorous or unbiased approaches available (Dicks et al., 2017).

The Collaboration for Environmental Evidence (www.environmentalevi dence.org) and the Conservation Evidence project (www.conservationevi dence.com) aim to address the needs of conservation practitioners and policy-makers with more rigorous methods of knowledge synthesis, namely systematic reviews, systematic maps (Collaboration for Environmental Evidence, 2013; see also Chapter 7) and subject-wide evidence syntheses (Sutherland et al., 2019b; see also Chapter 4). They do so by actively involving stakeholders in the selection of topics to synthesise and the collation and subsequent evaluation of the evidence found (Dicks et al., 2016; Haddaway et al., 2017).

To evaluate the overall success of this alignment effort, we recently asked how well evidence collated by the Conservation Evidence project on the subject of sustainable food production matched the priority knowledge needs of decision-makers. Five independent exercises (Pretty et al., 2010; Dicks et al., 2013a, 2013b; Ingram et al., 2013; Jones et al., 2014), involving 240 people from across business, practice, policy-making and academia, had generated 286 priority questions faced by decision-makers. We sorted these into five categories, following the Driver–Pressure–State–Impact–Response (DPSIR) framework (Maxim et al., 2009). This conceptual framework describes interactions between society and the environment in a way that is meaningful for policy. Social and economic developments (Driving Forces, D) exert Pressures (P) on the environment and, as a consequence, the State (S) of the environment changes. This leads to Impacts (I) on ecosystems, human health and society, which may elicit a societal Response (R) that feeds back on D, S or I. We added a category for questions about underlying science that did not fit the DPSIR categories (Figure 8.1).

Of all the priority questions, 189 (66%) were about responses (R), which are the focus of the Conservation Evidence project. Evidence had already been summarised that could help answer 35 of these questions (12% overall; Smith et al., 2015; Sutherland et al., 2019a).

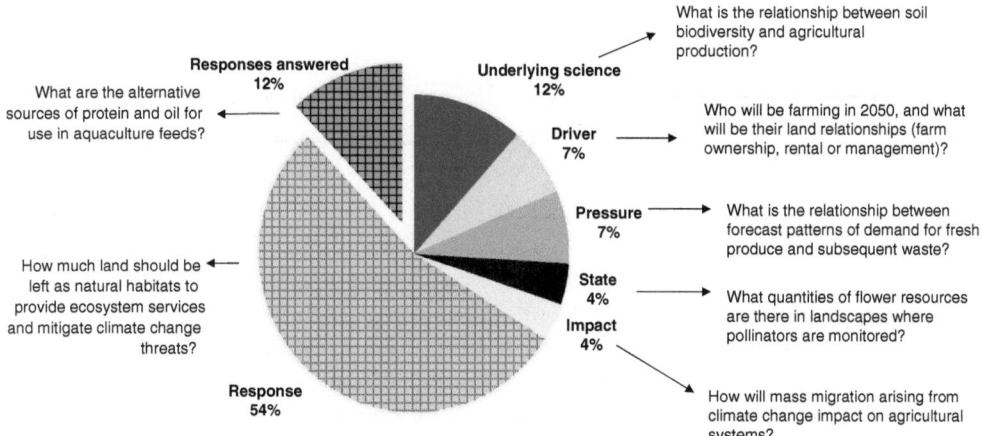

**Figure 8.1** Categorisation of 286 priority questions identified by stakeholders as relevant to sustainable food production (Pretty et al., 2010; Dicks et al., 2013a, 2013b; Ingram et al., 2013; Jones et al., 2014) according to the Driver–Pressure–State–Impact–Response framework. Examples of questions are provided for each category. The extracted segment represents questions already answered by evidence summaries provided by the Conservation Evidence project.

In a similar vein, Cook et al. (2013a) investigated the contribution of systematic reviews to conservation decision-making, finding that 35% of the 43 reviews considered practical on-the-ground management, while most addressed interventions relevant to policy. Cook et al. (2013a) argued that the benefits for conservation could be significantly enhanced by increasing the number of systematic reviews focused on questions of direct management relevance.

These two analyses show there is some alignment between high-quality evidence synthesis methods and the needs of conservation practitioners and policy-makers, but it could be improved. Below, we provide a series of examples of mechanisms to enhance this alignment at a range of scales.

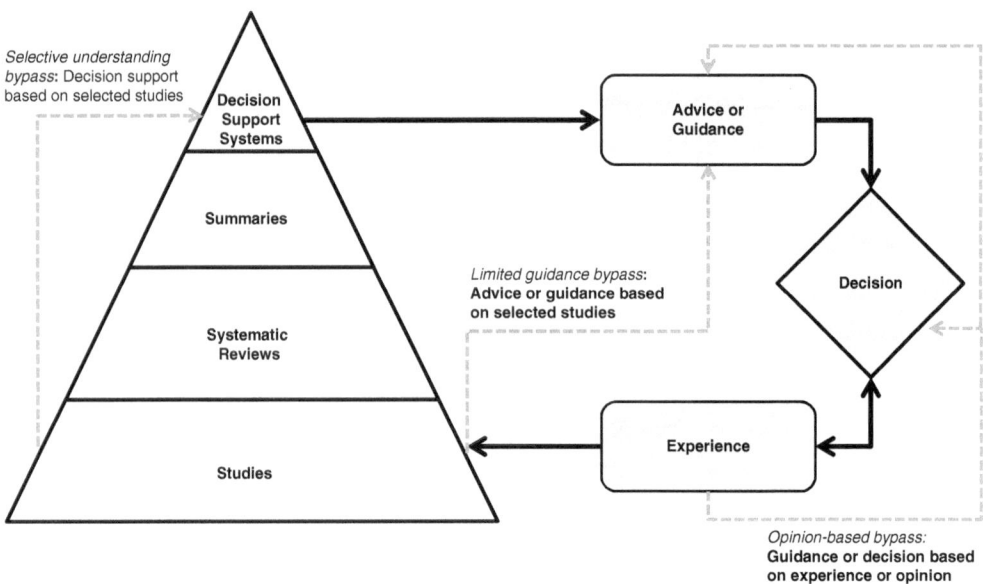

**Figure 8.2** A schematic showing how scientific information could support environmental decision-making (Dicks et al., 2014). The triangle on the left shows an evidence hierarchy, in which summaries, such as those produced by the Conservation Evidence project, integrate evidence from across studies and systematic reviews, and form the basis for information flowing into decision support systems. In these circumstances, environmental decisions (shown by the 'Decision' diamond on the right) are based on the best-available evidence, combined with the expertise and local knowledge of the practitioner or policy-maker (described by the 'Experience' box). Dashed lines illustrate bypass routes currently taken to inform environmental decisions.

## 8.3 Decision support systems

Decision support systems are tools designed to assist decision-makers, for example, by visually or numerically illustrating different possible outcomes to a question, or leading users through logical decision steps (Dicks et al., 2014). Often software-based, they represent a link between relevant science and decision-making (Dicks et al., 2014; Figure 8.2). Decision support systems are useful for incorporating evidence into decisions related to a specific question that has been widely and repeatedly addressed. It is also important that the evidence can be converted into simple numerical or visual formats.

There are many decision support tools available covering various aspects of environmental science. For instance, Zasada et al. (2017) identified 60 research projects funded between 2002 and 2013 under the European Commission's 6th and 7th Framework Programmes that had developed decision support tools for landscape and environmental management. Of these, only 61% still existed in 2014, and only half were updated after the projects that developed them ended, although this seems a pre-requisite for ongoing use. The uptake of decision support systems depends on a range of factors, including ease of use, performance, whether they are recommended by peers and the level of marketing (Rose et al., 2016). Uptake can be enhanced by ensuring that users are closely involved in the conception and design of the tools (Rose et al., 2018).

While decision support systems are often designed by researchers as a way of incorporating scientific knowledge into practice, most are based on one particular model, study or approach to a scientific question and represent a 'bypass' of the evidence hierarchy (Figure 8.2 and see Dicks et al., 2014). There are only a few examples where they represent the best-available scientific knowledge, based on rigorous synthesis of evidence.

One such decision support tool is the online biodiversity metric incorporated into the Cool Farm Tool (available at www.coolfarmtool.org), which provides scores for the likely benefits for biodiversity of a range of farm management actions. The actions that are included are selected according to a combination of expert judgement and assessments of summarised evidence conducted by the Conservation Evidence project. Each farm management action is assigned scores reflecting the benefit for overall biodiversity, and also for 11 species groups (e.g. woodland birds, beneficial invertebrates), weighted according to the evidence. Actions that are strongly supported by the evidence provided by the Conservation Evidence syntheses (Sutherland et al., 2019a) are scored more highly than those for which effectiveness is not known.

Another example is the set of greenhouse gas emission calculators used in agriculture to support mitigation by changing farm management. These tools incorporate models of greenhouse gas emissions and carbon storage according to vegetation type and farming practice (Richards et al., 2016). These calculators combine empirical models with emission factors collated by the Intergovernmental Panel on Climate Change (see 'National and International Scientific Assessments'). Although the outputs from these tools are only as good as the data that they are based on, new information can be added to improve their performance as it becomes available. For example, Richards et al. (2016) demonstrated that two widely used software tools tend to over-estimate emissions from smallholder farms in tropical environments, but suggest that this is probably due to a systematic bias in literature, with most data coming from temperate regions, rather than bias in the models themselves. As empirical data are included from a wider range of environments, more accurate disaggregated emissions factors will become available for different parts of the world. If the decision support systems are maintained and updated, this new knowledge will directly influence decision-making at farm level.

## 8.4 Active knowledge exchange mechanisms

Active knowledge exchange mechanisms are the most diverse alignment mechanism of the three considered in this chapter. Our concept is similar to that of 'boundary organisations' identified by some other authors (Guston, 2001; Cook et al., 2013b), in that they operate in both scientific and practical spheres, but retain distinct lines of accountability to both groups. They can take a variety of institutional forms, from a dedicated, self-funded or government-funded organisation to a network of people working together across organisations (see also Chapter 13).

The reputation of such a body depends on its ability to produce or broker knowledge that is salient, credible and legitimate (Cash et al., 2003; Sarkki et al., 2015) while maintaining transparency. Credibility refers to the scientific adequacy of the technical evidence and arguments. Salience is the relevance of the brokered knowledge to the needs of decision-makers. Legitimacy reflects the perceptions that the production of information has been respectful of stakeholders' divergent values and beliefs, unbiased in its conduct and fair in its treatment of views and interests. Achieving all these values requires adequate attention to governance from the outset.

Here, we provide examples of knowledge exchange mechanisms operating at a subnational scale, related to a particular environmental issue or landscape (Wadden Sea case study); at a national or international scale but restricted to environmental science (EKLIPSE mechanism); and at a national or international scale ranging across all scientific knowledge (European Scientific Advice Mechanism, and UK Parliamentary Office of Science and Technology).

### 8.4.1 Management of the Wadden Sea

At a subnational scale, van Enst et al. (2016) provided a detailed case study of three contrasting knowledge exchange mechanisms that have been important in aligning scientific evidence with policy and management decisions around the Wadden Sea, a shallow estuarine sea in the Netherlands. Competing cockle-fishing, gas extraction and biodiversity conservation interests generate continuous debate over the scientific knowledge, and the strategic use or misuse of such knowledge has played a pivotal role in disputes (Floor et al., 2013). Knowledge exchange mechanisms were devised to improve the transparent use of evidence. Two of the knowledge exchange mechanisms were government-funded: the Wadden Academy, a science-led organisation that oversees monitoring and data-gathering, and the Netherlands Commission for Environmental Assessment, which produces official reports. The third, IMSA Amsterdam, is a commercial think-tank and consultancy, focused on mediating between stakeholders, science and policy. These three organisations worked together to improve the salience, credibility and legitimacy of the scientific knowledge that was available, allowing it to be influential in decision-making related to the cockle-fishery and gas-exploitation controversies. Their efforts ultimately reduced conflict and improved environmental outcomes for the Wadden Sea, for example by enabling more sustainable fishing methods to be adopted (van der Molen et al., 2015; van der Molen, 2018).

### 8.4.2 The EKLIPSE mechanism

Knowledge exchange mechanisms focused on one environmental issue can develop deep, long-term relationships between a core set of stakeholders and researchers. When operating across many different issues at national or international scale, relationships with experts and other stakeholders are generally short-term and must continually be re-established as the topic of interest to policy changes. One possible approach to this is provided by the EKLIPSE mechanism (Watt et al., 2018; www.eklipse-mechanism.eu), which engages relevant actors from science, policy and society to identify evidence relevant to European policy. EKLIPSE accepts requests for knowledge synthesis on specific issues from policy-makers and other societal actors. A wide network of knowledge-holders can respond to the request, often through the formation of an expert working group (Wyborn et al., 2018). To give an example, the European Commission requested scientific knowledge on how to evaluate nature-based solutions (solutions inspired and supported by nature) for their ability to enhance sustainability in cities. In response, EKLIPSE convened a pan-European expert group to conduct a rapid evidence assessment and build a framework for evaluating the costs and benefits of nature-based solutions. This was disseminated as a policy report and an open-access scientific paper (Raymond et al., 2017).

### 8.4.3  The European Scientific Advice Mechanism and UK Parliamentary Office of Science and Technology

At a larger scale, knowledge exchange mechanisms can provide an interface between science and policy across all scientific issues. Usually these are national or international, such as the UK Parliamentary Office for Science and Technology (POST; Norton, 1997) and the European Union Scientific Advice Mechanism (ec.europa.eu/research/sam/index.cfm). At this level, knowledge exchange mechanisms have tended to settle on one particular way of doing things that works. At the POST, for instance, a Board selects subjects for briefing notes, known as POSTnotes, from among ideas gathered from a range of sources, including parliamentarians, the public and other stakeholders (www.parliament.uk/post). POSTnotes are generally researched through a series of interviews with key experts. Almost 600 POSTnotes have been published since 1989, on subjects ranging from the psychological health of military personnel to new plant-breeding technologies. All are freely available online and held in the House of Commons library.

The European Union Scientific Advice Mechanism, on the other hand, responds to requests for advice from the 'College of European Commissioners' through a group of government-appointed scientific advisers. It delivers evidence review reports on specific issues, drawing on a network of expertise from more than 100 European scientific academies in over 40 countries (e.g. The Royal Society in the UK, Hungarian Academy of Sciences). For both it and POST, adherence to a clearly defined process is a way of building credibility and assuring transparency. However, it does not necessarily provide the flexibility to address the diversity of issues and problems faced by environmental policy decision-makers.

To summarise, active knowledge exchange mechanisms can have a range of scales, formats and institutional arrangements. This plurality is the best approach to linking science and policy in decision-making contexts, where different types of questions continually arise.

### 8.5  National and international scientific assessments

A longer-term approach to aligning evidence synthesis with conservation policy decisions involves governments or international bodies mandating large-scale, scientific assessments in broad areas of strong policy interest. Examples include the assessment reports conducted by the Intergovernmental Panel on Climate Change (IPCC; www.ipcc.ch), Intergovernmental Science Policy Platform on Biodiversity and Ecosystem Services (IPBES; www.ipbes.net) and Millennium Ecosystem Assessment (www.millenniumassessment.org; see Chapter 16 for further details of mechanism and function of the Millennium Ecosystem Assessment and the IPBES science–policy platform). These global assessments involve hundreds or even thousands of scientists around the world, including indigenous and local knowledge-holders in the case of IPBES (Sutherland et al., 2014; see also Chapter 16).

Generally, governments define the scope of the assessment and identify or nominate a set of experts to conduct it (IPCC, 2015). The nominated experts form working groups and develop report texts, which are subject to extensive, transparent review, first by other experts and then by governments. Following review, the report texts are converted into concise summary documents (usually called 'Summary for Policy-makers'), the final text of which is agreed by governments. Each statement in the summary document must be traceable back to the full scientific report and, from there, to individual pieces of research or sources of knowledge. Through this process, science and policy influence one another in a two-way exchange of knowledge over very large temporal and spatial scales.

The IPCC, which has been active for almost three decades, has built a strong reputation for providing an overview of climate science across a range of disciplines, from geophysics to economics. There are now clear links from the scientific understanding of human-induced climate change and its impacts to policies controlling greenhouse gas emissions at national and international levels. Most recently, the Paris Climate Agreement of December 2015 is a global accord under which nations have made pledges and set emissions targets to keep global temperature rise below 2°C (Clemencon, 2016; Tobin et al., 2018). A large quantity of scientific research underlies these policy pledges, which would likely not have happened, or not have been so extensive, without the IPCC assessment process. Forty-five different global climate models are now being used together to link levels of greenhouse gas emissions to long-term global temperature rise under different emissions scenarios (Collins et al., 2013). There is also a plethora of analyses and modelling connecting economic activity to greenhouse gas emissions (e.g. Vandyck et al., 2016) and threshold temperate rises with specific impacts on environments, economies and human well-being (IPCC, 2014).

The Millennium Ecosystem Assessment (2005) was the first global evaluation of the status of ecosystems, and developed the ecosystem services framework for understanding how nature can benefit people. The ecosystem services concept originated in the academic world (Potschin & Haines-Young, 2016), but the Millennium Ecosystem Assessment formalised the thinking, providing a conceptual framework and nomenclature for ecosystem services. Since its publication, a growing number of countries have conducted their own national ecosystem assessments (Schröter et al., 2016) and the policy ground is being set for their results to be used in national natural-capital accounting. Both Aichi Biodiversity Target 2 from the Convention on Biological Diversity's Strategy Plan 2011–2020 (Convention

Table 8.1 *A summary of the costs associated with three mechanisms to align evidence synthesis with policy and practice in the environmental field, compared to the costs of individual evidence synthesis methods*

| Activity | When to apply | Cost (£) |
|---|---|---|
| *Mechanisms to align evidence synthesis with the needs of policy and practice* | | |
| Decision support tools | Specific question, repeatedly addressed | 380,000–3.9 million per tool[1] |
| Knowledge exchange mechanisms | Many questions arising | 600,000 per year[2] |
| International assessments | One big, broad issue | ~3 million per year[3] |
| *Individual evidence synthesis methods* | | |
| Systematic review | Many studies address a single question | 19,000–190,000[1] |
| Subject-wide evidence synthesis | Multiple sources of relevant evidence exist | Initial cost: 45,000–480,000 Update cost: 20% of initial cost[1] |

[1]  Dicks et al. (2014);
[2]  Cost of the EKLIPSE mechanism;
[3]  www.ipcc.ch; www.ipbes.net.

on Biological Diversity, 2010) and Action 5 of the EU Biodiversity Strategy to 2020 (European Commission, 2011) call for biodiversity values to be incorporated into national accounting.

Large-scale assessments are most effective at aligning scientific evidence with decisions when there is a broad issue of strong political interest, such as climate change or biodiversity loss. The assessments are expensive (see Table 8.1), so there must be substantial political commitment and a source of funds over the relatively long term.

Given the obvious power of national and international scientific assessments to influence policy, it is now more important than ever to incorporate into them the transparent, unbiased repeatable methods that have been developed for evidence synthesis. Currently, the rigour and reliability of large-scale scientific assessments rely on extensive peer review, rather than systematic searching or careful elicitation methods that reduce bias. Evidence synthesis methods are usually not reported (with some exceptions, such as chapter 6 of the Intergovernmental Science Policy Platform on Biodiversity and Ecosystem Services pollination report; IPBES, 2016). However, such assessments are conducted over long timescales, with the IPCC, for example, producing a global assessment report every 5–10 years. With this amount of time and money available (see Table 8.1) there is a clear opportunity to develop rigorous processes of evidence synthesis within this framework. As a first step, we urge policy-makers and institutions involved in commissioning large-scale scientific assessments to require authors to report their underlying synthesis methods.

## 8.6 What does it all cost?

The cost of the alignment mechanisms outlined in this chapter varies considerably, both within and among the different activities (Table 8.1). These costs should be interpreted in the context of total spending on scientific research. For example, the budget of the European Commission's flagship scientific research programme, Horizon 2020, is approximately £8 billion per year.

The organisations that fund research and aspire to be evidence-informed already invest heavily in improving interactions between science, policy and practice. Unfortunately, they frequently fund expensive decision support systems that are not maintained or used a few years later (Zasada et al., 2017) and large-scale reviews or scientific assessments that do not follow clear protocols to reduce bias. The challenge in aligning evidence synthesis with decision-making is not to find the money, but to demand and enable improved rigour and continuity in activities that are already taking place.

No single mechanism will be best for aligning evidence with policy and practice in all contexts. Each has strengths and weaknesses, and can be applied in different circumstances and at different scales. International assessments have redirected policies and scientific endeavour on a very large scale, but would be unlikely to align specific scientific findings with conservation practice at smaller scales. At smaller scales, the potential of decision support systems to incorporate rigorously collated environmental evidence has hardly been tapped.

At every level, mechanisms to link synthesised evidence with policy and practice decisions need to be funded sufficiently to ensure salience, legitimacy, credibility and transparency. These linking mechanisms need access to methods of collating and communicating evidence that are well-developed, transparent and widely understood (Cook et al., 2017; Dicks et al., 2017) and are just as important as the research itself, if not more so.

## References

Cash, D. W., Clark W. C., Alcock, F., et al. 2003. Knowledge systems for sustainable development. *Proceedings of the National Academy of Sciences*, 100, 8086–8091.

Clemencon, R. 2016. The two sides of the Paris climate agreement: dismal failure or historic breakthrough? *Journal of Environment & Development*, 25, 3–24.

Collaboration for Environmental Evidence. 2013. *Guidelines for Systematic Review and Evidence Synthesis in Environmental Management. Version 4.2.* Available from

www.environmentalevidence.org
/Documents/Guidelines/Guidelines4.2.pdf.

Collins, M., Knutti, R., Arblaster, J., et al.
2013. Long-term climate change:
projections, commitments and
irreversibility. In: Stocker, T. F., Qin, D.,
Plattner, G.-K., et al., editors, *Climate
Change 2013: The Physical Science Basis.
Contribution of Working Group I to the Fifth
Assessment Report of the Intergovernmental
Panel on Climate Change.* Cambridge:
Cambridge University Press.

Convention on Biological Diversity. 2010.
Decision X/2. The Strategic Plan for
Biodiversity 2011–2020 and the Aichi
Biodiversity Targets. UNEP/CBD/COP/DEC/
X/2, 29 October 2010.

Cook, C. N., Mascia, M. B., Schwartz, M. W., et al.
2013b. Achieving conservation science that
bridges the knowledge-action boundary.
*Conservation Biology*, 27, 669–678.

Cook, C. N., Nichols, S. J., Webb, J. A., et al. 2017.
Simplifying the selection of evidence
synthesis methods to inform
environmental decisions: a guide for
decision-makers and scientists. *Biological
Conservation*, 213(Part A), 135–145.

Cook, C. N., Possingham, H. P. & Fuller, R. A.
2013a. Contribution of systematic reviews
to management decisions. *Conservation
Biology*, 27, 902–915.

Defra. 2014. *Supporting Document to the National
Pollinator Strategy: For Bees and Other
Pollinators in England.* London: Department
for Environment, Food and Rural Affairs.

Dicks, L. V., Abrahams, A., Atkinson, J., et al.
2013a. Identifying key knowledge needs
for evidence-based conservation of wild
insect pollinators: a collaborative
cross-sectoral exercise. *Insect Conservation
and Diversity*, 6, 435–446.

Dicks, L. V., Bardgett, R. D., Bell, J., et al. 2013b.
What do we need to know to enhance the
environmental sustainability of
agriculture? A prioritisation of knowledge
needs for the UK food system. *Sustainability*,
5, 3095–3115.

Dicks, L. V., Haddaway, N., Hernández-
Morcillo, M., et al. 2017. Knowledge
synthesis for environmental decisions:
an evaluation of existing methods, and
guidance for their selection, use and
development – a report from the
EKLIPSE project. EKLIPSE D3.1, Version
1.0.

Dicks, L. V., Walsh, J. & Sutherland, W. J. 2014.
Organising evidence for environmental
management decisions: a 4S hierarchy.
*Trends in Ecology & Evolution*, 29, 607–613.

Dicks, L. V., Wright, H. L., Ashpole, J. E., et al.
2016. What works in conservation? Using
expert assessment of summarised evidence
to identify practices that enhance natural
pest control in agriculture. *Biodiversity and
Conservation*, 25, 1383–1399.

European Commission. 2011. *Our Life Insurance,
Our Natural Capital: An EU Biodiversity
Strategy to 2020.* 3.5.2011 COM(2011) 244.
2011. Brussels: European Commission.

Floor, J. R., van Koppen, C. S. A. &
Lindeboom H. J. 2013. A review of science-
policy interactions in the Dutch Wadden
Sea – the cockle fishery and gas
exploitation controversies. *Journal of Sea
Research*, 82, 165–175.

Guston, D. H. 2001. Boundary organizations in
environmental policy and science: an
introduction. *Science, Technology, & Human
Values*, 26, 399–408.

Haddaway, N. R., Kohl, C., Rebelo da Silva, N.,
et al. 2017. A framework for stakeholder
engagement during systematic reviews
and maps in environmental management.
*Environmental Evidence*, 6, 11.

Ingram, J. S. I., Wright, H. L., Foster L., et al.
2013. Priority research questions for the UK
food system. *Food Security*, 5, 617–636.

IPBES. 2016. *The Assessment Report of the
Intergovernmental Science–Policy Platform on
Biodiversity and Ecosystem Services on
Pollinators, Pollination and Food Production.*
Bonn: IPBES.

IPCC. 2014. Summary for policy-makers. In:
Field, C. B., Barros, V. R., Dokken, D. J.,

et al., editors, *Climate Change 2014: Impacts, Adaptation, and Vulnerability. Part A: Global and Sectoral Aspects. Contribution of Working Group II to the Fifth Assessment Report of the Intergovernmental Panel on Climate Change* (pp. 1–32). Cambridge: Cambridge University Press.

IPCC. 2015. *IPCC Factsheet: How Does the IPCC Review Process Work?* IPCC Secretariat, Switzerland. Available from www.ipcc.ch /news_and_events/docs/factsheets/ FS_review_process.pdf (accessed 29 April 2018).

Jones, A. C., Mead, A., Kaiser, M. J., et al. 2014. Prioritization of knowledge needs for sustainable aquaculture: a national and global perspective. *Fish and Fisheries*, 16, 668–683.

Lidskog, R. 2014. Representing and regulating nature: boundary organisations, portable representations, and the science–policy interface. *Environmental Politics*, 23, 670–687.

Maxim, L., Spangenberg, J. H. & O'Connor, M. 2009. An analysis of risks for biodiversity under the DPSIR framework. *Ecological Economics*, 69, 12–23.

Millennium Ecosystem Assessment. 2005. *Ecosystems and Human Well-being: Synthesis.* Washington, DC: Island Press.

Norton, M. 1997. The UK Parliamentary Office of Science and Technology and its interaction with the OTA. *Technological Forecasting and Social Change*, 54, 215–231.

Potschin, M. & Haines-Young, R. 2016. Ecosystem services in the twenty-first century. In: Potschin, M., Haines-Young, R., Fish, R. & Turner, R.K., editors, *Routledge Handbook of Ecosystem Services* (pp. 1–9). London:Routledge.

Pretty, J., Sutherland, W. J., Ashby, J., et al. 2010. The top 100 questions of importance to the future of global agriculture. *International Journal of Agricultural Sustainability*, 8, 219–236.

Raymond, C. M., Frantzeskaki, N., Kabisch, N., et al. 2017. A framework for assessing and implementing the co-benefits of nature-based solutions in urban areas. *Environmental Science & Policy*, 77, 15–24.

Richards, M., Metzel, R., Chirinda, N., et al. 2016. Limits of agricultural greenhouse gas calculators to predict soil $N_2O$ and $CH_4$ fluxes in tropical agriculture. *Scientific Reports*, 6, 26279.

Rose, D. C., Parker, C., Fodey, J. O. E., et al. 2018. Involving stakeholders in agricultural decision support systems: improving user-centred design. *International Journal of Agricultural Management*, 6, 80–89.

Rose, D. C., Sutherland, W. J., Parker, C., et al. 2016. Decision support tools for agriculture: towards effective design and delivery. *Agricultural Systems*, 149, 165–174.

Sarkki, S., Tinch, R., Niemelä, J., et al. 2015. Adding iterativity to the credibility, relevance, legitimacy: a novel scheme to highlight dynamic aspects of science-policy interfaces. *Environmental Science and Policy*, 54, 505–512.

Schröter, M., Albert, C., Marques, A., et al. 2016. National ecosystem assessments in Europe: a review. *BioScience*, 66, 813–828.

Smith, R. K., Dicks, L.V. & Sutherland, W. J., 2015. *Scientific Evidence to Address Priority Knowledge Needs For Sustainable Agriculture.* Conservation Evidence resources. Available from www .conservationevidence.com/synopsis/down load/19 (accessed 28 April 2018).

Sutherland, W. J., Dicks L. V., Ockendon, N., et al. 2019a. *What Works in Conservation 2019.* Cambridge: Open Books Publishers.

Sutherland, W. J., Taylor, N. G., MacFarlane, D., et al. 2019b. Building a tool to overcome barriers in the research–implementation space: the Conservation Evidence database. *Biological Conservation*, 283, DOI:10.1016/j. biocon.2019.108199

Sutherland, W. J., Gardner, T. A., Haider L. J., et al. 2014. How can local and traditional knowledge be effectively incorporated into international assessments? *Oryx*, 48, 1–2.

Tobin, P., Schmidt, N. M., Tosun, J., et al. 2018. Mapping states Paris climate pledges: analysing targets and groups at COP 21. *Global Environmental Change – Human and Policy Dimensions*, 48, 11–21.

Vanbergen J., Heard, M. S., Breeze, T., et al. 2014. *Status and Value of Pollinators and Pollination Services*. London: Department for the Environment, Food and Rural Affairs.

van der Molen, F. 2018. How knowledge enables governance: the coproduction of environmental governance capacity. *Environmental Science & Policy*, 87, 18–25.

van der Molen, F., Puente-Rodríguez, D., Swart, J. A. A., et al. 2015. The coproduction of knowledge and policy in coastal governance: integrating mussel fisheries and nature restoration. *Ocean & Coastal Management*, 106, 49–60.

Vandyck, T., Keramidas, K., Saveyn, B., et al. 2016. A global stocktake of the Paris pledges: implications for energy systems and economy. *Global Environmental Change*, 41, 46–63.

van Enst, W. I., Runhaar, H. A. C. & Driessen, P. P. J. 2016. Boundary organisations and their strategies: three cases in the Wadden Sea. *Environmental Science & Policy*, 55, 416–423.

Watt, A. D., Ainsworth, G., Balian, E., et al. 2018. Building a mechanism for evidence-informed European policy on biodiversity and ecosystem services through engagement of knowledge holders. *Evidence and Policy*, http://doi.org/10.1332/174426418X15314036194114

Wood, A. M., Kaptoge, S., Butterworth, A. S., et al. 2018. Risk thresholds for alcohol consumption: combined analysis of individual-participant data for 599 912 current drinkers in 83 prospective studies. *The Lancet*, 391, 1513–1523.

Wyborn, C., Louder, E., Harrison, J., et al. 2018. Understanding the impacts of research synthesis. *Environmental Science & Policy*, 86, 72–84.

Young, J. C., Waylen, K., Sarkki, S., et al. 2014. Improving science–policy dialogue to meet the challenges of biodiversity conservation: having conversations rather than talking at one-another. *Biodiversity and Conservation*, 23, 387–404.

Zasada, I., Piorr, A., Novo, P., et al. 2017. What do we know about decision support systems for landscape and environmental management? A review and expert survey within EU research projects. *Environmental Modelling & Software*, 98, 63–74.

# PART II

# Influencing and making decisions

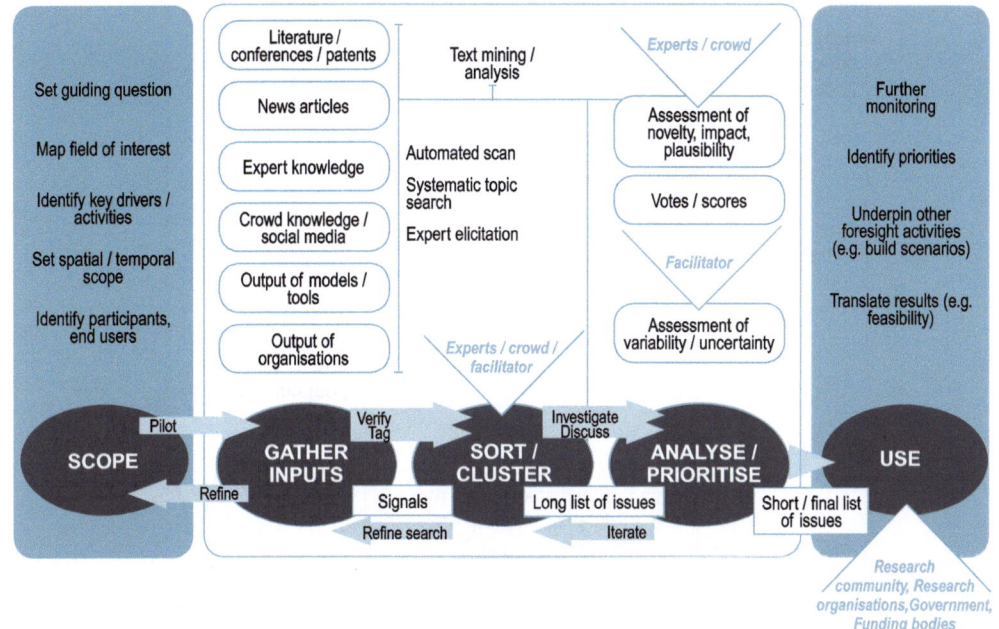

**Figure 3.1** General framework for horizon scanning, reflecting the key steps in the procedure (ovals), inputs and products (rounded rectangles), key outputs (rectangles), actors and end users (triangles), and activities and methods (floating text). Process adapted from Amanatidou et al. (2012). (A black and white version of this figure will appear in some formats.)

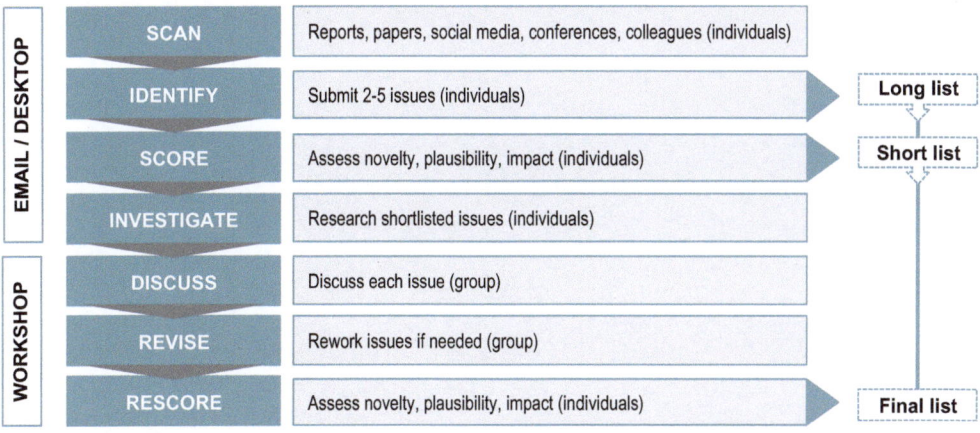

**Figure 3.2** The Delphi-style horizon-scanning approach often used in conservation (Sutherland et al., 2011). Figure reproduced from Wintle et al. (2017), published under the Creative Commons Attribution 4.0 Licence. (A black and white version of this figure will appear in some formats.)

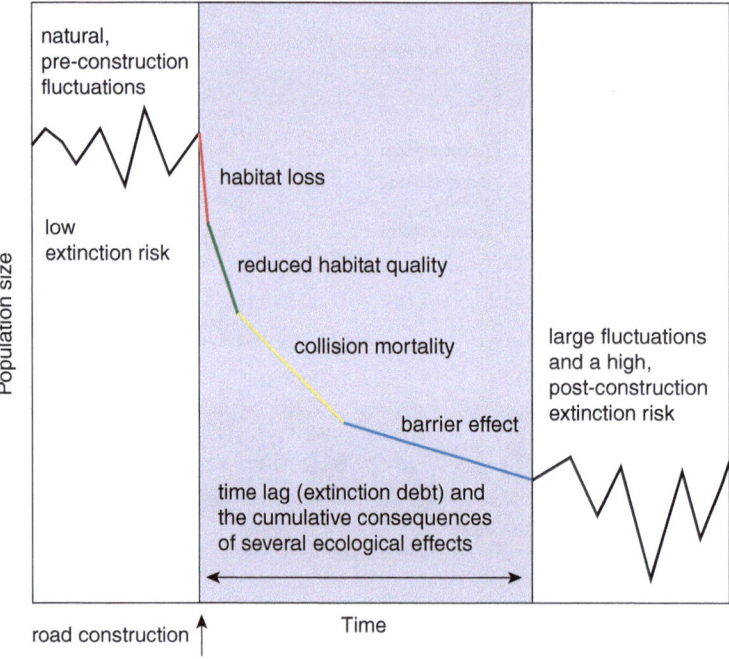

**Figure 4.1** The multiple causes of bat population reduction by road construction and the delayed response (extinction debt). Adapted from Forman et al. (2003). (A black and white version of this figure will appear in some formats.)

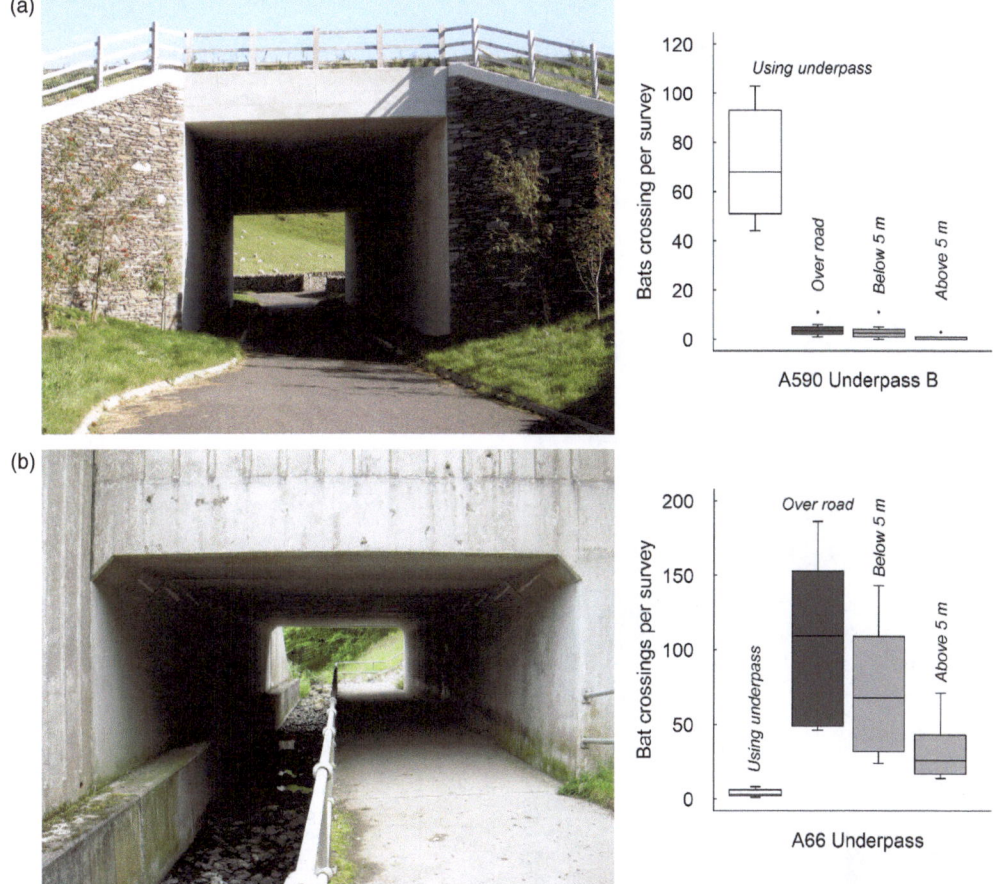

**Figure 4.2** Two underpasses found to vary in effectiveness in guiding bats safely under roads. (a) An effective underpass on the A590, Cumbria, UK; (b) an ineffective underpass on the A66, Cumbria, UK. Boxplots show the number of bats crossing per survey using the underpass and crossing over the road above at safe and unsafe heights (above and below 5 m, traffic height). The variable success of underpasses underlines the need to understand the details of conservation interventions; in this example, the location of the underpasses impacted on how effective they were. From Berthinussen and Altringham (2012b). (A black and white version of this figure will appear in some formats.)

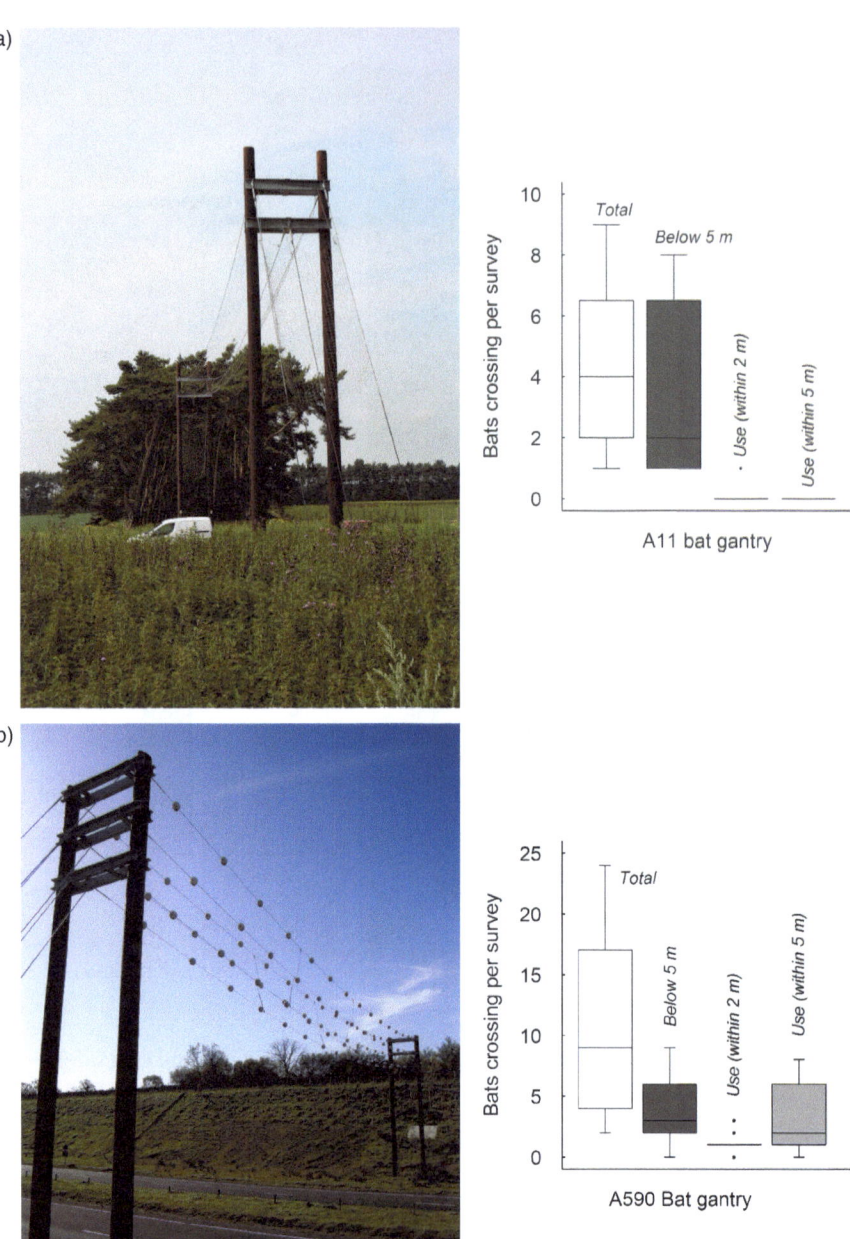

**Figure 4.3** Two bat gantry designs: (a) wire mesh design on the A11, Norfolk, UK; (b) wire and ball design on the A590, Cumbria, UK. Boxplots show the results of surveys carried out to test the effectiveness of the gantries in guiding bats safely over the road. Data were recorded for the total number of bats crossing per survey, the numbers crossing at unsafe heights (below 5 m, traffic height) and the numbers using the gantry according to two definitions of 'use' (flying within either 2 m or 5 m of the wires above traffic height). The bat gantry story neatly demonstrates the need to test conservation interventions before rolling them out on a wide scale. From Berthinussen and Altringham (2012b, 2015). (A black and white version of this figure will appear in some formats.)

**Figure 5.1** Using the Unmatched Count technique to ask about illegal bushmeat hunting in the Ugalla Wildlife Reserve, Tanzania. Picture by Paulo Wilfred. (A black and white version of this figure will appear in some formats.)

**Figure 5.2** Paulo Wilfred and his research assistant recording an illegal meat smoking rack in Ugalla Wildlife Reserve. (A black and white version of this figure will appear in some formats.)

**Figure 5.4** Hans Cosmas Ngoteya (second from right) setting up a beehive with local youths, as an alternative livelihood project. (A black and white version of this figure will appear in some formats.)

**Figure 5.7** WCS Indonesia team members measuring guitarfish at Tanjung Luar port. Photo provided by WCS-Indonesia. (A black and white version of this figure will appear in some formats.)

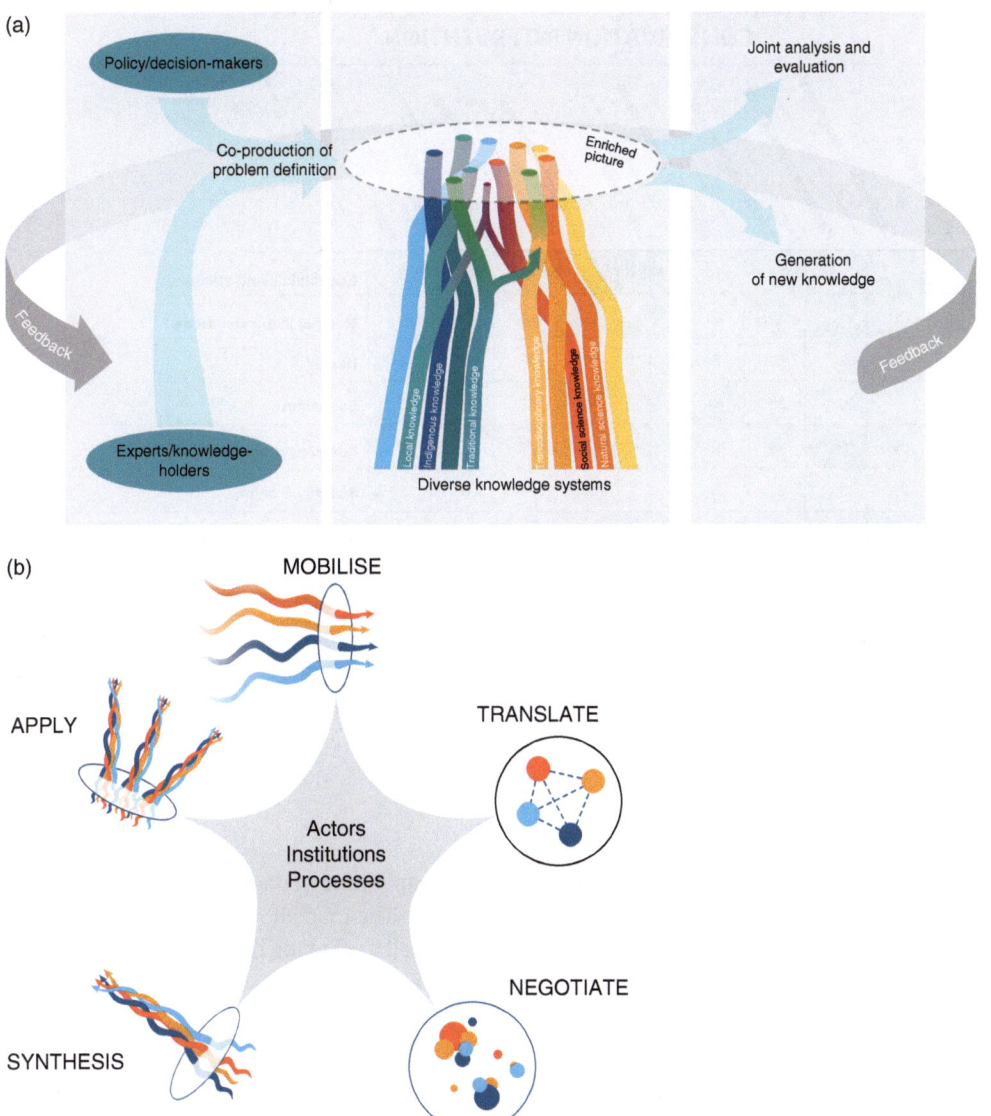

**Figure 6.1** The Multiple Evidence Base approach in action. (a) The three phases of a MEB approach: joint problem formulation, generating an enriched picture with contribution from multiple sources of evidence and joint analysis and evaluation of knowledge (Tengö et al., 2014). (b) Actors, institutions and processes are at the core of the five tasks required for successful collaboration across diverse knowledge systems. The different colours of the lines and dots in parts (a) and (b) represent different knowledge systems, or streams of knowledge within knowledge systems (Tengö et al., 2017). (A black and white version of this figure will appear in some formats.)

# CONSERVATION INTERVENTION

| Area protection | Land/water management | Resource management | Species management | Education & awareness | Law & policy | Livelihood incentives | Ext. capacity building | Sustainable use | Other | HUMAN WELL-BEING |
|---|---|---|---|---|---|---|---|---|---|---|
| 247 | 213 | 278 | 91 | 34 | 149 | 248 | 80 | 22 | 9 | Economic living standards |
| 158 | 151 | 185 | 52 | 20 | 99 | 119 | 28) | 16 | 5 | Material living standards |
| 22 | 16 | 24 | 6 | 6 | 12 | 17 | 4 | 0 | 3 | Health |
| 49 | 43 | 68 | 23 | 41 | 30 | 56 | 17 | 5 | 5 | Education |
| 102 | 105 | 140 | 45 | 18 | 75 | 89 | 47 | 13 | 2 | Social relations |
| 45 | 29 | 33 | 21 | 5 | 19 | 16 | 13 | 3 | 1 | Security & safety |
| 133 | 162 | 202 | 58 | 31 | 134 | 109 | 54 | 16 | 6 | Governance & empowerment |
| 36 | 37 | 23 | 19 | 15 | 24 | 47 | 25 | 10 | 3 | Subjective well-being |
| 21 | 17 | 10 | 11 | 5 | 6 | 21 | 8 | 2 | 3 | Culture/spirituality |
| 1 | 3 | 3 | 1 | 1 | 2 | 2 | 1 | 0 | 0 | Freedom of choice/action |
| 0 | 0 | 4 | 2 | 0 | 2 | 3 | 0 | 1 | 0 | Other |

NO. OF STUDIES

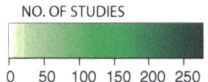

0   50   100  150  200  250

**Figure 7.1** An example of an evidence 'heat map' linking conservation interventions with human well-being outcomes. The map allows the user to assess the evidence base for gaps and gluts as well as clicking on each box to further examine the relevant studies (after McKinnon et al., 2016). (A black and white version of this figure will appear in some formats.)

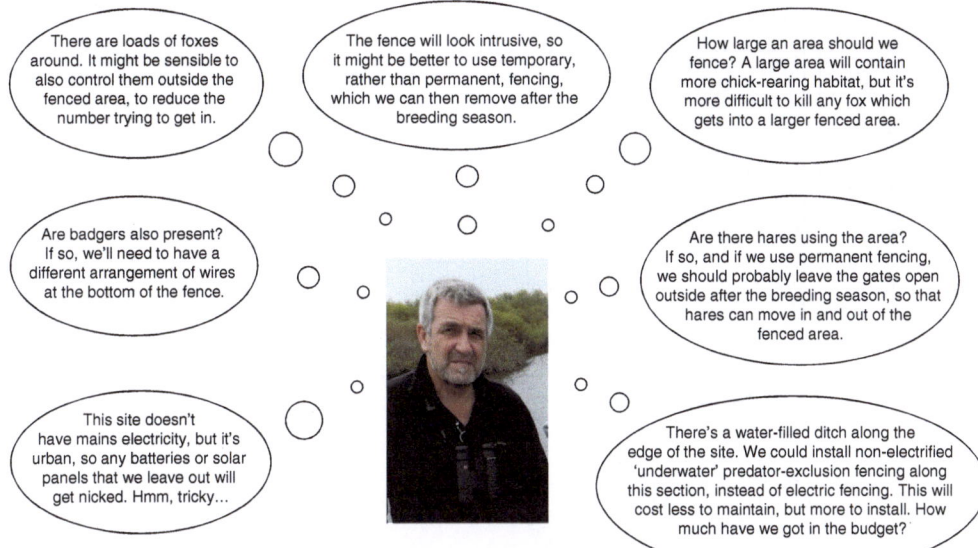

**Figure 9.1** Decision-making at sites often involves taking account of a range of site-specific factors. Here, an ecological adviser ponders over details of the design of predator-exclusion fencing used to protect ground-nesting waders. Photo by Malcolm Ausden. (A black and white version of this figure will appear in some formats.)

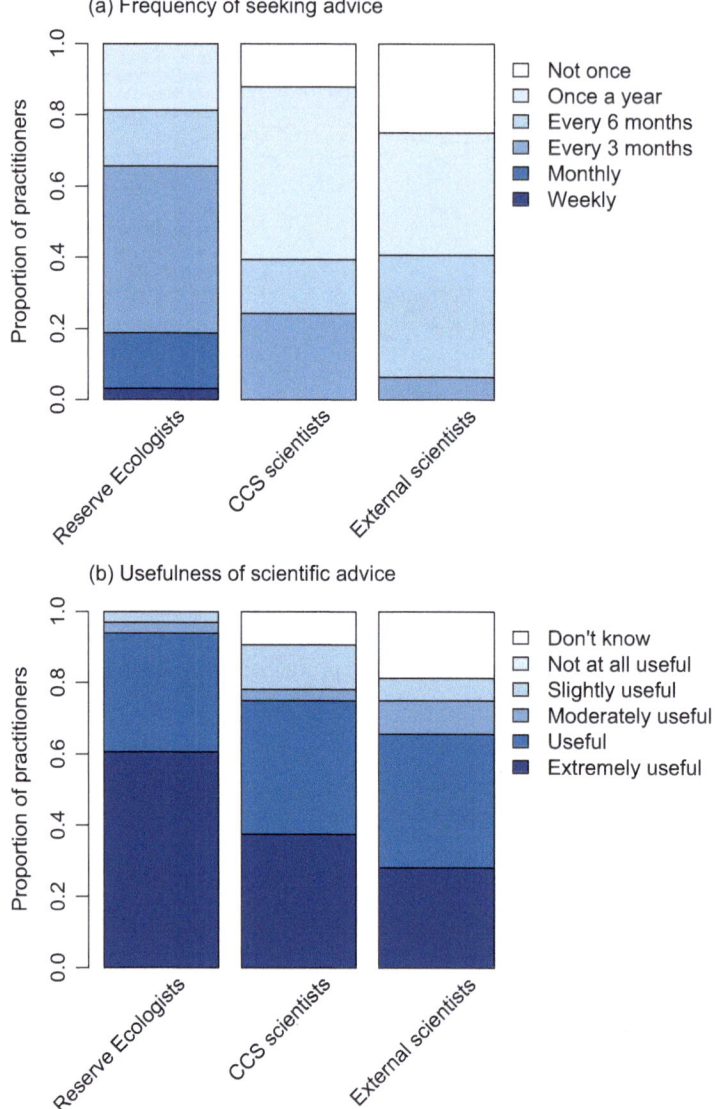

**Figure 9.2** The frequency with which 36 RSPB practitioners (mainly site managers and conservation officers) seek scientific advice from Reserve Ecologists (in-house ecological advisers), Centre for Conservation Scientists (CCS, in-house conservation scientists) and external scientists, and their perceived usefulness of this scientific advice from each source. There was a 78% response rate (46 practitioners were invited to participate) and survey methods are described in Walsh (2015; Chapter 4). (A black and white version of this figure will appear in some formats.)

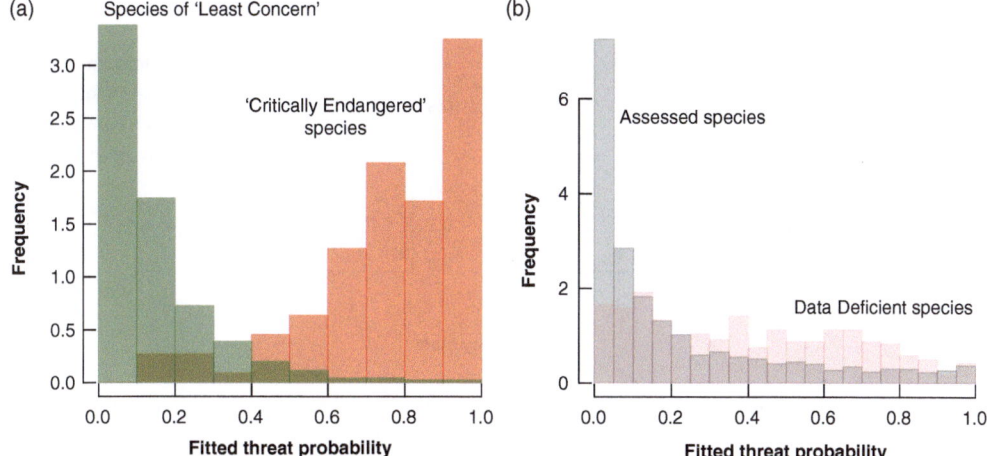

**Figure 11.1** The importance of dealing with uncertainty in conservation assessments. We used models to generate threat probabilities for mammals. (a) These probabilities do an effective job of distinguishing species that are Least Concern (green bars) from those that are Critically Endangered (orange bars); (b) our models were used to predict threat probabilities for species that were Data Deficient (DD) (pink bars) compared to species that were assessed (grey bars) (i.e. to reduce uncertainty in assessment). (A black and white version of this figure will appear in some formats.)

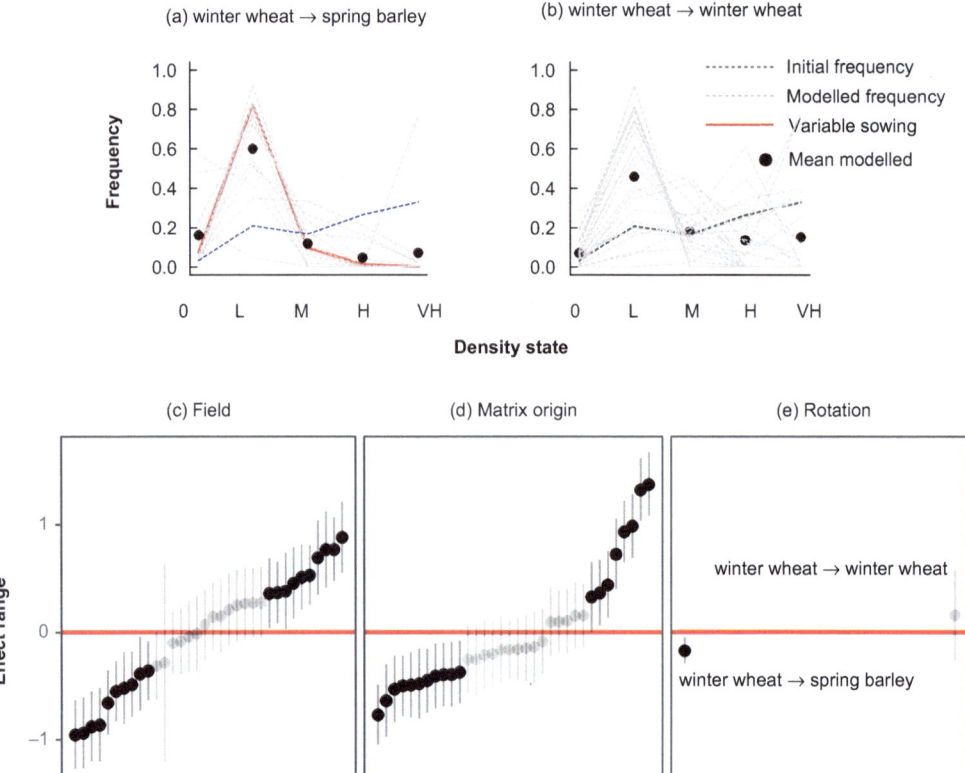

**Figure 11.2** Uncertainty and benchmarking in weed control. (a,b) Predicted responses of populations of the weed *Alopecurus myosuroides* to rotational management. The initial frequency of weeds at each sowing density was the same in each case (dashed · blue line). Each grey line represents a matrix generated from a different field following two forms of management. (a) What *would have been* the density (0 = zero, L = low, M = medium, H = high and VH = very high) of an average field *had* it been planted with spring barley. This is compared with (b) the predicted response from maintaining winter wheat. The red line in (a) represents a single field that was managed with variable sowing densities. Figures (c–e) compare the observed effect of management with difference sources of background variation to disentangle the uncertainty in management. We generated models for each field: 22 in winter wheat and 12 rotated from winter wheat to spring barley, and their results are presented in rank order. The effect range is the estimate of the random effect for each field, location or rotation. (A black and white version of this figure will appear in some formats.)

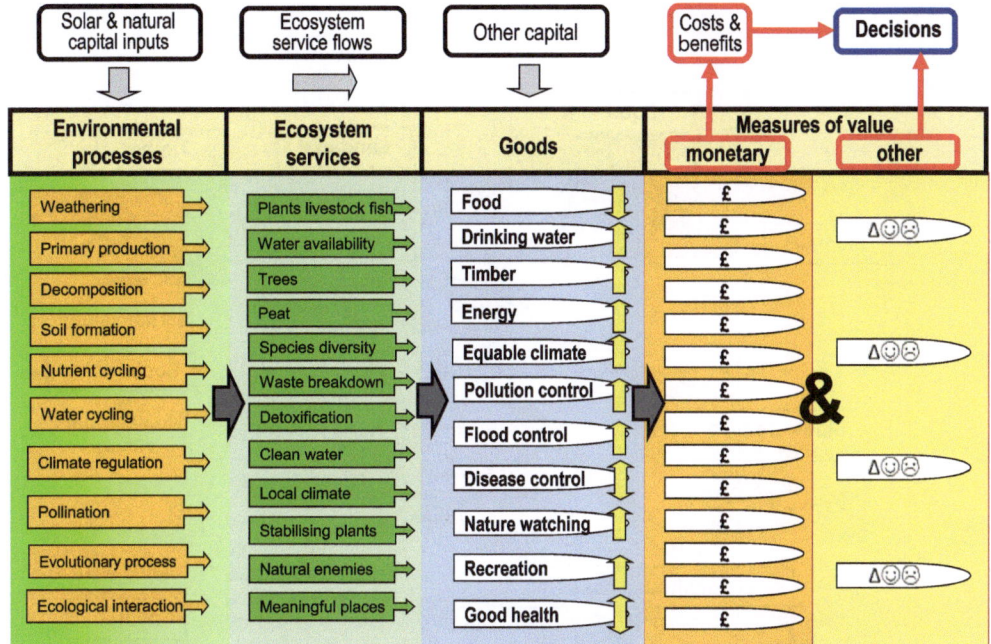

**Figure 12.1** Decision-making and the environment: from natural capital to decisions. The yellow arrows illustrate the multiple effects typical of a change in natural capital, in this case those arising from an investment to establish woodland on a currently farmed area. (A black and white version of this figure will appear in some formats.)

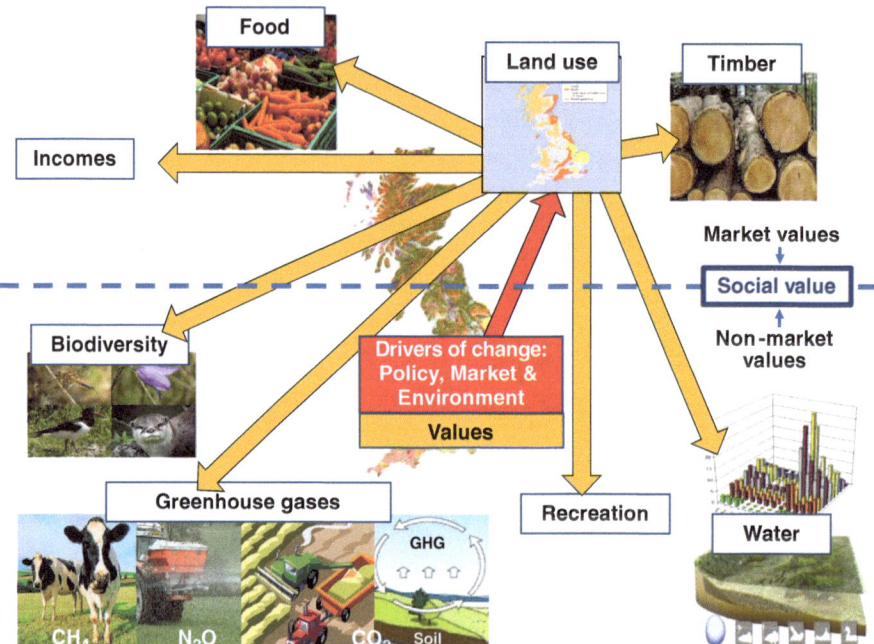

**Figure 12.2** The drivers, consequences and values of land-use change, associated with agricultural land use in Great Britain and incorporated within the conceptual framework of the National Ecosystem Assessment (Mace et al., 2011). (A black and white version of this figure will appear in some formats.)

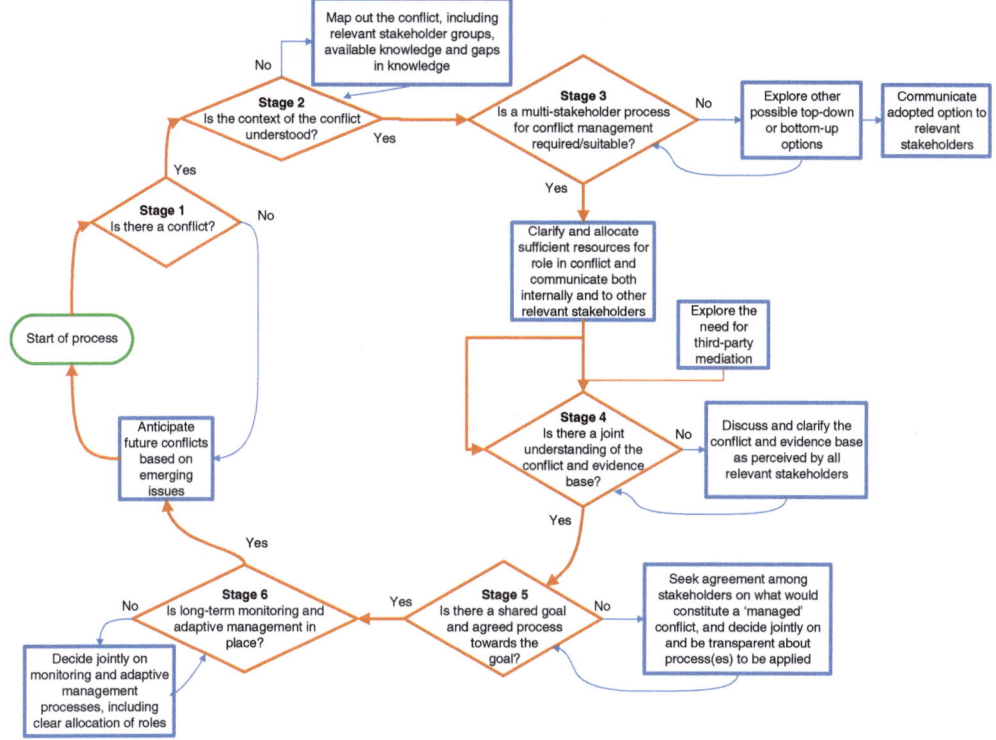

**Figure 14.1** Stepwise approach aimed at enabling decision-makers to identify, manage and monitor conservation conflicts. Diamond shapes indicate the six key decision stages. Squares state what needs to happen to go from one decision stage to the next. Adapted from Young et al. (2016a). (A black and white version of this figure will appear in some formats.)

**Figure 15.1** The 20 Aichi Biodiversity Targets. Image: Copyright BIP/SCBD. (A black and white version of this figure will appear in some formats.)

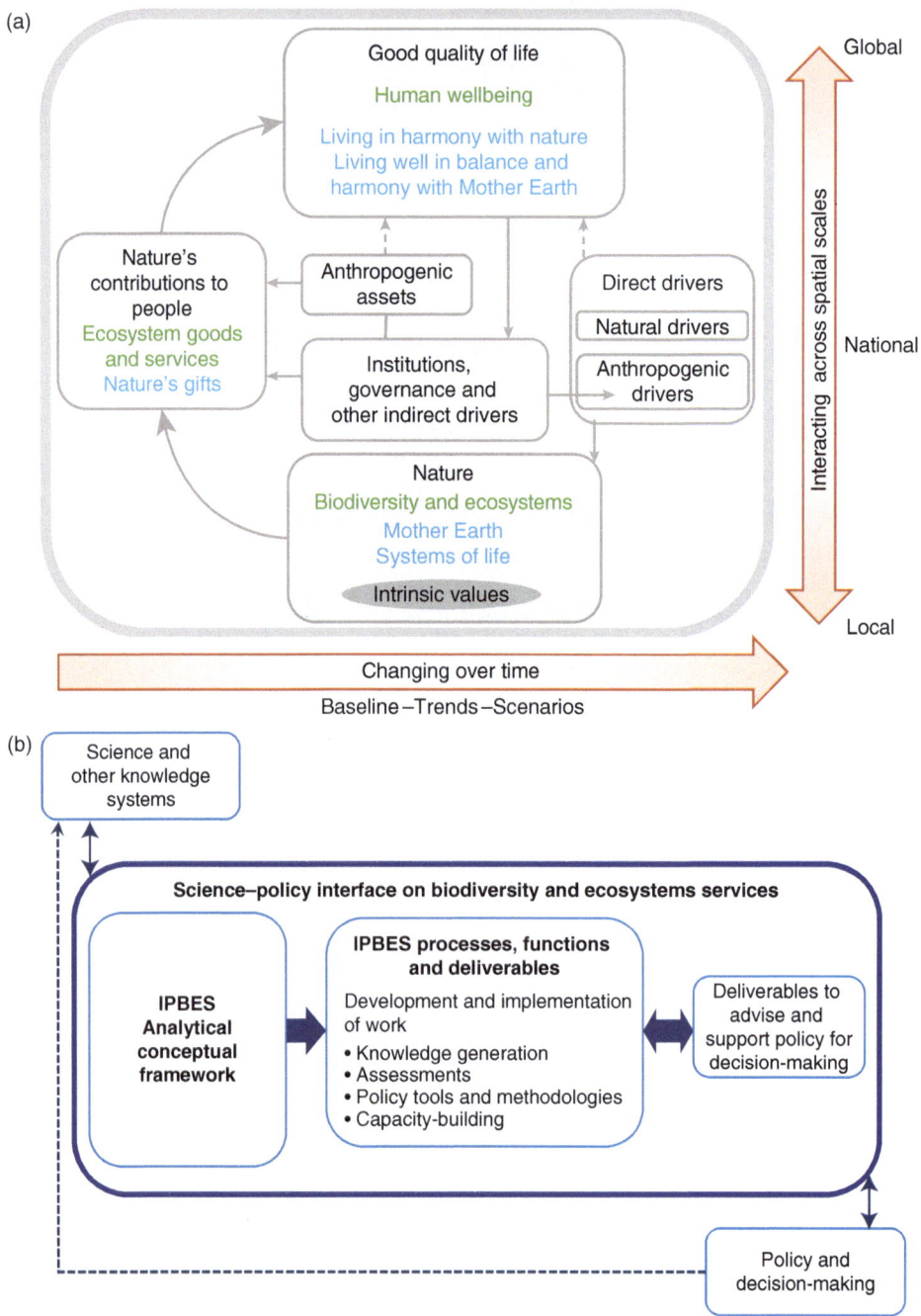

**Figure 15.2** (a) IPBES operational model of the Platform (adapted from IPBES, 2014), (b) analytical conceptual framework of assessments (adapted from Díaz et al., 2015). (A black and white version of this figure will appear in some formats.)

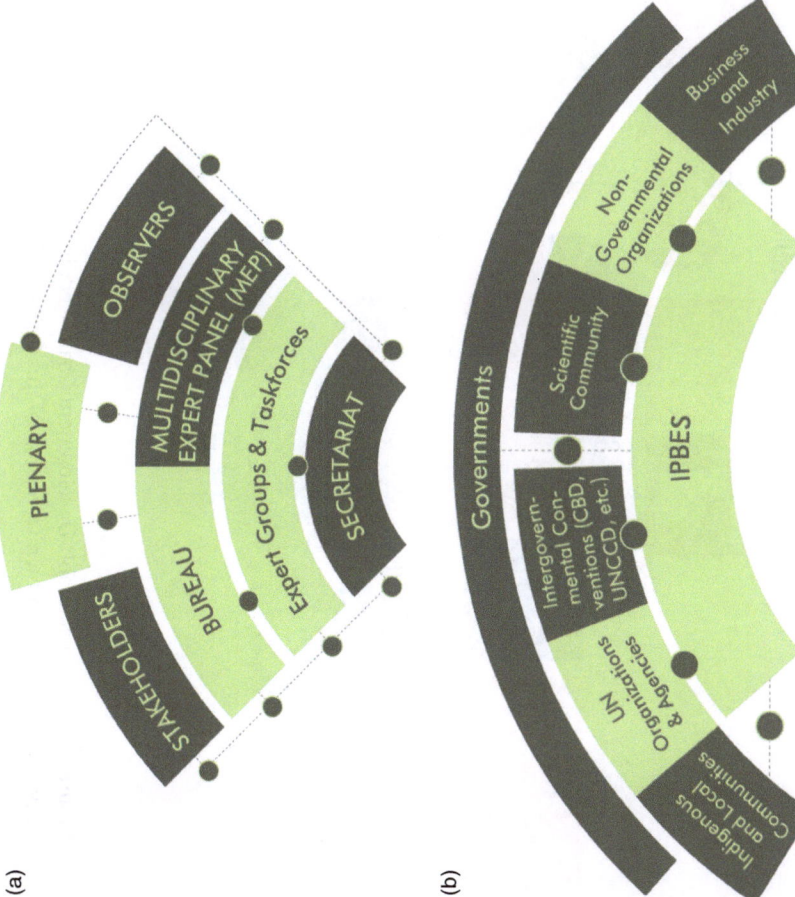

(a)

(b)

**Figure 15.3** Structures of IPBES (a) science–policy platform, (b) intergovernmental plenary (IPBES, 2018b). (A black and white version of this figure will appear in some formats.)

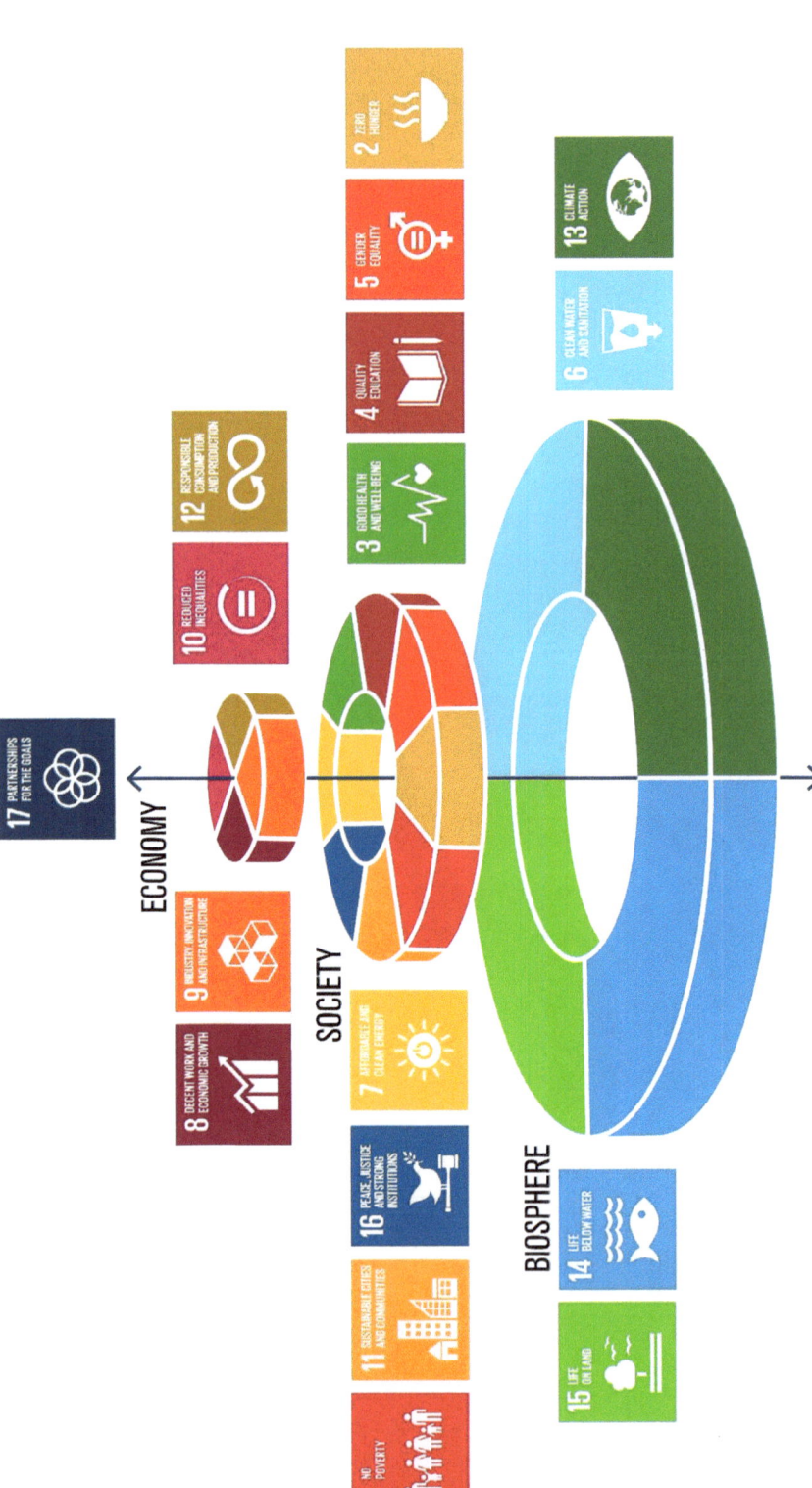

**Figure 15.4** The Sustainable Development Goals 'wedding cake' (source/credit: Azote Images for Stockholm Resilience Centre, Stockholm University). (A black and white version of this figure will appear in some formats.)

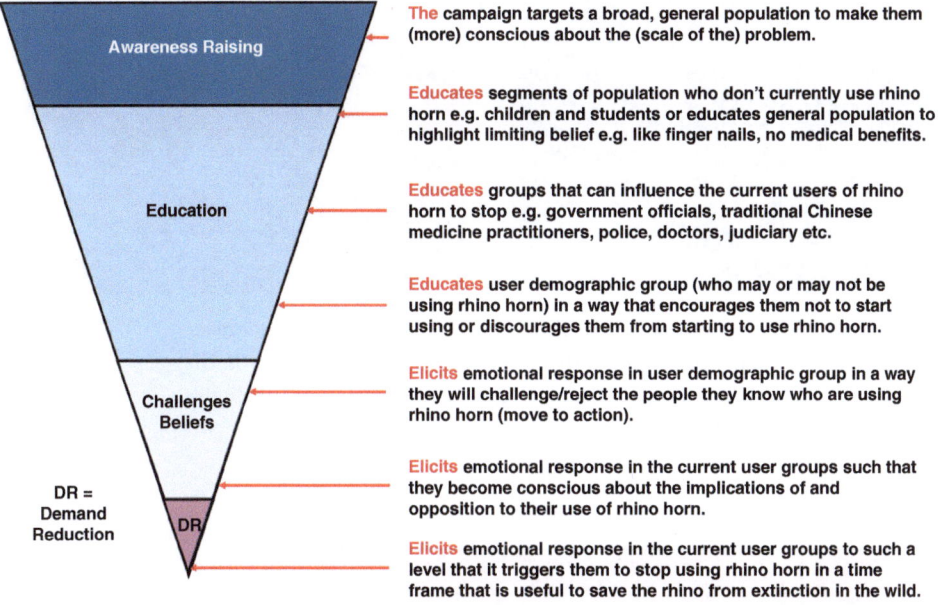

**Basic Test to Differentiate Demand Reduction from Awareness Raising and Education**

Awareness Raising

The campaign targets a broad, general population to make them (more) conscious about the (scale of the) problem.

Educates segments of population who don't currently use rhino horn e.g. children and students or educates general population to highlight limiting belief e.g. like finger nails, no medical benefits.

Education

Educates groups that can influence the current users of rhino horn to stop e.g. government officials, traditional Chinese medicine practitioners, police, doctors, judiciary etc.

Educates user demographic group (who may or may not be using rhino horn) in a way that encourages them not to start using or discourages them from starting to use rhino horn.

Challenges Beliefs

Elicits emotional response in user demographic group in a way they will challenge/reject the people they know who are using rhino horn (move to action).

Elicits emotional response in the current user groups such that they become conscious about the implications of and opposition to their use of rhino horn.

DR = Demand Reduction

DR

Elicits emotional response in the current user groups to such a level that it triggers them to stop using rhino horn in a time frame that is useful to save the rhino from extinction in the wild.

**Figure 17.1** Model showing differences between behaviour-change and awareness-raising campaigns developed by Nature Needs More Ltd for its Breaking The Brand RhiNo Campaign (Breaking The Brand, 2016). (A black and white version of this figure will appear in some formats.)

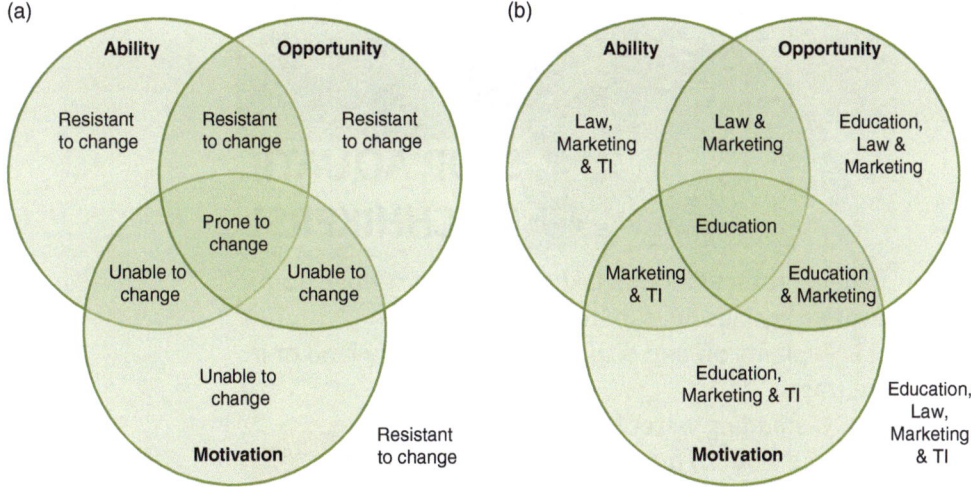

**Figure 19.1** Diagram showing how a person's ability, opportunity and/or motivation determines (a) whether they are prone, unable or resistant to change and (b) the appropriateness of the four different behaviour change approaches of education, law, marketing and technical intervention (TI) under these different conditions (adapted from Rothschild, 2000; Santos et al., 2011). (A black and white version of this figure will appear in some formats.)

(a) 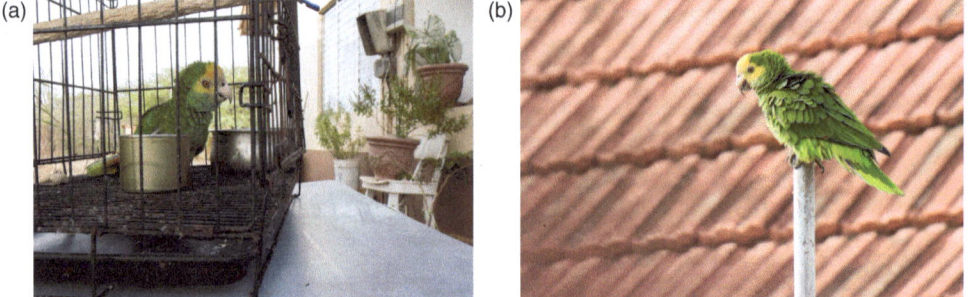 (b)

**Figure 19.2** The lora or yellow-shouldered Amazon parrot (*Amazona barbadensis*) that was the focus of a social marketing campaign on the Caribbean island of Bonaire. (A black and white version of this figure will appear in some formats.)

**Protect Your Waters**

**STOP AQUATIC HITCHHIKERS!**

To help protect Florida's lakes, rivers and springs from invasive species, be sure to:
**CLEAN** plants off and dispose of them on dry land or in the trash.
**DRAIN** standing water from your boat.
**DRY** your boat to prevent the transport of aquatic hitchhikers.

For more information, visit:
www.plants.ifas.ufl.edu or www.protectyourwaters.net

**Figure 19.3** Promotional material encouraging boat owners in the Greater Yellowstone Area to adopt practices that will reduce the spread of invasive species. (A black and white version of this figure will appear in some formats.)

# CHAPTER NINE

# The use of evidence in decision-making by practitioners

MALCOLM AUSDEN

*Reserves Ecology, Royal Society for the Protection of Birds*

and

JESSICA C. WALSH

*Monash University*

## 9.1 Introduction

Conservation practitioners are usually tasked with a very diverse set of activities within their job. A typical week for a reserve manager might involve managing staff, volunteers, contractors and budgets; liaising with people both within and outside of their organisation; dealing with health and safety and other legal obligations; taking part in a range of meetings; and replying to numerous emails about a wide range of topics. If the site is heavily visited, there will invariably be many tasks regarding visitors. In addition, practitioners also have to decide how best to manage their site for conservation.

In this chapter, we describe the processes that organisations and practitioners use to make conservation decisions, the trade-offs between resources spent monitoring and carrying out conservation management, and the types of information practitioners use to inform these decisions. We then discuss ways to ensure that decisions at sites are based on good evidence. We combine literature and theory on what constitutes best practice for reserve management with our practical experience. While our examples are focused on conservation land management at the site level, these frameworks and processes are generally applicable to decision-making in many other conservation contexts.

## 9.2 Types of conservation decisions made by practitioners

Decisions about the conservation management of sites are often complex. There are several reasons for this. First, many types of habitat management aim to achieve multiple objectives, and these will differ between sites. For example, a fire regime might aim to prevent an area of grassland from succeeding to scrub, while also aiming to maintain or increase plant species richness and provide a continuity of suitable conditions for particular bird species. Habitat management can also involve using different techniques in combination. For example, a wetland might be managed using a combination of livestock grazing and water-level control, while an area of dwarf-shrub heath might be managed through a combination of grazing, cutting and burning. Good management of sites, therefore, rarely involves simply implementing 'off-the-shelf' conservation actions. Furthermore, even where a single technique is used to benefit a single species (or group of species), practitioners usually still need to tailor the details of how it is implemented to the specific circumstances at their site.

Finally, decisions can also involve trade-offs between ecological, social and economic factors, and there may also be great uncertainty about the risks and benefits of each option. Meanwhile, practitioners are often working with limited resources, the scientific evidence may be conflicting, multiple decision-makers and stakeholders might have different preferences and opinions, and people inherently often do not make rational decisions.

## 9.3 Decision-making processes used by conservation organisations

The conservation management of nature reserves and other protected areas is usually the product of several levels of decision-making: strategic-level decisions, site-level management planning and what we will call 'day-to-day decision-making' by practitioners. Decisions taken at each of these levels are influenced by the decision-making process, the people involved in decisions and the evidence used to inform them (Table 9.1).

Decisions at the strategic level focus on the overall aims of the reserve network in which individual reserves sit, as well as the formulation of policies within which these reserves operate. An example of a policy might be an organisation's approach to allowing wildlife hunting on its land, including the range of acceptable methods allowed. Strategic decisions are discussed elsewhere and we will not focus on these here (Margules & Pressey, 2000; Pressey et al., 2007; Wilson et al., 2009).

Site-level management planning processes (or site action planning) help practitioners develop objectives for reserves and identify the management actions needed to achieve them. For example, they might help decide the aims of managing a wetland, the desired water-level regime and proportions of swamp and open water, and the frequency of cutting the swamp vegetation needed to achieve these. These processes are also used to decide what monitoring is needed to determine whether the actions are achieving these objectives, or to detect other important changes, particularly those that might trigger management actions.

**Table 9.1** *A summary of factors that influence conservation management decisions at different levels*

| Components of a decision | Strategic-level decisions | Site-level planning decisions | Day-to-day decisions |
|---|---|---|---|
| Frequency, context and potential consequences | These are made infrequently to set long-term, overarching aims and objectives for a network of sites, and policies within which sites operate. They require high-level planning and foresight, because the consequences of strategic decisions are high. They set the context within which site-level decisions are made | These commonly occur on a five-year cycle, but may be reviewed more frequently. They determine which management actions to implement to achieve agreed goals and objectives for individual sites. This planning stage is crucial, because it provides the context within which day-to-day management decisions are made | These often need to be made quickly, with the details of decisions often important in determining whether or not conservation actions will be successful |
| Decision-support tools and planning processes | Frameworks and methods to assist with strategic decisions include prioritisation decision support tools, horizon-scanning exercises, discussions or structured expert elicitation | Adaptive management/ management planning processes which include decision theory, multi-criteria decision analysis, structured decision-making, risk analysis and evidence-based decision-making | Usually none |
| Decision-makers involved | Strategic directors and managers, and sometimes funding bodies, policy-makers, boards of governors. Scientists may also be involved | Varies greatly depending on the organisation, but in addition to practitioners, their line managers, scientists, advisers, specialists and other | Site-based practitioners and, if they are unfamiliar with the technique, then also through discussion with fellow practitioners and advisers |

**Table 9.1** (*cont.*)

| Components of a decision | Strategic-level decisions | Site-level planning decisions | Day-to-day decisions |
| --- | --- | --- | --- |
| | | stakeholders can be involved, together with other stakeholders | |
| Information used | Informed by the strategic objectives and vision of the organisation, as well as government policy and law. Ecological, economic, social and political factors would be considered | Information about the conservation status of species and habitats, threats, effectiveness and costs of management actions, along with social and economic factors, objectives of the protected area network, organisational policies and available resources | Personal experience, colleague's advice or a quick internet search would often be the basis of day-to-day decisions |

Finally, the actions agreed through the site-level planning decisions are implemented via 'day-to-day decision-making' by practitioners. For example, a practitioner wanting to install boxes to provide roosting habitat for bats would need to decide which trees would be suitable, and at what height and orientation on the tree the boxes would be most effective.

### 9.3.1 Site-level management planning processes

Management planning processes and frameworks help practitioners make conservation decisions and ensure that the decisions made are based on logic. We provide two examples of organisations' management planning procedures in Box 9.1.

The procedures used by different organisations to set priorities and create management plans vary according to differences in their organisational structure, objectives and culture. However, in our experience, effective management and decision-making systems include the following six features.

1. They involve a range of people who, collectively, possess the expertise and knowledge needed to make well-informed decisions. They include site-

---

**Box 9.1. Examples of management planning processes used by different conservation organisations**

**Royal Society for the Protection of Birds (RSPB):** *a land-owning, science-based conservation non-governmental organisation (NGO) in the UK, whose 215 reserves comprise mainly intensively managed cultural landscapes.*

The overall aims for the RSPB's nature reserve network are set out in its Reserves Strategy,[1] which is usually reviewed every five years. The strategy lists the particular species and habitats that the network aims to benefit, together with, for example, how the organisation aims to use the network to help people connect with nature. This strategy therefore sets the context within which the objectives of individual reserves are made.

Each RSPB reserve has a management plan, based on a standard template. This plan is 'owned' by the site's practitioners, but its preparation involves a meeting with key individuals to agree on the long-term vision and objectives for the site, together with subsequent discussions. These key individuals are the practitioner's line manager, an ecological adviser, a land agent and, if required, other scientists and specialists. Preparation of the plan can also include discussions with members of the local community.

Each management plan contains the reserve's long-term vision, objectives, management and monitoring actions and five-year work programme. The Features–Attributes–Factors framework is used to decide these actions (Box 9.2). The draft management plan is checked and approved at both regional and national levels of the organisation and, if the site is a nationally designated site for protection, also by the relevant statutory agency.

Each reserve reports the progress towards achieving its management objectives annually and this report is audited by ecological advisers. An annual site-based meeting is also held at all key sites, involving site-based staff, their manager and an ecological adviser to help resolve any outstanding issues and plan work for the following year. Sites that are failing to make good progress are discussed with regional and national staff and a plan is developed to resolve any issues.

**New Zealand Department of Conservation (DOC):** *A government agency responsible for the conservation and management of native species, ecosystems and a third of the land in New Zealand.*

Conservation management in New Zealand is guided by the New Zealand Biodiversity Strategy and Action Plan[2] and the draft Threatened Species

---

[1]  http://ww2.rspb.org.uk/Images/rspbreserves2012_tcm9-326414.pdf
[2]  www.doc.govt.nz/nature/biodiversity/nz-biodiversity-strategy-and-action-plan/

## Box 9.1. (cont.)

Strategy,[3] which are produced by DOC. This is in addition to management plans for broader landscape management issues, National Parks, site-based management prescriptions for ecosystems and species[4] and Threatened Species Recovery Plans.[5] An annual '5-year Statement of Intent' sets out the longer-term directions for the DOC, as well as the management actions to be undertaken that year.

These plans are written variously by managers, policy staff, scientists and operations staff within the organisation, in partnership with Tangata whenua (NZ's indigenous people) and in consultation with the public, private land-owners, relevant agencies and organisations. Collectively these plans outline objectives, targets or goals (often quantitative), time-bound management actions, research priorities and monitoring activities. They inform annual operational work programmes and provide the basis for output and outcome monitoring and annual reporting.[6]

The planning process for DOC ecosystems and threatened species management focuses on producing specific, consistent and transparent action-based work projects in priority order to best meet agreed outcome-based objectives. Some of these outcome objectives include condition of ecosystems and long-term persistence of threatened species. Projects list the actions required to mitigate key pressures at sites. These projects are embedded directly into the Department's Business Planning software, and when budgets are agreed the approved projects are simply 'activated' in the software and are then available for operations staff to work on. Key elements include having stable, overarching, outcome-based objectives; having standardised database entry of prescriptions that feed directly into the Department's business planning processes; and having the ability to identify the most cost-effective set of prescriptions based on different priorities. Research, monitoring and evaluation of management are built into the planning and decision-making processes through DOC's Biodiversity Monitoring and Reporting System. This system helps to identify changes and monitor success.

*Section written jointly with Richard Maloney, Department of Conservation, New Zealand.*

---

[3] www.doc.govt.nz/get-involved/have-your-say/all-consultations/2017/draft-threatened-species-strategy-consultation/draft-threatened-species-strategy/

[4] www.doc.govt.nz/about-us/our-role/managing-conservation/natural-heritage-management/identifying-conservation-priorities/

[5] www.doc.govt.nz/about-us/science-publications/series/threatened-species-recovery-plans/

[6] www.doc.govt.nz/about-us/our-role/corporate-publications/annual-reports-archive/

based practitioners, their line managers and other advisers, scientists, experts and other stakeholders.

2. They involve an explicit process that helps identify appropriate actions. A variety of frameworks are used in management planning to help aid decision-making. We describe two examples in Box 9.2. Other methods used to help practitioners identify solutions for complex environmental problems include structured decision-making (Gregory et al., 2012), multi-criteria decision-making (Davis et al., 2003) and risk analysis (Pollard et al., 2008).

3. Practitioners are involved in the decision-making and have 'ownership' of the final management actions. There are many examples of site management plans that have been produced by consultants and other people not involved in managing the site, which just sit on shelves gathering dust. Practitioners typically have a lot to do, and want to focus on managing their sites. Therefore, decision-making frameworks need to be as straightforward and unbureaucratic as possible, while still ensuring that decisions are the result of a logical process.

4. Decisions should be underpinned by good scientific evidence. Evidence-based decision-making involves the integration of scientific research, expertise and local knowledge (Sutherland et al., 2004; Walsh, 2015). Scientific evidence can be obtained from scientific studies, reviews, summaries of evidence, decision support tools or advice from scientific advisors. In cases where evidence and data are limited, all available knowledge, including expertise and opinion, can be used for initial management decisions. This should be accompanied with monitoring, evaluation and experimentation where possible to learn and generate the required evidence.

5. The contents of the site management plan are checked and 'signed off' by colleagues who are involved in producing it. This ensures that standards are maintained, and that the contents of the management plan are sensible, feasible and consistent with regional, national and in some cases international priorities. It also helps to ensure 'buy-in' from relevant people in the rest of the organisation, some of whom might be involved in allocating resources for the site.

6. They include a process for evaluating and reviewing whether the site is achieving its objectives and, if not, helps identifies what to change. This process is a key component of adaptive management (Runge, 2011; Westgate et al., 2013; Murphy & Weiland, 2014), which has been adopted in principle by many conservation organisations and agencies. However, research suggests that successful implementation of adaptive management remains elusive in many projects (Keith et al., 2011; McFadden et al., 2011).

---

**Box 9.2. Examples of two frameworks used in site-based decision-making**

**Pressure–State–Response.** This framework has been widely used to develop environmental indicators, e.g. by Birdlife International for monitoring Important Bird Areas (Organisation for Economic Co-operation and Development, 1993; Birdlife International, 2006). It identifies negative pressures on habitats and species at a site; the state these habitats and populations are in; and what responses are required to reduce, or prevent, the impacts of these pressures.

For example, for an area of forest the *pressures* might be illegal logging and hunting; it might define the *state* of the forest in terms of its extent and population abundance of key species, while the *response* or interventions might be changes in conservation designation or protection and other projects aimed at preventing illegal logging and hunting.

**Features–Attributes–Factors.** This is the UK government's framework for identifying actions to carry out in protected areas (JNCC, 2004). The first step is to identify the important conservation *features* at the site. These features can be species, assemblages of species, habitats or, more rarely, processes.

The second step is to identify the best measures of condition of these features, and to set targets (or target ranges) for them. These measures of condition are called *attributes*. Commonly used attributes for a species will be its population size and productivity. Attributes for a habitat might include measures of its structure and of the abundance of positive or negative indicator species.

The final step is to identify the main *factors* that are thought to determine whether a feature's attribute will achieve its target condition and to set targets (or target ranges) for these factors. For a species, factors that might affect whether it attains its target population size could include levels of illegal persecution or its food supply. For a habitat, factors that might affect whether it attains its target condition might include levels of nutrient run-off and the management regime.

---

### 9.3.2 Day-to-day decision-making

To implement actions agreed in a site's management plan, practitioners still need to make frequent decisions about the details of the interventions. Consider this example about protecting the nests of ground-nesting waders in the UK. The scientific evidence shows that predator-exclusion fencing can

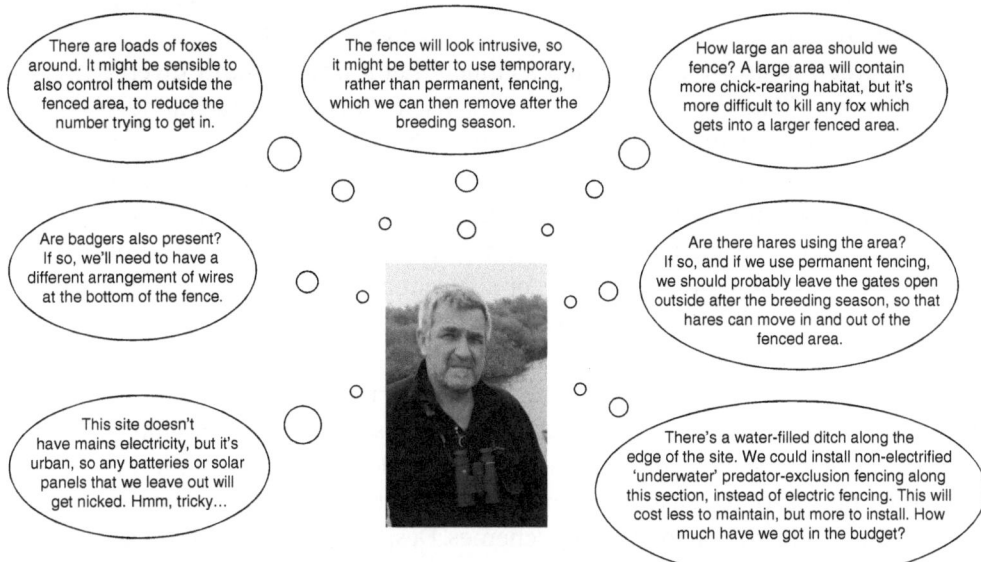

**Figure 9.1** Decision-making at sites often involves taking account of a range of site-specific factors. Here, an ecological adviser ponders over details of the design of predator-exclusion fencing used to protect ground-nesting waders. Photo by Malcolm Ausden. (A black and white version of this figure will appear in some formats. For the colour version, please refer to the plate section.)

be used to increase the nest survival and overall breeding success of ground-nesting waders (Sutherland et al., 2018) and the site's management plan includes an action to install predator-exclusion fencing. However, practitioners still need to consider many minute details before installing the fencing, to address local circumstances and try to maximise the effectiveness of the fencing (Figure 9.1).

When making decisions about the details of site management, a practitioner or their adviser will usually have a mental image of what they consider to be ideal habitat for a particular species or set of species. They will then compare the habitat present at a site with this ideal state and, based on a combination of past experience and other information, identify what they think needs to take place. This process will typically involve a visual assessment of the site, together with information from surveys and monitoring, the presence and population trends of key species, and their own and others' experience of the impacts of management actions in the past and at other sites.

## 9.4 Monitoring information used in decision-making

The resources that practitioners have available for monitoring (i.e. staff time and money) usually come from the same 'pot' as those used for carrying out conservation work. Therefore, practitioners must make a trade-off decision. They need to conduct sufficient monitoring to reliably inform whether actions are having their desired effect, but not so much that it unnecessarily diverts resources away from the conservation work itself. Similarly, practitioners need to target surveillance efforts to detecting changes that, if they occur, would trigger conservation action. This is a different approach to that of a conservation scientist, who may be interested in investigating the underlying mechanisms causing a change, the effectiveness of an action (or set of actions) and in disentangling the effects of different actions. To do this would usually involve replicates and controls, and detailed monitoring sufficient for the results to be published.

These trade-offs are important to get right, because monitoring and surveillance can be expensive. For example, on the RSPB's reserve network, monitoring, one-off surveys and surveillance are pared down to the minimum considered necessary to reliably inform management and contribute data to a small number of national monitoring schemes. Despite this, they still cost an estimated 7% of the total costs of maintaining this reserve network.

The type and quality of data collected during monitoring depends on the management question. At the one extreme, detailed monitoring is not needed to determine whether cutting grass reduces its height. At the other extreme, considerable resources can be required to determine levels of predation, or changes in the botanical composition of species-rich grassland. Practitioners and their advisers may invest more resources into monitoring if they are using a novel technique, applying a standard technique in a novel situation, if there is a high level of uncertainty about the results, or if the results are difficult to observe visually. The results would then ideally feedback into the planning processes to inform future decisions, and also be written up and disseminated to other practitioners.

## 9.5 Information used by practitioners to inform decision-making

Multiple studies have investigated the types of information used by practitioners from the UK, South Africa, Australia, Brazil and the USA, their level of access, and which sources they find most useful (Pullin et al., 2004; Pullin & Knight, 2005; Cook et al., 2010, 2012; Seavy & Howell, 2010; Bayliss & Randall, 2011; Young & Van Aarde, 2011; Matzek et al., 2014; Walsh, 2015; Giehl et al., 2017). These have shown that practitioners use a wide range of sources to inform conservation management decisions, with 'personal experience' the most common source of information usually reported. For example, practitioners from government and non-government agencies in the UK and South Africa said they use personal experience, monitoring data and advice from scientific advisors and managers most frequently when making management decisions (Walsh, 2015). Management plans, policy documents and decision support tools were less-frequently used. In contrast, scientific papers and unpublished research were rarely used directly to inform decisions (Walsh, 2015).

However, given the complexity of the types of decisions that practitioners make, we need to be cautious in concluding, from the results of simplified surveys, that most conservation decisions are based on personal experience, rather than scientific evidence.

First, as described in Section 9.3, practitioners' decisions usually, but not always (see Pullin et al., 2004), take place *within the context* of 'higher-level' decisions, which have involved different people and thereby been based on different sources of information, potentially including scientific evidence.

Second, as described in Section 9.2, conservation management often involves the use of a combination of methods to benefit a wide range of species, tailored to specific circumstances at a site. Therefore, even if the decision to undertake an action (or set of actions) is underpinned by scientific evidence, the details of how best to implement it will usually require an additional 'layer' of personal experience and ecological 'nous' and expertise.

Third, 'personal experience' in any case consists of a mixture and accumulation of experiential and scientific knowledge, which is difficult to disentangle. An experienced practitioner may have read a relevant scientific paper a decade ago, or been informed of best practice that was itself based on scientific evidence. However, having since carried out the same or similar management activity for many years, they may now consider their source of information to be 'personal experience'.

Scientific and ecological advisors provide an important link between science and practice by giving practitioners direct advice and bite-size information chunks of up-to-date, relevant scientific research. There is clear evidence of the value of advisers in increasing the effectiveness of conservation actions (Ingram, 2008; Ewen et al., 2013). While a scientist will typically have in-depth knowledge of a particular subject area, a good ecological adviser will have a broader range of knowledge and experience of conservation management across multiple sites. Most importantly, good ecological advisers will have the ability to translate the results of science into practical management advice, which will involve their experience of the use of similar management actions at other sites.

On RSPB reserves, practitioners place a higher value on the advice given by dedicated ecological advisers than on advice provided by scientists, although the latter is still highly valued (Figure 9.2). The full role of these ecological advisers entails:

- providing ecological advice to practitioners, through the management planning process, project teams and other ad-hoc means;
- 'signing off' all important ecological decisions made on these reserves;

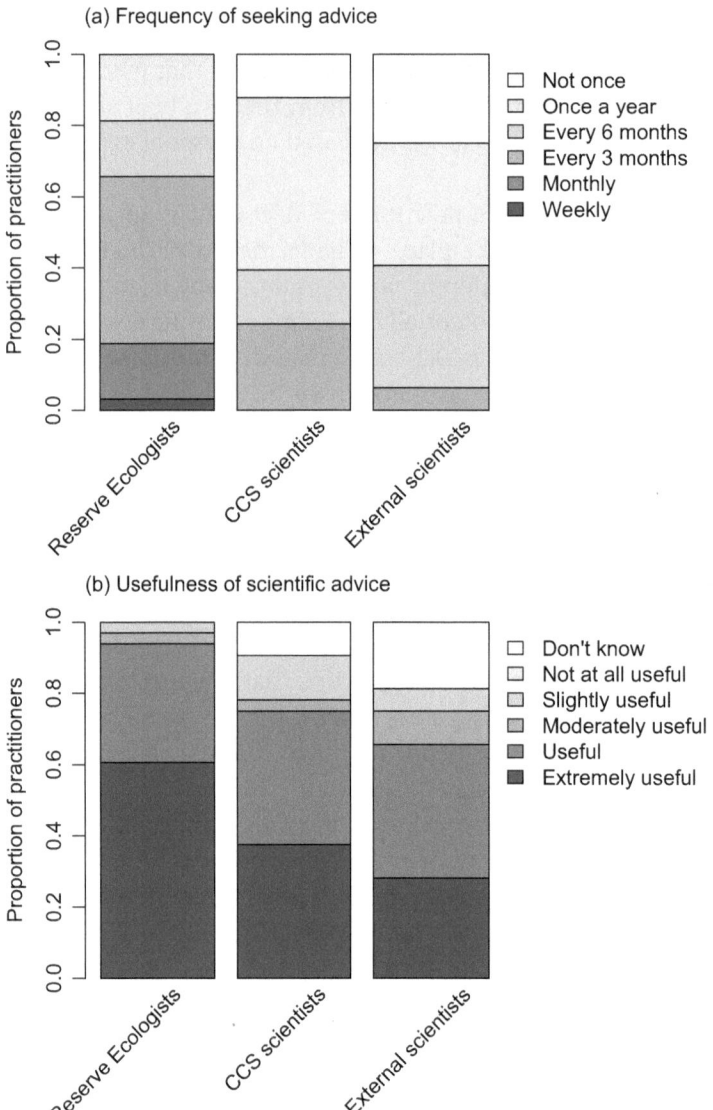

**Figure 9.2** The frequency with which 36 RSPB practitioners (mainly site managers and conservation officers) seek scientific advice from Reserve Ecologists (in-house ecological advisers), Centre for Conservation Scientists (CCS, in-house conservation scientists) and external scientists, and their perceived usefulness of this scientific advice from each source. There was a 78% response rate (46 practitioners were invited to participate) and survey methods are described in Walsh (2015; Chapter 4). (A black and white version of this figure will appear in some formats. For the colour version, please refer to the plate section.)

- annually auditing the effectiveness of reserve management; and
- developing and encouraging the use of best practice, both within and outside the organisation.

These advisers need to have credibility with practitioners, many of whom will have a more detailed knowledge of, and close emotional attachment to, the land on which advice is being given. Similar advisers also have a critical role within government agencies, which provide grants to landowners through agri-environment and land management schemes.

The cost of providing these services by the dedicated ecological advisers at the RSPB is about 4% of the total costs of managing the reserve network. Therefore, if the provision of this advice increases the cost-effectiveness of reserve management by more than 4%, employing these advisers is a good use of conservation resources.

## 9.6 How important is it to use scientific evidence in decision-making?

There is an underlying assumption that decisions based on scientific evidence are more effective than those based solely on personal experience. However, there is little evidence in the conservation field to support the assumption that scientific evidence improves conservation outcomes. In the medical field, however, there are several examples where medical procedures and drugs that were once considered 'best practice' were found to be ineffective or caused severe unintended consequences once the scientific evidence had been collated and synthesised (Sackett & Rosenberg, 1995; Morris et al., 2011).

The best evidence demonstrating the impact of using scientific evidence for conservation decisions comes from a study that measured practitioners' likelihood of using different methods of reducing predation on birds before and after providing them with a summary of scientific evidence about the efficacy of each intervention (Walsh et al., 2014). After reading the summarised scientific information, each participant was asked whether they were more or less likely to use each intervention. On average, practitioners changed their likelihood of using 46% of the interventions shown. Practitioners were more likely to use effective interventions after reading the evidence and less likely to use ineffective actions, suggesting access to the summarised scientific evidence could improve some conservation decisions. Even so, most participants said they would continue using their existing method(s), which they still considered to be the best solution for their set of circumstances (Walsh et al., 2014).

The importance of scientific evidence will vary according to the type of decisions being made. For example, we would hope that a practitioner would check the latest scientific evidence on the best way to control a newly arrived, invasive non-native species. We would not, though, expect an experienced wetland manager to check the scientific literature every time they make a decision about manipulating water levels, although it would still be valuable for them to keep up-to-date with the results of new research. This might be via scientific summaries, magazines, or other information sources that synthesise new research into an accessible format, through meetings with relevant societies and by talking with scientific advisors.

We also suspect that the extent to which resources are wasted on implementing ineffective, non-evidence–based interventions varies greatly in different situations. In the case of widely adopted management interventions carried out by science-based organisations with good systems of planning and adaptive management, most interventions are likely to be *underpinned* by good evidence, but with actions tailored with personal experience to suit the site's specific circumstances, and achieve its often complex objectives.

On the other hand, it is possible that ineffective interventions are implemented more frequently where there is less access to scientific advisers and the results of published science (e.g. Giehl et al., 2017). Another situation where ineffective interventions may also be more widespread is where a developer and their consultants put in place compensatory or offsetting measures that enable them to proceed with development, but have little or no interest in whether these measures prove effective (e.g. Harper & Quigley, 2005; Chapter 4). The consequences and wasted resources of ineffective interventions will be amplified if they are integrated into policy, and widely applied through standardised prescriptions, as occurred when the scientific evidence was not consulted while designing some European Union agri-environment scheme prescriptions (Dicks et al., 2014).

## 9.7 Ways to increase the use of scientific evidence in decision-making

Despite the infrequent direct use of scientific papers by most practitioners, and the perceived low level of usefulness of scientific papers in informing decision-making, it is striking that practitioners typically value advice given to them by scientists (Walsh, 2015). Therefore, any lack of evidence-based decision-making in conservation is clearly not driven by practitioners' aversion to the use of scientific evidence.

However, there are a number of barriers to the use of scientific papers by practitioners (Walsh et al., 2019). Only a small proportion of papers published in ecological journals contain information that is useful for practitioners (Matzek et al., 2014), while the results described in many papers are often fairly incomprehensible to most people outside of academia, often due to the complex statistical techniques used. In addition, many scientific papers are unavailable to practitioners due to publishers' paywalls, although the increase in open-access journals will help with this (Fuller et al., 2014). Therefore, given that scientific papers on their own are unlikely to bridge the gap between science and practice, this leaves two complementary approaches. The first is increasing the synthesis and translation of scientific research into more easily accessible, practical information. The second is ensuring that decision-making processes involve advisors and scientists who help interpret the science and ensure that decisions are based on sound evidence.

Systematic reviews published through the Collaboration for Environmental Evidence are considered the most robust, unbiased level of evidence (Chapter 7). While systematic reviews are invaluable in informing medical practice and are becoming more popular in environmental management, they are often of limited use to conservation practitioners, because their conclusions are usually too generic (Cook et al., 2013). To return to our previous example, a meta-analysis might conclude that predator-exclusion fences usually increase nesting success of a range of bird species, across a range of habitats (e.g. Smith et al., 2011). However, practitioners are unlikely to be interested in their effect across a range of species and habitats. Instead, they will usually be more interested in knowing how to maximise the effectiveness of fencing at protecting a particular species against a specific predator, or set of predators, under similar conditions to those which occur at their site (see Figure 9.1). Because of this, summaries of scientific research that evaluate the success of more specific actions may be of greater use to practitioners. Examples of these include Conservation Evidence synopses (www.conservationevidence.com/synopsis/) and 'What Works in Conservation' (Sutherland et al., 2018).

In addition to the use of evidence summaries, in our experience, the most favoured forms of communication about the effectiveness of conservation actions by practitioners are: one-to-one advice; practical management workshops; practical management handbooks and case studies; visiting sites where the interventions have been implemented; and discussions with fellow practitioners who have practical experience of implementing the technique.

In conclusion, we suggest five key requirements to delivering effective conservation interventions at a site. These are:

- ensuring there are sufficient resources;
- ensuring good decision-making, planning processes and adaptive management are in place, and that these involve people who have relevant expertise;
- employing skilled ecological advisors who can keep up-to-date with the relevant scientific and other literature, spread best practice and who are able to advise practitioners on site-specific solutions based on a combination of science and experience;

- developing projects and collaborations with in-house conservation scientists and universities; and
- employing skilled and knowledgeable practitioners who care about the effectiveness of what they are doing, keep up-to-date with accessible forms of information and who are subsequently able to make informed ecological decisions on a day-to-day basis (as well as being able to do a myriad of other things).

## References

Bayliss, H. R. & Randall, N. P. 2011. Science for action: perceptions of the role of research in invasive species management. *In Practice: Bulletin of Ecology and Environmental Management*, 72, 14–15.

BirdLife International. 2006. *Monitoring Important Bird Areas: A Global Framework*. Version 1.2. Cambridge: BirdLife International.

Cook, C. N., Carter, R. W. B., Fuller, R. A., et al. 2012. Managers consider multiple lines of evidence important for biodiversity management decisions. *Journal of Environmental Management*, 113, 341–346.

Cook, C. N., Hockings, M. & Carter, R. W. 2010. Conservation in the dark? The information used to support management decisions. *Frontiers in Ecology and the Environment*, 8, 181–186.

Cook, C. N., Possingham, H. P. & Fuller, R. A. 2013. Contribution of systematic reviews to management decisions. *Conservation Biology*, 27, 902–915.

Davis, F. W., Stoms, D. M., Costello, C. J., et al. 2003. *A Framework for Setting Land Conservation Priorities using Multi-criteria Scoring and an Optimal Fund Allocation Strategy*. Santa Barbara, CA: National Centre for Ecological Analysis and Synthesis, University of California, Santa Barbara.

Dicks, L. V., Hodge, I., Randall, N. P., et al. 2014. A transparent process for 'evidence-informed' policy making. *Conservation Letters*, 7, 119–125.

Ewen, J. G., Adams, L. & Renwick, R. 2013. New Zealand Species Recovery Groups and their role in evidence-based conservation. *Journal of Applied Ecology*, 50, 281–285.

Fuller, R. A., Lee, J. R. & Watson, J. E. M. 2014. Achieving open access to conservation science. *Conservation Biology*, 28, 1550–1557.

Giehl, E. L. H., Moretti, M., Walsh, J. C., et al. 2017. Scientific evidence and potential barriers in the management of Brazilian protected areas. *PLoS ONE*, 12, e0169917.

Gregory, R., Failing, L., Harstone, M., et al. 2012. *Structured Decision Making: a Practical Guide to Environmental Decision Choices*. Chichester: Wiley-Blackwell.

Harper, D. J. & Quigley, J. T. 2005. No net loss of fish habitat: a review and analysis of habitat compensation in Canada. *Environmental Management*, 36, 343–355.

Ingram, J. 2008. Are farmers in England equipped to meet the knowledge challenge of sustainable soil management? An analysis of farmer and advisor views. *Journal of Environmental Management*, 86, 214–228.

JNCC. 2004. *Common Standards Monitoring. Introduction to the Guidance Manual*. JNCC. Available from http://jncc.defra.gov.uk/pdf/CSM_introduction.pdf

Keith, D. A., Martin, T. G., McDonald-Madden, E., et al. 2011. Uncertainty and adaptive management for biodiversity conservation. *Biological Conservation*, 144, 1175–1178.

Margules, C. R. & Pressey, R. L. 2000. Systematic conservation planning. *Nature*, 405, 243–253.

Matzek, V., Covino, J., Funk, J. L. et al. 2014. Closing the knowing–doing gap in invasive plant management: accessibility and interdisciplinarity of scientific research. *Conservation Letters*, 7, 208–215.

McFadden, J. E., Hiller, T. L. & Tyre, A. J. 2011. Evaluating the efficacy of adaptive management approaches: is there a formula for success? *Journal of Environmental Management*, 92, 1354–1359.

Morris, Z. S., Wooding, S. & Grant, J. 2011. The answer is 17 years, what is the question: understanding time lags in translational research. *Journal of the Royal Society of Medicine*, 104, 510–520.

Murphy, D. D. & Weiland, P. S. 2014. Science and structured decision making: fulfilling the promise of adaptive management for imperiled species. *Journal of Environmental Studies and Sciences*, 4, 200–207.

Organisation for Economic Co-operation and Development. 1993. *OECD Core Set of Indicators for Environmental Performance Reviews: A synthesis report by the Group on the State of the Environment*. Environment Monographs No. 83. Paris: OECD.

Pollard, S. J. T., Davies, G. J., Coley, F., et al. 2008. Better environmental decision making – recent progress and future trends. *Science of the Total Environment*, 400, 20–31.

Pressey, R. L., Cabeza, M., Watts, M. E., et al. 2007. Conservation planning in a changing world. *Trends in Ecology & Evolution*, 22, 583–592.

Pullin, A. S. & Knight, T. M. 2005. Assessing conservation management's evidence base: a survey of management-plan compilers in the United Kingdom and Australia. *Conservation Biology*, 19, 1989–1996.

Pullin, A. S., Knight, T. M., Stone, D. A., et al. 2004. Do conservation managers use scientific evidence to support their decision-making? *Biological Conservation*, 119, 245–252.

Runge, M. C. 2011. An introduction to adaptive management for threatened and endangered species. *Journal of Fish and Wildlife Management*, 2, 220–233.

Sackett, D. L. & Rosenberg, W. M. C. 1995. On the need for evidence-based medicine. *Journal of Public Health Medicine*, 17, 330–334.

Seavy, N. E. & Howell, C. A. 2010. How can we improve information delivery to support conservation and restoration decisions? *Biodiversity and Conservation*, 19, 1261–1267.

Smith, R. K., Pullin, A. S., Stewart, G. B., et al. 2011. Is nest predator exclusion an effective strategy for enhancing bird populations? *Biological Conservation*, 144, 1–10.

Sutherland, W. J., Dicks, L. V., Ockendon, N., et al. 2018. *What Works in Conservation 2018*, 2nd edition. Cambridge: Open Book Publishers.

Sutherland, W. J., Pullin, A. S., Dolman, P. M., et al. 2004. The need for evidence-based conservation. *Trends in Ecology & Evolution*, 19, 305–308.

Walsh, J. C. 2015. Barriers and solutions to implementing evidence-based conservation. PhD thesis, University of Cambridge.

Walsh, J. C., Dicks, L. V. & Sutherland, W. J. 2014. The effect of scientific evidence on conservation practitioners' management decisions. *Conservation Biology*, 29, 88–98.

Walsh, J. C., Dicks, L. V., Raymond, C. M. & Sutherland, W. J. 2019. A typology of barriers and enablers of scientific evidence use in conservation practice. *Journal of Environmental Management*, 250, 109481.

Westgate, M. J., Likens, G. E. & Lindenmayer, D. B. 2013. Adaptive management of biological systems: a review. *Conservation Biology*, 158, 128–139.

Wilson, K. A., Cabeza, M. & Klein, C. J. 2009. Fundamental concepts of spatial conservation prioritization. In: Moilanen, A., Wilson, K. A. & Possingham, H. P., editors, *Spatial Conservation Prioritization*. Oxford: Oxford University Press.

Young, K. D. & Van Aarde, R. J. 2011. Science and elephant management decisions in South Africa. *Biological Conservation*, 144, 876–885.

# CHAPTER TEN

# Effective engagement of conservation scientists with decision-makers

DAVID C. ROSE
*University of Reading*
MEGAN C. EVANS
*University of Queensland*
and
REBECCA M. JARVIS
*Auckland University of Technology
and Sydney Institute of Marine Science*

## 10.1 Introduction

This chapter offers advice on how the conservation science community can effectively engage with decision-makers. The rationales for why we, as scientists, need to do this have been widely discussed in the literature. Often, the reasons offered are normative, pragmatic, or instrumental (de Vente et al., 2016); in other words, there is a belief that engaging with decision-makers leads to better-informed, more acceptable decisions. Indeed, better engagement may lead to the greater uptake of evidence for conservation decisions, something which some scholars argue is a priority for effective management (e.g. Sutherland & Wordley, 2017; Gardner et al., 2018).

Engagement with decision-makers of all types is needed because scientific evidence rarely influences policy and practice in a straightforward way; rather, evidence is considered as one part of a 'messy' decision-making progress alongside other forms of knowledge, interests, beliefs, pragmatics and other factors (Lawton, 2007; Adams & Sandbrook, 2013; Rose, 2014a; Young et al., 2014; Evans et al., 2017). This is particularly true in the case of complex problems, such as biodiversity conservation, where the science is often uncertain, solutions are not readily apparent and the implementation of conservation interventions affects a range of stakeholders with different values and interests (Jarvis et al., 2015a; Maron et al., 2016; Alford & Head, 2017; Rose, 2018). Appreciating and understanding this complexity is a necessary step for scientists who wish to learn how they can most effectively engage with and influence conservation decision-making (Toomey et al., 2016; Evans et al., 2017; Chapter 2).

Effective engagement with decision-makers can facilitate the use of scientific evidence in decision-making, while building support for interventions that are to be implemented on the ground (de Vente, 2016; Bodin, 2017; Roux et al., 2017). Indeed, there has recently been renewed calls for a 'new kind of science' (Keeler et al., 2017) that is more democratic and inclusive, and explicitly recognises the need to engage stakeholders in the production and utilisation of scientific knowledge (Enquist et al., 2017; Hallett et al., 2017; Wall et al., 2017).

We define engagement as the process by which decision-makers and other stakeholders (including scientists) influence how and what decisions are made. Engagement is a key component of doing conservation work, since conservation decisions will always affect, or be affected, by people (Kareiva & Marvier, 2007; Kothari et al., 2013). Poorly conducted engagement, however, has the potential to lead to detrimental outcomes (Young et al., 2013; Bodin, 2017; Reed et al., 2017), for example by failing to include all decision-makers in a representative, valued way, or by reinforcing existing power imbalances and inequality (e.g. Chambers, 1997; Brockington, 2007).

So, what does 'effective' engagement look like? Communication is unsurprisingly a fundamental component. Differences in organisational culture, incentives and language can make it difficult for decision-makers and scientists to understand one another (Caplan, 1979; Head, 2015; Newman et al., 2016) and this can lead to scientific evidence being mismatched with the needs of policy-makers and practitioners (Jarvis, 2015). Many other studies in conservation have noted that academic science is not always immediately relevant for practitioners (see Walsh et al., 2015). Difficulties in communication include science being presented in jargonistic, unusable formats (Walsh et al., 2015; Marshall et al., 2017;), the lack of open access publishing (Arlettaz et al., 2010), communicating only in one language (Amano et al., 2016) and poorly communicated policy demands (Neßhöver et al., 2016). Overall, Farwig et al. (2017) found that major differences in workflows, background and objectives create a 'research–implementation gap' (Cook et al., 2013; Jarvis et al., 2015a), which is difficult to bridge. Rose et al. (2018a) found agreement on the major barriers to the use of evidence in conservation policy among policy-makers, scientists and practitioners, but noted that solutions needed to be implemented.

However, effective engagement is not simply a matter of improving communication (Cash et al., 2002; Evans et al., 2017). Knowledge is inevitably co-produced (Miller & Wyborn, 2018) by multiple groups of people through an iterative process of knowledge exchange, mutual learning, negotiation and adaptation (Cash et al., 2002; Wyborn, 2015). While scientists cannot change the fact that scientific evidence is (necessarily) just one input into conservation decision-making, through effective engagement it is possible to influence how and what knowledge (and decisions) are co-produced (Miller & Wyborn, 2018).

Although it is impossible to construct a framework for good engagement that will work in all contexts (de Vente et al., 2016; Bodin, 2017; Reed et al., 2017) common principles of effective engagement include trust, reciprocity, respect, transparency, clear benefits to participants, co-learning and identifying all necessary decision-makers (see Table 10.1; de Vente et al., 2016; Enquist et al., 2017; Reed et al., 2017; Roux et al., 2017; Sterling et al., 2017). Engagement processes should be sensitive to cultural context and power relations and seek to disrupt existing inequalities, rather than reinforce them (Reed et al., 2017; Sterling et al., 2017).

In this chapter, we seek to illustrate common principles of effective engagement using several case studies. We first describe in more detail who the decision-makers in conservation are and how to ensure they are all identified and effectively engaged in a particular context. Next, we outline four case studies that provide examples of good engagement: the development of environmental offsets policy in Australia; community engagement in carnivore conservation in Costa Rica; participatory marine spatial planning in New Zealand; and the development of a code of conduct for marine conservation globally between researchers and NGOs. We conclude by providing 10 'top tips' for scientists engaging with decision-makers, by drawing on the literature, aforementioned case studies and our own experiences.

## 10.2  Who are decision-makers in conservation?

Conservation decisions are made by various individuals and organisations at different levels of governance (Newell et al., 2012; Evans et al., 2017). Throughout this chapter we use 'decision-makers' as an umbrella term to refer to the multiple groups that are involved in conservation policy and practice. The decision-makers involved in a particular conservation issue will vary, as will the local cultures, priorities, knowledge types, values and workflows. Engagement with decision-makers is more likely to be effective if scientists first work to gain an understanding of who may affect or be affected by conservation decisions in a particular context (Waylen et al., 2010; Enquist et al., 2017; Sterling et al., 2017).

It cannot be assumed that good practice for working with one type of decision-maker is transferable to working with another (de Vente et al., 2016; Reed et al., 2017). For example, it is likely that the most appropriate approaches will differ between a government policy-maker, an NGO practitioner, an academic researcher, a farmer and a local resident. Decision-makers will use varying language, hold particular, and personal, worldviews and be part of different decision-making cultures (Blicharska & Grandin, 2015).

Before engaging, a representative list of key decision-makers needs to be created. Reed et al. (2009) argue that three stages of stakeholder analysis are required at the start of collaborative forms of engagement: (1) identify all key actors; (2) differentiate between them by working to understand individual workflows, values, cultures and interests; and (3) understand relationships between actors, to help build alliances or prevent conflict (see also Colvin et al., 2016). A range of methods can be used to map influential decision-makers (see Reed et al., 2017 for a typology), including interviews, focus groups, Q-methodology, community workshops and the Delphi technique (Amit & Jacobson, 2018; Mukherjee et al., 2018; Nyumba et al., 2018; Young et al., 2018). Such techniques can help to identify key decision-makers, elucidate how different individuals use and value their land, understand their views on conservation and manage differences between groups.

There is also heterogeneity within groups of decision-makers. For example, in the context of tropical reforestation, Lazos-Chavero et al. (2016) noted that cattle ranchers vary by their age, herd size and educational background. It proved important to engage with a representative group of cattle ranchers because the workflows and priorities of farmers varied with farm size and this influenced uptake of management practices. Indeed, the literature details many such cases where knowledge exchange with practitioners or the public was ineffective because groups were assumed to be homogeneous (Chilvers & Kearnes, 2016). Taking account of intra-group hetereogeneity as well as inter-group variance thus adds an extra challenge to collaborative processes.

## 10.3 Case studies of good engagement

Many good examples of effective engagement in conservation exist in the literature from terrestrial (Fraser et al., 2006), freshwater (Nel et al., 2016) and marine systems (Granek & Brown, 2005). The nature of these successes varies from fostering an increased interest in conservation or natural resource management among local communities (e.g. Granek & Brown, 2005; Fraser et al., 2006; Roux et al., 2017) to traditional knowledge being valued alongside scientific information and fostering inclusivity and trust (Granek & Brown, 2005) to the formation of better decisions (Fraser et al., 2006; Nel et al., 2016).

Here, we highlight four case studies where engagement with decision-makers has helped conservation. They present examples of engagement with different types of decision-maker: first with government policy-makers, second with stakeholders at the community level, third with multiple stakeholders at a regional level, and fourth with multiple stakeholders at a global level.

### 10.3.1 Case Study 1: Engaging with policy-makers – development of the Australian Environmental Offsets Policy

In 2012, Australian academic researchers formulated a calculation-based approach that set a new standard for determining environmental offset requirements. In collaboration with federal policy-makers in the Australian Department of the Environment, the calculation approach was developed into a tool for making fair and robust decisions about offsets. This became the Offsets Assessment Guide, which underpins the Australian Environmental Offsets Policy (2012) (see www.environment.gov.au/system/files/resources/12630bb4-2c10-4c8e-815f-2d7862bf87e7/files/offsets-policy_2.pdf) and remains the tool for determining offsets for significant impacts on more than 1800 threatened species and ecological communities in Australia (Gibbons et al., 2015; Miller et al., 2015). This collaborative effort between academics and policy-makers was enabled by long-term, effective relationships, significant government investment in research specifically to improve environmental decision-making,[1] support of senior executive members of the department and a decade of scientific research led by the research team and many colleagues.

Environmental offsets are routinely used as a tool to compensate for unavoidable impacts on biodiversity as a result of development activities such as mining, urban development and agricultural expansion (Maron et al., 2016). In Australia, offsets have been used as conditions of development approval by state and federal governments since the early 2000s (Maron et al., 2015; Evans, 2016). Regulatory decisions under Australia's federal environmental law was guided by a draft policy from 2007 onwards, but stakeholder dissatisfaction with this framework led to a policy review and development of a new draft environmental offsets policy in 2011 (Miller et al., 2015).

Stakeholder consultation led by the federal Department of the Environment indicated broad stakeholder agreement with the new draft policy principles, but also a clear desire for a scientifically robust framework for estimating offset requirements (Miller et al., 2015). The Department then approached academic researchers to develop an offset calculation framework that would enable impacts on threatened species and ecological communities to be adequately and effectively compensated, give effect to the policy principles and be accessible and easy-to-use for all stakeholders (Miller et al., 2015).

The development of the Offsets Assessment Guide was highly collaborative and iterative. Each major revision of the calculation framework produced by the academic researchers was tested by federal government operations staff to ensure ease of use, applicability to a range of decision contexts and adherence to the policy principles. This process of co-design enabled mutual learning and fostered a shared understanding of the different constraints and incentives that policy-makers and academic researchers work under. There was intense negotiation, compromise and robust debate. The researchers had to operate

---

[1] Specifically, through partnerships with the Australian Government's Commonwealth Environmental Research Facilities (CERF) program (2004–2008), National Environmental Research Program (NERP, 2011–2015) and National Environmental Science Programme (NESP, 2016–2020).

within a much shorter timeframe than is normally permitted in academia and learned to appreciate the government decision processes and ministerial requirements. The Department of the Environment recognised the need for the collaboration to result in academic publications for the researchers, and publication of work in the academic literature was considered a priority (Miller et al., 2015).

The research outcomes have now shaped environmental offsetting around the world (IUCN, 2016; Maseyk et al., 2016; Cowie et al., 2018). The researchers continue to work with governments, industry, local communities and international convening bodies to boost public and policy-maker capacity to engage with environmental offsets. The final independent report to the Australian Government on the $154 M National Environmental Research Program highlighted this work as one of the Program's most important impacts (Spencer et al., 2014):

> The Offsets Calculator has provided a useful tool to improve the efficiency and effectiveness of regulating development under the EPBC Act by assessing the suitability of offset proposals and assisting with planning and estimating future offset requirements ... The department credits the standing, expertise and assistance of the NERP Environmental Decisions Hub in building stakeholder understanding, trust and acceptance of the offsets policy and calculator, including by industry, NGOs and the jurisdictions. Stakeholder acceptance is crucial to its successful adoption and implementation of this policy.

### 10.3.2 Case Study 2: Engaging local communities – co-existence with large carnivores in Costa Rica

Amit and Jacobson (2018) present an example of community engagement in a project designed to facilitate co-existence between large carnivores (jaguars and pumas) and people in Costa Rica. Through the use of a group decision-making technique based on the Delphi process (see Mukherjee et al., 2015), they engaged 133 members of seven communities, as well as 25 multidisciplinary experts from government, NGOs and academic science. Four decision-making rounds were undertaken.

Round one: community representatives were identified by using a database of ranches with the potential for big cat attacks on livestock. After selecting two ranchers and two community leaders from each of seven 'attack hotspots', further participants were identified in consultation with them. At a workshop held at the University of Costa Rica, these local representatives were used to define the project agenda, to identify the major problems, and to brainstorm potential solutions. Draft solutions to incentivise co-existence were developed.

Rounds two and three: the draft incentives were reviewed through online questionnaires sent to a panel of multi-disciplinary experts (NGOs, academics, government). The draft list of incentives was iteratively developed based on the opinions of these experts.

Round four: a workshop was held with the communities in each of the seven 'attack hotspots'. They had an average duration of three hours and were conducted by five facilitators at venues such as schools and community halls. Through anonymous voting, and a satisfaction questionnaire, the study team were able to test for consensus, and the willingness of participants to pilot particular incentives.

Detailed results and other methodological information are presented in the original paper (Amit & Jacobson, 2018). The authors claim that their structured, bottom-up communication process stimulated social learning in a trusting, transparent, collaborative environment. Although one community declined to take part in future research, citing a lack of information provided in the process, the study team argued that the list of incentives for co-existence was able to integrate issues of governance, equity and social norms. As a result, support for the incentives, and for working in a transdisciplinary way, was strengthened in many of the communities.

### 10.3.3 Case Study 3: Engagement of multiple stakeholders and decision-makers at a regional level – the Sea Change – Tai Timu Tai Pari marine spatial planning process

In 2000 the Hauraki Gulf Marine Park (HGMP) was established to recognise the national significance of the Hauraki Gulf/Tīkapa Moana (also known as Te Moananui-ā-Toi) in New Zealand. While a number of management plans were developed over the years to mitigate key threats in the HGMP, they were never implemented. This lack of implementation was due to a lack of stakeholder involvement, weak governance and ineffective management (Hauraki Gulf Forum, 2011, 2014).

In response, Sea Change – Tai Timu Tai Pari was developed in 2013 as a new marine conservation and spatial planning process for the region. In contrast to previous planning efforts, Sea Change – Tai Timu Tai Pari was created as a collaborative, stakeholder-led, co-governance process to design, develop and action a new plan for the HGMP. A Stakeholder Working Group and a number of issues-based roundtables were established to navigate the co-development of the plan in consultation with mana whenua iwi and hapū (indigenous Māori tribes and sub-tribes), technical experts, local communities and stakeholders across a range of issues and priorities. This work was supported and assisted by five key partner agencies, including the Hauraki Gulf Forum, Waikato Regional Council, Auckland Council, the Ministry of Primary Industries and the Department of Conservation. In addition, conservation scientists were invited to collaborate with Sea Change – Tai Timu Tai Pari to develop participatory tools and approaches to enhance public and stakeholder engagement, while incorporating local knowledge and diverse values, views and priorities into the planning process (Jarvis et al., 2015b, 2016; Jarvis, 2016). The final plan was released in April 2017 (Sea Change – Tai Timu Tai Pari, 2017).

Effective engagement and collaboration was seen as critical for the Sea Change – Tai Timu Tai Pari process and the development of the plan. This highly collaborative approach required negotiation, perseverance and sacrifice, in addition to the vision and commitment offered by those involved. While some work is already underway, the next step of the plan will be broad implementation across all goals and key principles. Strong and effective co-governance will be key to continuing engagement and effective implementation. There are high hopes that mana whenua iwi and hapū, communities, agencies and government will continue to work together to protect and conserve the future of the HGMP, support healthy and prosperous communities and safeguard this precious natural resource.

### 10.3.4 Case Study 4: Engagement of researchers, practitioners and NGOs at a global level – developing a code of conduct for marine conservation

As marine conservation gathers pace around the globe to achieve our conservation targets and the Sustainable Development Goals, there is a risk that these efforts fail to engage stakeholders and local people effectively. As a result, some actions taken may undermine the rights, dignity and freedoms of local people by not considering their needs or involving them in conservation processes. In response, a code of conduct (COC) was developed to provide a social baseline for how marine conservation should be undertaken, while raising the profile of effective engagement practices and the need for community and stakeholder involvement (Bennett et al., 2017a).

The COC was developed to promote fair governance and decision-making, support social justice and promote transparency and accountability in our marine conservation actions. This includes principles of human rights, indigenous rights and food security, as well as ensuring that marine conservation is carried out in a fair, inclusive way that supports local people. The COC has the potential to have wide-ranging impacts in the way scientists and practitioners undertake marine conservation to ensure it is socially just and environmentally effective.

The lead authors of the proposed code of conduct undertook an initial scoping review and prepared an initial list of principles for discussion with the broader marine conservation community (Bennett et al., 2017a). Next, they convened a meeting with a diverse group of leading experts in marine conservation at the IUCN 2016 World Conservation Congress in Hawaii to

Table 10.1 *Key factors for effective engagement identified in five selected studies*

| Paper title and reference | Context | Key principles for good engagement |
|---|---|---|
| How does the context and design of participatory decision-making processes affect their outcomes? Evidence from sustainable land management in global drylands (de Vente et al., 2016) | Sustainable land management in global drylands | 1. Select participants carefully<br>2. Make participation easy<br>3. Build trust<br>4. Give participants relevant information<br>5. Give participants decision-making power<br>6. Utilise professional facilitators<br>7. Make a long-term commitment<br>8. Flexible language, location and design to the participants |
| Foundations of translational ecology (Enquist et al., 2017) | Considers what a new 'translational ecology' looks like – i.e. ecology that is trans-disciplinary and inclusive of stakeholders beyond academia | 1. Pursue co-production of knowledge<br>2. Ensure meaningful engagement with diverse stakeholders<br>3. Make a long-term commitment<br>4. Listen and respect views<br>5. Ensure everyone can contribute<br>6. Have a clear purpose for the engagement exercise |
| A theory of participation: what makes stakeholder and public engagement in environmental management work? (Reed et al., 2017) | Narrative literature search (multiple contexts) | 1. Understand local context<br>2. Include all stakeholders in a transparent and representative way<br>3. Ensure equal participation for all<br>4. Match levels of engagement with aims and strength of values held (longer engagement needed to change core beliefs) |
| Trans-disciplinary research for systemic change: who to learn with, what to learn about and how to learn (Roux et al., 2017) | Contemporary conservation issues in South Africa | 1. Make a long-term commitment<br>2. Use bridging agents or knowledge brokers to improve communication between groups<br>3. Researchers need to present as co-learners, not 'dominant masters'<br>4. Use mixed paradigm research designs<br>5. Be conscious of bias, e.g. self-selection, perceived superiority of scientific knowledge, attraction of simple solutions to complex problems |

**Table 10.1** *(cont.)*

| Paper title and reference | Context | Key principles for good engagement |
|---|---|---|
| Assessing the evidence for stakeholder engagement in biodiversity conservation (Sterling et al., 2017) | Literature review (multiple contexts) | 1. Ensure stakeholders can contribute meaningfully to process<br>2. Ensure transparency<br>3. Build trust<br>4. Recognise the values of stakeholders<br>5. Understand why stakeholders want to engage<br>6. Harness stakeholder champions<br>7. Make a long-term, trusting commitment<br>8. Incorporate local and traditional knowledge<br>9. Appreciate and respect local cultural context<br>10. Manage stakeholder relationships flexibly |

debate what is considered acceptable and unacceptable in marine conservation with researchers and practitioners from universities, non-profit organisations and government agencies from around the world. The final list of principles was agreed after several rounds of iterations with the authors and workshop participants, incorporating a thorough review of peer-reviewed literature, conservation policies and procedures and foundational policy documents.

The COC (Bennett et al., 2017a) was the result of this collaborative process and was communicated in a wide variety of formats to different media around the world, presented to policy-makers and discussed at high-level meetings, such as the United Nations Ocean Conference in June 2017. As a result, the COC has already been adopted as guiding principles for the Global Environment Facility Blue Carbon Project (GEF, 2017), with partners and beneficiaries that include the United Nations, 40 NGOs and a number of academic institutions, practitioners and members of the scientific community. The objective is for all Blue Carbon Projects to be developed following the COC by 2020. Engagement and discussion around the application of COC more broadly is ongoing. The goal is to establish the COC as a clearly articulated and comprehensive set of social standards to guide marine conservation actions at multiple scales and ensure that marine conservation goals are met through effective engagement, fair decision-making, accountability and inclusive participatory processes.

## 10.4 Ten tips for achieving good engagement

There have been few attempts to derive general principles of effective engagement from examples implemented in practice (Nguyen et al., 2017; Reed et al., 2017), as environmental management is such a context-specific endeavour (de Vente et al., 2016). As such, Reed et al. (2009) suggested that approaches to engagement should be flexible, adaptive and iterative based on local circumstances. With this in mind, we highlight 10 tips based on the case studies, the literature and our own experience (see also Table 10.1 for key factors identified in five other studies).

### 1. Know who you need to talk to

This important theme of inclusivity is commonplace in the literature (see Table 10.1). All relevant decision-makers need to be engaged with, or vital knowledge may be missed or unnecessary conflicts created (e.g. de Vente et al., 2016; Enquist et al., 2017; Lazos-Chaveros et al., 2016; Reed et al., 2017). The composition of key decision-makers will always vary with context and may depend on the specific impact that is sought, but robust stakeholder analyses should be conducted before commencement of work (Reed et al., 2009; de Vente et al., 2016). If time or resources are short, then decision-makers may be classified by the extent to which they are affected by a conservation issue (Reed et al., 2009), as Amit and Jacobson (2018) did by identifying 'predator attack hotspots'.

Once decision-makers are identified and engaged with, scientists should seek to differentiate between different groups and understand relationships between them. Part of this process can be an attempt to understand their workflows, their values and culture and even the constraints under which they work. Once groups have been differentiated, then different styles of engagement and conflict management might be needed to work with each (Blicharska & Grandin, 2015). Furthermore, an appreciation and understanding of political, social and cultural context is always useful (Sterling et al., 2017).

### 2. Engage early, with clearly defined aims

Decision-maker engagement must have a clear purpose in order for all participants to work together towards a clear goal and outcome (Enquist et al., 2017). Involving decision-makers at an early stage of a project may provide ownership of a project to local communities, building support, legitimacy, and trust, as well as leading to the production of relevant, 'use-inspired', or 'actionable' knowledge (Wall et al., 2017). The need for local community-led engagement was illustrated by the examples of human-carnivore co-existence in Costa Rica (Amit & Jacobson, 2018), marine conservation in New Zealand (Jarvis, 2015; Jarvis et al., 2015) and the biodiversity offsetting project stimulated by the Australian Department of the Environment (Miller et al., 2015).

### 3. Decision-makers should find it easy to engage

Participation for all decision-makers must be easy (de Vente et al., 2016). For example, meetings should be held in a convenient place for all and project timescales should consider the busy and varied workflows of all decision-makers involved, so as not to disincentivise engagement. Language should also be geared towards participants, and thus a common language and understanding should be developed wherever possible (Amano et al., 2016; de Vente et al., 2016). While we do not necessarily condone offering financial incentives for attendance, researchers could at least consider what the relative advantage of engagement is for decision-makers (what do different decision-makers gain from being part of the process?) and cover costs at the very least (particularly where poorer communities are being involved).

### 4. Embrace and include multiple knowledge(s), perspectives and worldviews

Engagement with decision-makers must be meaningful, and the perspectives and opinions of all stakeholders must be genuinely valued throughout the process (see all studies in Table 10.1). Participation should not merely be tokenistic. The first step towards this is humility on the part of researchers, which fosters a genuine sense to learn from others, while also accepting and appreciating that science is just one input into policy and practical processes. In their study of co-management in South African freshwater ecosystems, Roux et al. (2017) warn against perceived scientific authority, and an attitude that bemoans some decisions made by policy-makers and other stakeholders as irrational if they are not 'evidence-based'. The second step is to find ways of integrating multiple knowledge types into a project, including lay and indigenous knowledges, and local experiential knowledges, and ultimately fostering respect and understanding across different values and motivations (Sterling et al., 2017). The final step is to be able to reflect on your own values and motivations as a conservationist and be prepared to learn from those held by others (Bodin, 2017).

If these steps are followed, it is more likely that a truly collaborative spirit of cooperation will be achieved, which will help to build common understanding of an issue. This will not always mean that everyone agrees, but it will still be possible for all participants to understand each other's point of view. Such a collaborative spirit has been shown to help a range of conservation projects, including in the case studies highlighted above.

### 5. Think hard about power

As researchers, we must do more than simply speak truth to the most obvious powers-that-be (Chambers, 1997); rather, we should seek to understand how communities work as thoroughly as possible, something that may require long-term engagement (e.g. using ethnography). Lazos-Chavero et al. (2016) found that paying attention to gender, generational and power disparities in a given region was essential to the success of tropical reforestation schemes. Furthermore, Kleiber et al. (2015) showed that including women in the management of fisheries is essential for conservation success because a significant proportion of fishers are women (something that had often been ignored in previous studies). Thus, ensuring that all stakeholders have equal decision-making power is important for effective engagement. This also includes the balance of power between the stakeholders and the researchers themselves.

## 6. Build mutual trust

This theme is just about universally accepted in the literature and needs little explanation (see Table 10.1). Without mutual trust, transparency and respect, then engagement exercises with decision-makers are doomed to failure. Although Lacey et al. (2018) warn against too much trust (e.g. because this could lead to facts being accepted on 'blind faith'), it is logical to expect that relationships built on trust will yield better results. This is because participants will feel valued and able to challenge the opinion of others. Good practices for building trust include respecting participant confidentiality, following through on promises and committing to long-term engagement if it has been offered.

## 7. Good facilitation is key

Engagement processes need to have good facilitators (de Vente et al., 2016). As illustrated by guides on how to conduct participatory methods, such as focus groups (Nyumba et al., 2018), the facilitator plays a key role in managing group dynamics, encouraging stakeholder input and building trust. A good facilitator will be aware of potential sensitivities within the group (Gibbons et al., 2008) and be able to skilfully avoid and manage conflict, which is so important for a healthy engagement process (Amit & Jacobson, 2018; Chapter 14). In controversial cases in particular, which are not unusual when dealing with the complex problem of biodiversity loss, the potential for conflict is more pronounced.

## 8. Learn new skills for good engagement

Good engagement and facilitation is helped if the individual is a good communicator. As individuals, it will become increasingly important to develop a range of different skills (as per Jackson et al., 2017) and be able to communicate differently with different people. In doing so, it is important to recognise that conservation can greatly benefit from better use of qualitative methods that improve communication, enhance engagement and give voice to others (Mukherjee et al., 2018). However, it may not be possible for individuals to learn all the different skills key for good engagement themselves. Therefore the development of truly inter- and trans-disciplinary teams could be one approach to bring all the necessary tools and skills together and co-design research that properly integrates the natural and social sciences (Bennett et al., 2017b, 2017c) while engaging with stakeholders from the outset and throughout conservation processes (Reed et al., 2017). Where scientists feel unable to facilitate engagement processes effectively, much of the literature suggests using knowledge brokers (alternatively called boundary spanners or bridging agents; Cvitanovic et al., 2015; de Vente et al., 2016; Roux et al., 2017; Bednarek et al. 2018). These individuals have the skills to bridge the gap between varying backgrounds, cultures, interests and languages.

## 9. You don't have to reinvent the wheel – consider making use of existing spaces and opportunities

In conservation, there are several good schemes that encourage scientists to engage better with decision-makers across research, policy and practice (see Elliott et al., 2018 for a global database of 650 conservation capacity initiatives). Such schemes have been developed to reflect requirements for the foundational skills necessary for good engagement while also providing existing opportunities for conservationists to develop their own capacity for effective communication, interpersonal interaction and boundary-crossing. By making use of such schemes, conservation scientists can develop their engagement skills while also being able to better adapt to the changing needs of conservation.

An additional point worthy of consideration is whether conservation researchers make the most of existing informal spaces of engagement to harness the views of decision-makers. Chilvers et al. (2017) criticise engagement processes for usually being established on the terms of researchers. In other words, groups of stakeholders are assembled to talk about an issue that is framed and defined by researchers or policymakers, such as through public forums (see Chilvers & Kearnes, 2016). Very rarely do we seek to 'listen in' on existing spaces of public participation (e.g. in the village hall, in the pub, on social media) to see what people are concerned about. Could the same criticism be levelled at conservation engagement exercises? Do we seek to assemble groups of decision-makers to discuss conservation issues that we have already framed, rather than asking, for example, local communities to devise the questions of interest (see tip 4)? We suggest that it is important to consider these questions in order that engagement exercises are led by communities, rather than done to them.

## 10. Don't give up!

The need for long-term engagement is commonly highlighted in the literature (see Table 10.1). One important aspect to take from our recommendations is that they will not always yield immediate, tangible rewards, but this should not be the sole aim of practising good engagement. Rather, ongoing, long-term engagement can lead to a change in the overall policy framing of problems and solutions (Rose et al., 2017), something which can occur diffusely over long timescales (Owens, 2015). Reed et al. (2017) argue that engagement in controversial issues, where people hold deep core values, will need to be more long term (de Vente, 2016; Roux et al., 2017). It can take some time to build the trust and confidence for stakeholders to contribute, and continued engagement after implementation is usually required for conservation projects (Lazos-Chavero et al., 2016). So it is vital not to give up; as Amit and Jacobson (2018) argue, 'participatory decision-making has an inherent phase of struggle and frustration', which is perfectly normal. Sterling et al. (2017) further describe knowledge co-production as a 'slow' process because it requires long-term committed engagement from all sides.

However, it is also important to note that flexibility of process is also key (Sterling et al., 2017). When inviting decision-makers to contribute to a project, the outcome might be different to the one that the researcher envisaged. Indeed, because you are incorporating multiple values and perspectives into decision-making, the unexpected may be the norm. Most importantly, expect the unexpected and don't give up!

We acknowledge that it is not easy for conservation scientists to initiate and manage collaborative research projects, particularly those that work with a variety of stakeholder groups outside of academia. There are certainly challenges in achieving the new kind of science that Keeler et al. (2017) envisaged (or in embracing the 'post-normal' reality, see Colloff et al., 2017; Rose, 2018), which would be more inclusive of people beyond academia. This includes practical difficulties (e.g. time, money) of engaging decision-makers (Sutherland et al., 2017), as well as the challenge for conservation scientists of developing the skills needed to engage with people, a task for which many of us are not traditionally trained (Jackson et al., 2017). Furthermore, being actively involved with decision-makers might not be something that appeals to individual conservation scientists. Although the boundaries between science, policy and practice are fluid (Rose, 2014b; Toomey et al., 2016), scientists sometimes worry about moving beyond their comfort zone. Yet, if there is a scientific discipline in which advocacy is easier to do, then it should be mission-driven conservation biology (Soulé, 1985; Rose et al., 2018b).

Ultimately, achieving effective engagement and conservation impact may mean changing the way conservationists work, including those housed in universities and research institutions. One significant challenge is for academic conservation scientists to find the time, motivation and support to engage decision-makers (Chapin, 2017; Keeler et al., 2017; Littell et al., 2017). Often, academics are not rewarded adequately for producing tangible impacts (Jarvis et al., 2015; Tyler, 2017), and activities focused on delivering these impacts are still widely sidelined in favour of career-enhancing academic publication. However, there is no real reason why impact cannot be better incentivised, and new opportunities developed to explore the different ways we can better navigate science, policy and practice. Why, for example, cannot academic departments have dedicated policy teams to highlight policy demand and to foster collaboration with decision-makers? A new kind of conservation science could certainly be imagined, which would reward outreach and incentivise inter-, multi- and trans-disciplinary collaborative work. Where we are unable to invest the time to engage with decision-makers ourselves, we could make much better use of knowledge brokers or boundary spanners (Bednarek et al., 2018).

### 10.5 Acknowledgement

This research was supported by the Australian Government's National Environmental Science Program through the Threatened Species Recovery Hub (MCE).

## References

Adams, W. M. & Sandbrook, C. 2013. Conservation, evidence and policy. *Oryx*, 47, 329–335.

Alford, J. & Head, B. W. 2017. Wicked and less wicked problems: a typology and a contingency framework. *Policy and Society*, 36, 397–413.

Amano T., González-Varo J. P. & Sutherland, W. J. 2016. Languages are still a major barrier to global science. *PLoS Biology*, 14(12), e2000933.

Amit, R. & Jacobson, S. K. 2018. Participatory development of incentives to coexist with jaguars and pumas. *Conservation Biology*, 32, 938–948. https://doi.org/10.1111/cobi .13082

Arlettaz, R., Schaub, M., Fournier, J., *et al.* 2010. From publications to public actions: when conservation biologists bridge the gap between research and implementation. *BioScience*, 60, 835–842.

Bednarek, A. T., Wyborn, C., Cvitanovic, C., et al. 2018. Boundary spanning at the science–policy interface: the practitioners' perspectives. *Sustainability Science*, 13, 1175–1183. https://doi.org/10.1007 /s11625-018-0550-9

Bennett, N. J., Roth, R., Klain, S. C., et al. 2017b. Conservation social science: understanding and integrating human dimensions to improve conservation. *Biological Conservation*, 205, 93–108.

Bennett, N. J., Roth, R., Klain, S. C., et al. 2017c. Mainstreaming the social sciences in conservation. *Conservation Biology*, 31, 56–66.

Bennett, N. J., Teh, L., Ota, P., et al. 2017a. An appeal for a code of conduct for marine conservation. *Marine Policy*, 81, 411–418.

Blicharska, M. & Grandin, U. 2015. Why protect biodiversity? Perspectives of conservation professionals in Poland. *International Journal of Biodiversity Science Ecosystem Services & Management*, 11, 349–362.

Bodin, O. 2017. Collaborative environmental governance: achieving collective action in social-ecological systems. *Science*, 357 (6352).

Brockington, D. 2007. Forests, community conservation, and local government performance: the village forest reserves of Tanzania. *Society and Natural Resources*, 20, 835–848.

Caplan, N. 1979. The two-communities theory and knowledge utilization. *American Behavioral Scientist*, 22, 459–470.

Cash, D., Clark, W. C., Alcock, F., et al. 2002. Salience, credibility, legitimacy and boundaries: linking research, assessment and decision making. KSG Working Paper Series RWP02-046. http://dx.doi.org/10 .2139/ssrn.372280

Chambers, R. 1997. *Whose Reality Counts? Putting the First Last.* London: ITDG Publishing.

Chapin, F. S. 2017. Now is the time for translational ecology. *Frontiers in Ecology and the Environment*, 15, 539.

Chilvers, J. & Kearnes, M. 2016. *Remaking Participation. Science, Environment and Emerging Publics.* Abingdon: Routledge.

Chilvers, J., Pallett, H. & Hargreaves, T. 2017. Public engagement with energy: broadening evidence, policy and practice. Briefing note to the UK Energy Research Centre. Available from www.ukerc.ac.uk/ publications/public-engagement-with-energy.html

Colloff, M. J., Lavorel, S., van Kerkhoff, L. E., et al. 2017. Transforming conservation science and practice for a postnormal world. *Conservation Biology*, 31, 1008–1017.

Colvin, R. M., Witt, G. B. & Lacey, J. 2016. Approaches to identifying stakeholders in environmental management: insights from practitioners to go beyond the "usual suspects". *Land Use Policy*, 52, 266–276.

Cook, C. N., Mascia, M. B., Schwartz, M. W., et al. 2013 Achieving conservation science that bridges the knowledge–action boundary. *Conservation Biology*, 27, 669–678.

Cowie, A. L., Orr, B. J., Castillo Sanchez, V. M., et al. 2018. Land in balance: the scientific conceptual framework for Land Degradation Neutrality. *Environmental Science and Policy*, 79, 25–35.

Cvitanovic, C., Hobday, A. J., van Kerkhoff, L., et al. 2015. Improving knowledge exchange among scientists and decision-makers to facilitate the adaptive governance of marine resources: a review of knowledge and research needs. *Ocean and Coastal Management*, 112, 25–35.

de Vente, J., Reed, M. S., Stringer, L. C., et al. 2016. How does the context and design of participatory decision making processes affect their outcomes? Evidence from sustainable land management in global drylands. *Ecology and Society*, 21(2), 24.

Elliott, L., Ryan, M. & Wyborn, C. 2018. Global patterns in conservation capacity development. *Biological Conservation*, 221, 261–269.

Enquist, C. A. F., Jackson, S. T., Garfin, G. M., et al. 2017. Foundations of translational ecology. *Frontiers in Ecology and the Environment*, 15, 541–550.

Evans, M. C. 2016. Deforestation in Australia: drivers, trends and policy responses. *Pacific Conservation Biology*, 22, 130–150.

Evans, M. C., Davila, F., Toomey, A., et al. 2017. Embrace complexity to improve conservation decision making. *Nature Ecology & Evolution*, 1, 1588.

Farwig, N., Ammer, C., Annighöfer, P., et al. 2017. Bridging science and practice in conservation: deficits and challenges from a research perspective. *Basic and Applied Ecology*, 24, 1–8.

Fraser, E. D. G., Dougill, A. J., Mabee, W. E., et al. 2006. Bottom up and top down: analysis of participatory processes for sustainability indicator identification as a pathway to community empowerment and sustainable environmental management. *Journal of Environmental Management*, 78, 114–127.

Gardner, C. J., Waeber, P. O., Razafindratsima, O. H., et al. 2018. Decision complacency and conservation planning. *Conservation Biology*, 32, 1469–1472. https://doi.org/10.1111/cobi.13124

Gibbons, P., Zammit, C., Youngentob, K., et al. 2008. Some practical suggestions for improving engagement between researchers and policy-makers in natural resource management. *Ecological Management and Restoration*, 9, 182–186.

Gibbons, P., Evans, M. C., Maron, M., et al. 2015. A loss–gain calculator for biodiversity offsets and the circumstances in which no net loss is feasible. *Conservation Letters*, 9, 252–259.

Global Environment Facility (GEF) Blue Carbon Project. 2017. Blue carbon code of conduct. Available from https://news.gefblueforests.org/blue-carbon-code-of-conduct

Granek, E. E. & Brown, M. A. 2005. Co-management approach to marine conservation in Mohéli, Comoros Islands. *Conservation Biology*, 19, 1724–1732.

Hallett, L. M., Morelli, T. L., Gerber, L. R., et al. 2017. Navigating translation ecology: creating opportunities for scientist participation. *Frontiers in Ecology and the Environment*, 15, 578–586.

Hauraki Gulf Forum. 2011. State of our gulf: Tikapa Moana – Hauraki Gulf state of the environment report 2011. New Zealand.

Hauraki Gulf Forum. 2014. State of our gulf: Tikapa Moana – Hauraki Gulf state of the environment report 2014. New Zealand.

Head, B. W. 2015. Relationships between policy academics and public servants: learning at a distance? *Australian Journal of Public Administration*, 74, 5–12.

IUCN. 2016. *IUCN Policy on Biodiversity Offsets*. Gland: IUCN.

Jackson, S. T., Garfin, G. M. & Enquist, C. A. F. 2017. Toward an effective practice of translational ecology. *Frontiers in Ecology and the Environment*, 15, 540.

Jarvis, R. M. 2015. Putting people back in the picture: a social research agenda for a social–ecological approach to conservation. PhD thesis, Auckland University of Technology.

Jarvis, R. M., Bollard Breen, B., Krägeloh, C. U., et al. 2015b. Citizen science and the power of public participation in marine spatial planning. *Marine Policy*, 57, 21–26.

Jarvis, R. M., Bollard Breen, B., Krägeloh, C. U., et al. 2016. Identifying diverse conservation values for place-based spatial planning using crowdsourced voluntary geographic information. *Society & Natural Resources*, 29, 603–616.

Jarvis, R. M., Borrelle, S. B., Bollard Breen, B., et al. 2015a. Conservation, mismatch and the research implementation gap. *Pacific Conservation Biology*, 21, 105–107.

Kareiva, P. & Marvier, M. 2007. Conservation for the people. *Scientific American*, 297, 50–57.

Keeler, B. L., Chaplin-Kramer, R., Guerry, A. D., et al. 2017. Society is ready for a new kind of science – is academia? *BioScience*, 67, 591–592.

Kleiber, D. L., Harris, L. M. & Vincent, A. 2015. Gender and small-scale fisheries: a case for counting women and beyond. *Fish and Fisheries*, 16, 547–562.

Kothari A., Camill, P. & Brown, J. 2013. Conservation as if people also mattered: policy and practice of community-based conservation. *Conservation and Society*, 11, 1–15.

Lacey, J., Howden, M., Cvitanovic, C., et al. 2018. Understanding and managing trust at the climate science–policy interface. *Nature Climate Change*, 8, 22–28.

Lawton, J. H. 2007. Ecology, politics and policy. *Journal of Applied Ecology*, 44, 465–474.

Lazos-Chavero, E., Zinda,J. Bennett-Curry, A., et al. 2016. Stakeholders and tropical reforestation: challenges, trade-offs, and strategies in dynamic environments. *Biotropica*, 48, 900–914.

Littell, J. S., Terrando, A. J. & Morelli, T. L. 2017. Balancing research and service to decision-makers. *Frontiers in Ecology and the Environment*, 15, 598.

Maron, M., Bull, J. W., Evans, M. C., et al. 2015. Locking in loss: baselines of decline in Australian biodiversity offset policies. *Biological Conservation*, 192, 504–512.

Maron, M., Ives, C. D., Kujala, H., et al. 2016. Taming a wicked problem: resolving controversies in biodiversity offsetting. *BioScience*, 66, 489–498.

Marshall, N., Adger, N., Attwood, S., et al. 2017. Empirically derived guidance for social scientists to influence environmental policy. *PLoS ONE*, 12(3), e0171950.

Maseyk, F., Barea, L., Stephens, R., et al. 2016. A disaggregated biodiversity offset accounting model to improve estimation of ecological equivalency and no net loss. *Biological Conservation*, 204, 322–332.

Miller, K. L., Trezise, J. A., Kraus, S., et al. 2015. The development of the Australian environmental offsets policy: from theory to practice. *Environmental Conservation*, 42, 306–314.

Miller, C. A. & Wyborn, C. 2018. Co-production in global sustainability: histories and theories. *Environmental Science & Policy*, https://doi.org/10.1016/j .envsci.2018.01.016

Mukherjee, N., Hugé, J., Sutherland, W. J., et al. 2015. The Dephi technique in ecology and biological conservation: applications and guidelines. *Methods in Ecology and Evolution*, 6, 1097–1109.

Mukherjee, N., Zabala, A., Hugé, J., et al. 2018. Comparison of techniques for eliciting views and judgements in decision-making, *Methods in Ecology and Evolution*, 9, 54–63.

Neßhöver, C., Vandewalle, M., Wittmer, H., et al. 2016. The Network of Knowledge approach – improving the science and society dialogue on biodiversity and ecosystem services in Europe. *Biodiversity and Conservation*, 25, 1215–1234.

Nel, J. L., Roux, D. J., Driver, A., et al. 2016. Knowledge co-production and boundary work to promote implementation of conservation plans. *Conservation Biology*, 30, 176–188.

Newell, P., Pattberg, P. & Schroeder, H. 2012. Multiactor governance and the environment. *Annual Review of Environment and Resources*, 37, 365–387.

Newman, J., Cherney, A. & Head, B. W. 2016. Do policy-makers use academic research? Reexamining the "Two Communities" theory of research utilization. *Public Administration Review*, 76, 24–32.

Nguyen, V. M., Young, N. & Cooke, S. J. 2017. A roadmap for knowledge exchange and mobilization research in conservation and natural resource management. *Conservation Biology*, 31, 789–798.

Nyumba, T. O., Wilson, K., Derrick, J., et al. 2018. The use of focus group discussion methodology: insights from two decades of application in conservation. *Methods in Ecology and Evolution*, 9, 20–32.

Owens, S. 2015. *Knowledge, Policy, and Expertise: The UK Royal Commission on Environmental Pollution 1970–2015*. Oxford: Oxford University Press.

Reed, M. S., Graves, A., Dandy, N., et al. 2009. Who's in and why? A typology of stakeholder analysis methods for natural resource management. *Journal of Environmental Management*, 90, 1933–1949.

Reed, M. S., Vella, S., Challies, E., et al. 2017. A theory of participation: what makes stakeholder and public engagement in environmental management work? *Restoration Ecology*, 26(S1), S7–S17.

Rose, D. C. 2014a. Five ways to enhance the impact of climate science. *Nature Climate Change*, 4, 522–524.

Rose, D. C. 2014b. Boundary work. *Nature Climate Change*, 4, 1038.

Rose, D. C. 2018. Avoiding a post-truth world: embracing post-normal conservation science. *Conservation and Society*, 16, 518–524.

Rose, D. C., Brotherton, P. M., Owens, S., et al. 2018a. Honest advocacy for nature: presenting a persuasive narrative for conservation. *Biodiversity and Conservation*, 27, 1703–1723.

Rose, D. C., Mukherjee, N., Simmons, B. I., et al. 2017. Policy windows for the environment: tips for improving the uptake of scientific knowledge. *Environmental Science and Policy*, http://dx.doi.org/10.1016/j.envsci.2017.07.013

Rose, D. C., Sutherland, W. J., Amano, T., et al. 2018b. The major barriers and their solutions for evidence-informed conservation policy. *Conservation Letters*, 11 (5), e12564. https://doi.org/10.1111/conl.12564

Roux, D. J., Nel, J. L., Cundill, G., et al. 2017. Transdisciplinary research for systemic change: who to learn with, what to learn about and how to learn. *Sustainability Science*, 12, 711–726.

Sea Change – Tai Timu Tai Pari. 2017. Sea Change – Tai Timu Tai Pari Hauraki Gulf Marine Spatial Plan. Available from www.seachange.org.nz/Read-the-Plan/

Spencer, C., McVay, P. & Sheridan, S. 2014. Evaluation of the National Environmental Research Program (NERP).

Soulé, M. E. 1985. What is conservation biology?, *BioScience*, 35, 727–734.

Sterling, E. J., Betley, E., Sigouin, A., et al. 2017. Assessing the evidence for stakeholder engagement in biodiversity conservation. *Biological Conservation*, 209, 159–171.

Sutherland, W. J., Shackelford, G. & Rose, D. C. 2017. Collaborating with communities: co-production or co-assessment? *Oryx*, 51, 569–570.

Sutherland, W. J. & Wordley, C. F. R. 2017. Evidence complacency hampers conservation. *Nature Ecology & Evolution*, 1, 1215–1216.

Toomey, A. H., Knight, A. T. & Barlow, J. 2016. Navigating the space between research and implementation in conservation: research–implementation spaces. *Conservation Letters*, 10, 619–625.

Tyler, C. 2017. Wanted: academics wise to the needs of government. *Nature*, 552, 7.

Wall, T. U., McNie, E. & Garfin, G. M. 2017. Use-inspired science: making science usable and useful to decision-makers. *Frontiers in Ecology and the Environment*, 15, 551–559.

Walsh, J. C., Dicks, L. V. & Sutherland, W. J. 2015. The effect of scientific evidence on conservation practitioners' management decisions. *Conservation Biology*, 29, 88–98.

Waylen, K. A., Fischer, A., McGowan, P. J. K., et al. 2010. Effect of local cultural context on the success of community-based conservation interventions. *Conservation Biology*, 24, 1119–1129.

Wyborn, C. A. 2015. Connecting knowledge with action through coproductive capacities: adaptive governance and connectivity conservation. *Ecology and Society*, 20, 11.

Young, J. C., Jordan, A., Searle, K. R., et al. 2013. Does stakeholder involvement really benefit biodiversity conservation? *Biological Conservation*, 158, 359–370.

Young, J. C., Rose, D. C., Mumby, H. S., et al. 2018. A methodological guide to using and reporting on interviews in conservation science research. *Methods in Ecology and Evolution*, 9, 10–19.

Young, J. C., Waylen, K. A., Sarkki, S., et al. 2014. Improving the science–policy dialogue to meet the challenges of biodiversity conservation: having conversations rather than talking at one-another. *Biodiversity and Conservation*, 23, 387–404.

# CHAPTER ELEVEN

# Conservation decisions in the face of uncertainty

ROBERT P. FRECKLETON
*University of Sheffield*

## 11.1 Introduction

Scientific evidence is fundamental to solving a suite of real-world issues and research is crucial in informing solutions to pressing issues such as climate change, food security, evolved resistance and land management (Thomas et al., 2004; Godfray et al., 2010; Hicks et al., 2018; Watson et al., 2018). This evidence takes a range of forms, including the results of small- and large-scale experiments (Firbank et al., 2003), meta-analyses (Johnson & Curtis, 2001; Batáry et al., 2011), systematic reviews (Pullin & Stewart, 2006) and predictive models (Taylor & Hastings, 2004; Stratonovitch et al., 2012). Decision-makers need to be able to choose between options using the best evidence available (Sutherland & Freckleton, 2012).

Unfortunately, ecological systems are enormously variable at just about every scale that we study them (Holling, 1973). This variability has numerous sources and, collectively, they contribute to what may be known as 'uncertainty'. In recognising the role of uncertainty, it is important to recognise that this may arise both as an intrinsic property of the system as well as a nuisance through inadequate data or observation. In terms of intrinsic sources, for example, *spatial* variability results from variations in conditions from place to place (Tilman & Karieva, 1997), while *temporal* variability similarly results from variations in systems through time (Huston, 1994). On the other hand, the measurements of the system may contain inaccuracies. For instance, *observational* variance is a consequence of our inability to perfectly measure systems, instead relying on sampling in order to build up a picture of the dynamical properties of the system (Dennis et al., 2006; Freckleton et al., 2006).

Addressing all types of variability and stochasticity is important in making decisions, and we need to recognise the different sources and how they contribute to uncertainty. Consider a simple example: imagine that we are attempting to implement a conservation measure to protect an organism and that a management intervention, $I$, may be an effective conservation action if implemented, and this yields a benefit, $b$. However, there is a cost, $c$, to implementing the action. If we know that the action is certain to work, there is a simple calculation: all other things being equal then, assuming they are measured in the same units, if $b > c$ then it would be worth performing $I$. If this is not true, then $I$ is not a favourable approach.

However, because variability is pervasive, the situation in conservation management is rarely so simple. We might not be certain that $I$ is always effective and instead suppose that we know that $I$ is effective with probability $p$; $p$ could have multiple interpretations depending on context. For example, in a spatially variable system, $I$ might be effective in a fraction $p$ of sites, but not in others: $p$ thus measures the spatial variance in outcomes. Alternatively, the evidence for $I$ being an effective strategy might be mixed, and therefore $p$ could measure some aspect of our belief that $I$ works.

When such uncertainty exists, the condition for a manager choosing to apply $I$ becomes $pb > c$. Note that typically $c$ should be known reasonably accurately as this will be costed in terms of the resources required to enact $I$. The benefit is now weighted by the uncertainty in efficacy of $I$. In terms of making correct management decisions, this simple condition suggests a number of interesting observations. First, as uncertainty increases (i.e. $p$ gets smaller) the likelihood of employing $I$ decreases. If $p$ measures spatial or temporal variability in outcomes, then this is sensible because if $I$ is less likely to work, so a manager should be less inclined to choose it. On the other hand, if $p$ measures a lack of knowledge of the effectiveness of $I$, then the inequality suggests conservatism: do not take action unless it is known that $I$ is effective with a high probability ($p > c/b$). If $p$ is measuring such uncertainty then the recommended action has nothing to do with the *actual* effectiveness of $I$. Being conservative thus results from ignorance.

A second significant behaviour occurs when both $p$ and $c$ are low: the likelihood of $I$ working is believed to be small but the cost is also small. In this case, employing $I$ may still be favoured by a manager if the benefit is very large and one might describe this as *superstitious* behaviour (i.e. doing something in the face of little evidence that it will work because the benefit is high and the cost is low). A large number of interventions possibly fall into this category.

Overall, this illustrative example demonstrates that the amount of uncertainty can contribute a great deal to the overall management outcome. In both of the hypothetical situations outlined above, the management applied, and consequent outcome, is suboptimal because it leads to biased impressions of the costs and benefits. Characterising uncertainty is thus vital.

## 11.2 Recognising types of uncertainty

The source of uncertainty is important and authors have proposed various approaches to classifying uncertainty in management. Regan et al. (2002) point out that many of the sources of variability leading to uncertainty described above may be termed *epistemic* (i.e. uncertainty in the system itself and its measurement). They also highlight a second source of uncertainty, namely *linguistic* uncertainty. This results from uncertainty in the language used to describe actions or systems, as well as resulting from the conveyance of information. As an example, in the UK there was a programme for government-hired shooters to exterminate ruddy ducks (*Oxyura jamaicensis*). During the cull, coot (*Fulica atra*), black-necked grebe (*Podiceps nigricollis*), common pochard (*Aythya ferina*) and common scoter (*Melanitta nigra*) individuals were also shot (Henderson, 2009). This resulted in part from inadequate communication with shooters (Henderson, 2009), who were not ornithologists and failed to distinguish between species. Consequently, there is a possibility of confusion, with procedures subsequently being developed to ensure that confusion is minimised. Although such uncertainty is undoubtedly important, I will concentrate on epistemic uncertainty *sensu* Regan et al. (2002), although some of the points made below could equally apply to a more inclusive definition.

Broadly speaking, it is useful to distinguish *intrinsic uncertainty* (analogous to the variance in model parameters in an ecological or statistical model) from *knowledge uncertainty* (by analogy with the measurement error or lack of data in a model). The reason for making the distinction between these two types of uncertainty is important: one is a property of the system itself, while the other is caused by a lack of understanding or data. The two are interactive, and this is perhaps the greatest challenge to making robust predictions in management. If the management outcomes are uncertain both in terms of intrinsic variability and knowledge then they will be largely unpredictable. In this circumstance, it is necessary to question the recommendations given, as well as to consider whether the approach to prediction is the correct one. Another option is to consider models that use an alternative more stable formulation (Taylor & Hastings, 2004; Freckleton et al., 2011).

## 11.3 Science versus practice: different perspectives on uncertainty

Scientists and practitioners have different perspectives, even if they are working on the same problem. The question of how to resolve this difference is a thorny one (Bradshaw & Borchers, 2000; Sutherland & Freckleton, 2012) and there is a pervasive perception of a science–policy gap (Bertuol-Garcia et al., 2018). Bradshaw and Borchers (2000) highlighted a series of ways in which the perspectives of science and practice may be misaligned. Of these there are two in which uncertainty plays a particularly important role.

### 11.3.1  Probabilistic, qualified evidence

In the introductory example above, the evidence for the effectiveness of a management intervention was measured as a probability. In terms of providing evidence, this is a routine way in which a scientist would express their recommendation. However, for implementing management, this can be problematic. For instance, telling a manager that there is a 70% chance that the intervention will work is only partly addressing the question of the manager, namely should they undertake the action or not? How is a manager to know whether their particular circumstances are likely to lead to them being in the 70% of cases in which the intervention works or in the 30% in which it fails?

In this context, the meaning of probabilities conveyed by scientists may not always be fully clear. Consider an everyday example. We might be told by a weather forecaster that there is a 50% chance of rain today. However, the meaning of that probability is not typically explained. Here are four interpretations.

(i)   It will either rain everywhere or nowhere: it could be one or other of these outcomes, for example, because it is not possible to predict the precise location of a weather system.
(ii)  It will rain for 50% of the time during the forecast period: for example, there are patchy rain clouds that are continually moving.
(iii) It will rain in 50% of places: for example, there are rain clouds cover 50% of the area that do not move.
(iv)  The forecaster is unable to tell us whether it will rain or not and is telling you to flip a coin.

The technical interpretation of a probability in a weather forecast is that this probability represents the fraction of times a given outcome (e.g. raining within a defined set of areas) occurs in a set of stochastic realisations. This definition, interestingly, can incorporate all four of the above interpretations. Nevertheless, the probability quoted is a form of *knowledge uncertainty* that has a very specific meaning: it is a measure of model uncertainty/variance.

This highlights a second aspect of scientific evidence that is problematic from the perspective of management, namely that scientific evidence is usually *qualified*. The statement 'there is a 50% chance of rain' from a scientific perspective should also be qualified by the statement 'across a set of simulations, given the assumption that the model is correct'. If the model is wrong then the prediction could be greatly different.

The task of a manager is to convert such evidence into action (i.e. the binary outcome of whether to act or not). As noted in the introduction, the decision then involves costs and benefits, defined in the widest sense and including values. To continue the hypothetical example, carrying an umbrella is low cost and high benefit, so a 50% chance of rain would render this a good strategy. On the other hand, a manager who is spraying a pesticide requires good conditions, and a 50% chance of rain would potentially carry an unacceptable risk that this costly action (in terms of fuel, time and chemicals) would fail.

## 11.3.2 General versus situational outcomes

The aim of science is typically to find answers that are as general and robust as possible. A scientist faced with evaluating the effectiveness of a management intervention will attempt to find whether there is evidence of its effectiveness, on average, and then probably focus on understanding the mechanisms that drive it. In contrast, a manager is faced with the task of managing a given site over a defined time period. There is a potential conflict between these perspectives, as the scientific perspective typically averages over variation arising from site-specific variations, whereas this is precisely the variation that a manager is focused on. For a scientist, the local variation at a specific site is essentially nuisance variance.

Although perhaps something of a caricature, there is undoubtedly a real problem in addressing these differences in perspectives. The situation is complicated by the difference in success measures for scientists and managers: scientists prove success by presenting results that are of interest to a wide range of others and that do not focus on specific instances (e.g. in scientific papers); managers measure success based on the state of their site. This difference in perspectives is reflected in the contrasting ways that scientists and managers treat uncertainty. From the science perspective the variation around the mean is a quantity that is to be minimised where possible; in contrast, a manager needs to know where their site sits with respect to this variation, and whether local circumstances render the overall average outcome pattern inapplicable.

## 11.4 Addressing uncertainty

In general, it is important that uncertainty is recognised and tackled to avoid common 'traps' (Millner-Gulland & Shea, 2017). These traps are varied, but include ignoring or not accounting for uncertainty, as well as focusing on irrelevant uncertainties and not clearly stating the objectives in framing problems (Millner-Gulland & Shea, 2017). Here I review three case studies, showing that there is a line of argument that ignores uncertainty and another that embraces it. In each case the value of conclusions, both for the scientist and the practitioner, require that uncertainty is fully evaluated.

### 11.4.1 Ignoring uncertainty should not be an option

One of the most important causes of uncertainty is lack of information. This is particularly an issue when information is lacking on rare and difficult-to-observe species, meaning that clade-wide conservation assessments are potentially compromised. The International Union for Conservation of Nature (IUCN) is an important organisation that collates data on the conservation status of species from a wide range of taxa into a set of threat states (Mace & Lande, 1991). This extensive and important exercise informs conservation strategies in a range of contexts (Rodrigues et al., 2006). The basis for the assessment is a five-point scale of threat status for wild extant species. Species are classified as Least Concern (LC), Near Threatened (NT), Vulnerable (VU), Endangered (EN) or Critically Endangered (CE). Extinct in the Wild and Extinct are categories of extinction beyond these five points, representing species loss.

The amount of data required to apply these criteria varies between taxa. In some cases the amount of information required is quite low. For example, the Nechisar nightjar (*Camprimulgus solala*) is classified as VU despite being known from only a single wing and a single sighting. On the other hand, for some groups (e.g. mammals and amphibians) the data requirements for the assignment of conservation status are more exacting. Those species for which sufficient information is not available are assigned a status termed Data Deficient (DD). The number of DD mammal species is a considerable fraction of the group (483 of 4186 species; i.e. >10%) of mammals studied by Jetz and Freckleton (2015).

Denoting species as DD is, effectively, a way of dealing with uncertainty. It is essentially the same as ignoring missing data in an analysis. This way of dealing with data uncertainty is, however, fraught with pitfalls, and a large literature exists on dealing with missing data and associated uncertainty (Nakagawa & Freckleton, 2008). It is well understood that non-randomness in the pattern of 'missingness' can yield highly misleading analyses.

In the case of conservation assessments, the concern with DD mammal species is that the factors that drive data deficiency are closely related to those that determine extinction threat. For instance, if species are difficult to observe it is likely to be because they only occur at low density in remote locations, or population trends are unknown because they are so rare. It is easy to see that this set of criteria could lead to species being ignored from conservation assessments even though they are threatened.

Jetz and Freckleton (2015) tested this hypothesis by applying a framework for phylo-spatial modelling of IUCN threats, then using this to predict the probability that DD species are threatened. Species that are DD are predicted to have much higher threat probabilities than those that have been assessed already (Figure 11.1). The fraction of threatened mammal species is therefore *underestimated* by the current system of assessment.

Interestingly, the same is not true of birds (Lee & Jetz, 2011), as a much smaller fraction of them are considered DD because a lower threshold of information is required to assess threat status. Thus, the recent taxonomic explosion that has

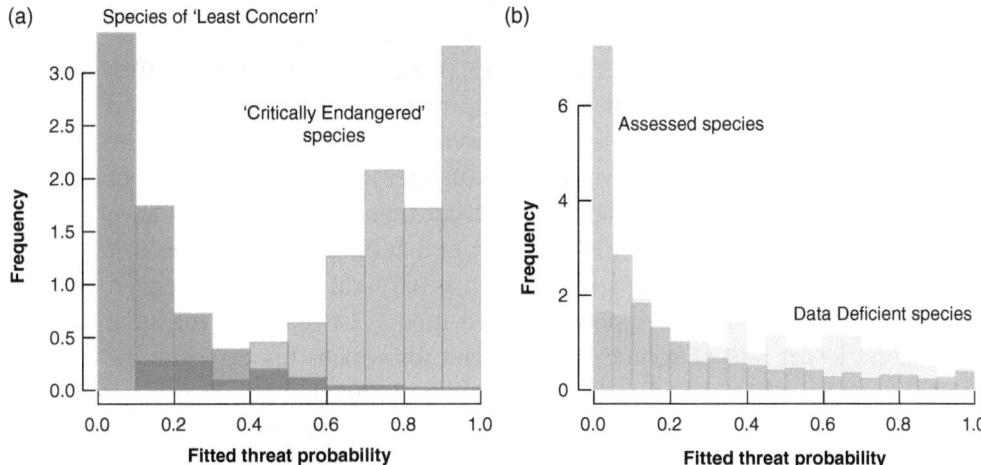

**Figure 11.1** The importance of dealing with uncertainty in conservation assessments. We used models to generate threat probabilities for mammals. (a) These probabilities do an effective job of distinguishing species that are Least Concern (green bars) from those that are Critically Endangered (orange bars); (b) our models were used to predict threat probabilities for species that were Data Deficient (DD) (pink bars) compared to species that were assessed (grey bars) (i.e. to reduce uncertainty in assessment). (A black and white version of this figure will appear in some formats. For the colour version, please refer to the plate section.)

led to the creation of 1000 new species of birds (del Hoyo et al., 2014, 2016) has not resulted in 1000 species being assigned to the DD category.

This example illustrates an important point about uncertainty that is relevant to conservation and management. Ignoring uncertainty by simply excluding cases where data are missing runs the risk of introducing bias and so, in general, should be addressed if at all possible (Millner-Gulland & Shea, 2017). In the introduction I noted that the likelihood of implementing an action is low, irrespective of its actual effectiveness, when there is great uncertainty associated with its effectiveness (i.e. the parameter $p$ is low). In this example, data-deficiency data result in no action being taken ($p$ is low because of uncertainty), although the evidence (Figure 11.1) is that the intervention (assigning status of 'threatened') is justified with high probability.

### 11.4.2 Providing more data/evidence

The preceding example highlights that, where possible, additional data should be used to plug gaps in knowledge. One of the ways that scientists tend to qualify conclusions (see Section 11.3.1) is to say that we cannot be confident because more data are required. As argued by Millner-Gulland and Shea (2017), this can prevent effective management-relevant advice being given.

The example from Jetz and Freckleton (2015) (see also Safi & Pettorelli, 2010; Bland et al., 2015) addressed this qualification by extracting as much information as possible out of the existing data using advanced statistical methods. There are a large range of techniques that have been used to infer missing data and it is not possible to review them here, except to point out that suitable methods have been developed (Nakagawa & Freckleton, 2008), or that the problem can be dealt with using flexible statistical frameworks, such as Bayesian modelling (Gelman et al., 1995). Another recent application used models to infer the maximal population growth rate of several shark species for which this demographic rate has not been otherwise estimated (e.g. Pardo et al., 2018).

In many cases, however, the bottom line is that sufficient data do not exist and there is no option but to collect more. Data are time-consuming and expensive to collect. Engaging in a programme of data collection will delay implementation and use up resources that could be targeted at on-the-ground management. Frequently there will not be resources available for data collection and hence the knowledge gap is never plugged.

On the assumption that more information could be obtained, a key question arises: will collecting more information improve management decisions (Maxwell et al., 2015)? Canessa et al. (2015) highlight a measure called the 'Value of Information' (VoI). This measure is the difference in outcome between the expected management action based only on whatever prior information was available, and action taken with new information provided (Yokota & Thompson, 2004; Canessa et al., 2015). They provide an example that is typical of many in conservation or land management. Imagine that a species of conservation concern occurs in one location within a protected area. The aim of conservation is to maximise the size of the population in the area over a specified time period. In order to meet this aim, one strategy could be to create a new population. However, imagine further that there is a chance that a disease could be present that would limit the effectiveness of the reintroduction. The VoI in this case reflects the change in estimated effectiveness that would be achieved by testing for the presence of disease before starting the reintroduction programme. Thus, a test might be performed and return a positive or negative result. Given a prior estimate of the prevalence of the disease, the difference between initial and updated estimates can be calculated using Bayesian updating. These differences then measure the VoI provided by conducting testing. This represents the possible improvement in decision-making through the removal of uncertainty.

### 11.4.3 Addressing uncertainty through benchmarking
A manager might apply a conservation intervention which, if the outcome is positive, leads to a question of whether the intervention should be used again,

or even recommended to another manager. Informal communication of outcomes of this sort are not unusual in land management (Henrich, 2001).

From a scientific perspective, this is not an acceptable way of proceeding unless appropriate controls and experimental design are used in the evaluation of the method. Furthermore, the intervention would ideally be evaluated at more than a single site. This reflects, of course, the tension between the situational and general perspectives of practitioners and scientists. There are pitfalls in both views. There is of course, no guarantee that if management appears to work at one site that it is not simply due to natural variation. Figure 11.2a gives an example of this from an agricultural case study. At one site a specific intervention was used and appeared to be successful. However, compared with the outcome on a set of farms that did not use the technique, there is no obviously large effect. On the other hand, if we are too picky about standards of evidence or data then there is a real danger that useful information will be discarded.

Developments such as evidence-based conservation promote the collation of evidence on the effectiveness of management (Sutherland, 2003; Sutherland et al., 2004; see also Chapter 4). The idea here is twofold. First, if the same management has been used in different places then, even if individual interventions do not meet the criteria of a randomised trial (as in Figure 11.2a), the collective body of evidence might be useful. Resources such as www.conservationevidence.com allow this work to be synthesised. Second, using systematic review approaches, it is possible to synthesise this information to provide answers to management problems (Pullin & Stewart, 2006; see also Chapter 7).

In the example shown in Figure 11.2a, a single manager implemented one management intervention. On its own this is not enough to determine effectiveness. However, if many people implement the same management then it may be possible to use non-intervention cases as a benchmark and compare the difference with those places where interventions were made. For example, Figures 11.2a and 11.2b show the distribution of weed population sizes in fields subject to intervention (Figure 11.2a) compared with those in which no intervention was made (Figure 11.2b). There is an apparent difference in outcome, but clearly with a high degree of variance. Modelling the data (Figure 11.2c–11.2e) reveals that, although there is an effect of the intervention (Figure 11.2e), there is also a high degree of variance resulting from the initial state (Figure 11.2c) or from the variation in population dynamics from field to field (Figure 11.2d). Consequently, the effect of management, although measurable (Figure 11.2e), is relatively small compared with the intrinsic variability of this system. In this example, the results in Figure 11.2c–e confirm the expectation that the specific management intervention *should* work, but they also confirm anecdotal local reports that the effectiveness of this approach is patchy, and suggest that frequently the positive effects observed may be attributable to other factors (the large negative effect sizes in Figure 11.2c and d).

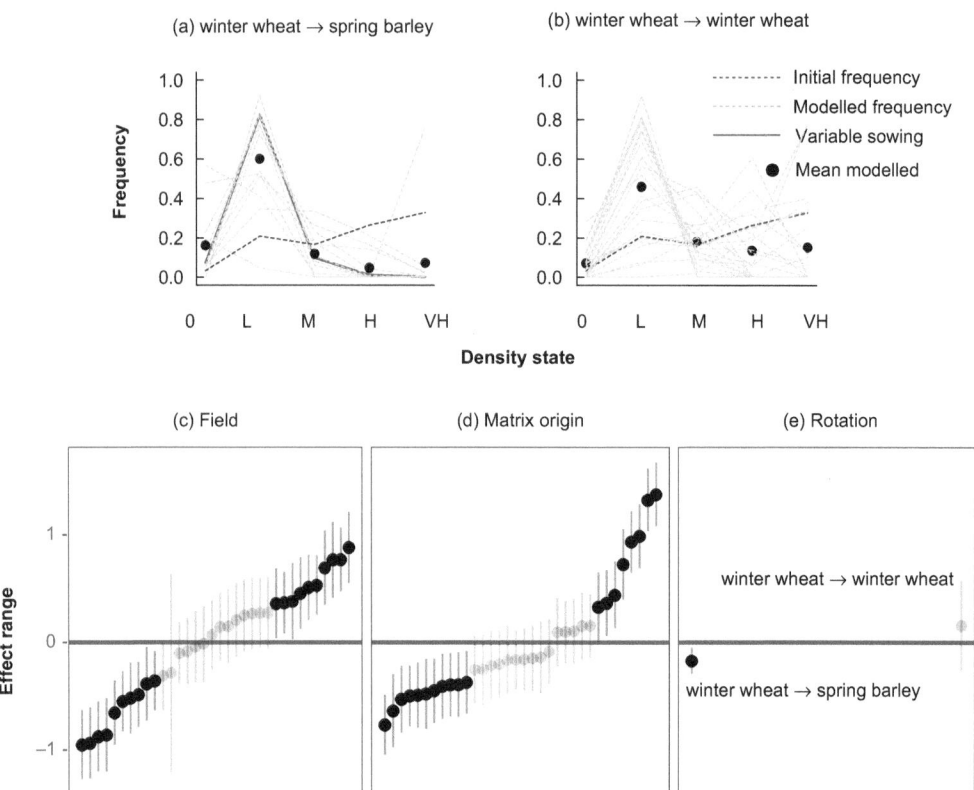

**Figure 11.2** Uncertainty and benchmarking in weed control. (a,b) Predicted responses of populations of the weed *Alopecurus myosuroides* to rotational management. The initial frequency of weeds at each sowing density was the same in each case (dashed blue line). Each grey line represents a matrix generated from a different field following two forms of management. (a) What *would have been* the density (0 = zero, L = low, M = medium, H = high and VH = very high) of an average field *had* it been planted with spring barley. This is compared with (b) the predicted response from maintaining winter wheat. The red line in (a) represents a single field that was managed with variable sowing densities. Figures (c–e) compare the observed effect of management with difference sources of background variation to disentangle the uncertainty in management. We generated models for each field: 22 in winter wheat and 12 rotated from winter wheat to spring barley, and their results are presented in rank order. The effect range is the estimate of the random effect for each field, location or rotation. (A black and white version of this figure will appear in some formats. For the colour version, please refer to the plate section.)

*Benchmarking* of this sort could be extremely valuable in aiding management decisions (Freckleton et al., 2018). Technological advances, such as widespread instrumentation of agricultural machinery, UAS technology (Paneque-Gálvez et al., 2014; Lambert et al., 2018) and remote sensing (Kerr & Ostrovsky, 2003; Turner et al., 2003) offer the possibility of widescale automated data collection at massive scales. When combined with ecological models, such data could provide a hitherto impossible resource for reducing uncertainty in predicting future management outcomes.

## References

Batáry, P., Báldi, A., Kleijn, D., et al. 2011. Landscape-moderated biodiversity effects of agri-environmental management: a meta-analysis. *Proceedings of the Royal Society B: Biological Sciences*, 278, 1894–1902.

Bertuol-Garcia, D., Morsello, C., El-Hani, C. N., et al. 2018. A conceptual framework for understanding the perspectives on the causes of the science–practice gap in ecology and conservation. *Biological Reviews*, 93, 1032–1055.

Bland, L. M., Orme, C. D. L., Bielby, J., et al. 2015. Cost-effective assessment of extinction risk with limited information. *Journal of Applied Ecology*, 52, 861–870.

Bradshaw, G. A. & Borchers, J. G. 2000. Uncertainty as information: narrowing the Science–policy gap. *Conservation Ecology*, 4, 7. [online] www.consecol.org/vol4/iss1/art7/.

Canessa, S., Guillera-Arroita, G., Lahoz-Monfort, J. J., et al. 2015. When do we need more data? A primer on calculating the value of information for applied ecologists. *Methods in Ecology and Evolution*, 6, 1219–1228.

del Hoyo, J., Collar, N. J., Christie, D. A., et al. 2014. *HBW and BirdLife International Illustrated Checklist of the Birds of the World. Volume 1: Non-passerines.* Barcelona: Lynx Edicions and BirdLife International.

del Hoyo, J., Collar, N. J., Christie, D. A., et al. 2016. *HBW and BirdLife International Illustrated Checklist of the Birds of the World. Volume 2: Passerines.* Barcelona: Lynx Edicions and BirdLife International.

Dennis, B., Ponciano, J. M., Lele, S., et al. 2006. Estimating density-dependence, process noise and observation error. *Ecological Monographs*, 76, 323–341.

Firbank, L. G., Heard, M. S., Woiwod, I. P., et al. 2003. An introduction to the Farm-Scale Evaluations of genetically modified herbicide-tolerant crops. *Journal of Applied Ecology*, 40, 2–16.

Freckleton, R. P., Hicks, H. L., Comont, D., et al. 2018. Measuring the effectiveness of management interventions at regional scales by integrating ecological monitoring and modelling. *Pest Management Science*, 74, 2287–2295.

Freckleton, R. P., Sutherland, W. J., Watkinson, A. R., et al. 2011. Density-structured models for plant population dynamics. *Amercian Naturalist*, 177, 1–17.

Freckleton, R. P., Watkinson, A. R., Green, R. E., et al. 2006. Census error and the detection of density-dependence. *Journal of Animal Ecology*, 75, 837–851.

Gelman, A., Carlin, J. B., Stern, H. S., et al. 1995. *Bayesian Data Analysis.* New York. NY: Chapman & Hall.

Godfray, H. C. J., Beddington, J. R., Crute, I. R., et al. 2010. Food security: the challenge of feeding 9 billion people. *Science*, 327, 812–818.

Henderson, I. 2009. Progress of the UK Ruddy Duck eradication programme. *British Birds*, 102, 680–690.

Henrich, J. 2001. Cultural transmission and the diffusion of innovations: adoption dynamics indicate that biased cultural transmission is the predominate force in behavioral change. *American Anthropologist*, 103, 992–1013.

Hicks, H. L., Comont, D., Coutts, S. R., et al. 2018. The factors driving evolved herbicide resistance at a national scale. *Nature Ecology and Evolution*, 2, 529-536.

Holling, C. S. 1973. Resilience and stability of ecological systems. *Annual Review of Ecology and Systematics*, 4, 1-23.

Huston, M. A. 1994. *Biological Diversity*. Cambridge: Cambridge University Press.

Jetz, W. & Freckleton, R. P. 2015. Towards a general framework for predicting threat status of data-deficient species from phylogenetic, spatial and environmental information. *Philosophical Transactions of the Royal Society B: Biological Sciences*, 370(1662), 20140016.

Johnson, D. W. & Curtis, P. S. 2001. Effects of forest management on soil C and N storage: meta analysis. *Forest Ecology and Management*, 140, 227-238.

Kerr, J. T. & Ostrovsky, M. 2003. From space to species: ecological applications for remote sensing. *Trends in Ecology & Evolution*, 18, 299-305.

Lambert, J. P. T., Hicks, H. L., Childs, D. Z., et al. 2018. Evaluating the potential of Unmanned Aerial Systems for mapping weeds at field scales: a case study with Alopecurus myosuroides. *Weed Research*, 58, 35-45.

Lee, T. M. & Jetz, W. 2011. Unravelling the structure of species extinction risk for predictive conservation science. *Proceedings of the Royal Society B: Biological Sciences*, 278, 1329.

Mace, G. M. & Lande, R. 1991. Assessing extinction threats: toward a reevaluation of IUCN Threatened species categories. *Conservation Biology*, 5, 148-157.

Maxwell, S. L., Rhodes, J. R., Runge, M. C., et al. 2015. How much is new information worth? Evaluating the financial benefit of resolving management uncertainty. *Journal of Applied Ecology*, 52, 12-20.

Millner-Gulland, E. J. & Shea, K. 2017. Embracing uncertainty in applied ecology. *Journal of Applied Ecology*, 54, 2063-2068.

Nakagawa, S. & Freckleton, R. P. 2008. Missing inaction: the dangers of ignoring missing data. *Trends in Ecology & Evolution*, 23, 592-596.

Paneque-Gálvez, J., McCall, M. K., Napoletano, B. M., et al. 2014. Small drones for community-based forest monitoring: an assessment of their feasibility and potential in tropical areas. *Forests*, 5, 1481-1507.

Pardo, S., Cooper, A. B., Reynolds, J. D., et al. 2018. Quantifying the known unknowns: estimating maximum intrinsic rate of population increase in the face of uncertainty. *ICES Journal of Marine Science*, 75, 953-963.

Pullin, A. S. & Stewart, G. B. 2006. Guidelines for systematic review in conservation and environmental management. *Conservation Biology*, 20, 1647-1656.

Regan, H. M., Colyvan, M. & Burgman, M. A. 2002. A taxonomy and treatment of uncertainty for ecology and conservation biology. *Ecological Applications*, 12, 618-628.

Rodrigues, A. S. L., Pilgrim, J. D., Lamoreux, J. F., et al. 2006. The value of the IUCN Red List for conservation. *Trends in Ecology & Evolution*, 21, 71-76.

Safi, K. & Pettorelli, N. 2010. Phylogenetic, spatial and environmental components of extinction risk in carnivores. *Global Ecology and Biogeography*, 19, 352-362.

Stratonovitch, P., Storkey, J. & Semenov, M. A. 2012. A process-based approach to modelling impacts of climate change on the damage niche of an agricultural weed. *Global Change Biology*, 18, 2071-2080.

Sutherland, W. 2003. Evidence-based conservation. *Conservation in Practice*, 4, 39-42.

Sutherland, W. J., Butchart, S. H. M., Connor, B., et al. 2018. A 2018 horizon scan of emerging issues for global conservation and biological diversity. *Trends in Ecology & Evolution*, 33, 47-58.

Sutherland, W. J. & Freckleton, R. P. 2012. Making predictive ecology more relevant to policy-makers and practitioners. *Philosophical Transactions of the Royal Society B: Biological Sciences*, 367, 322-330.

Sutherland, W. J., Pullin, A. S., Dolman, P. M., et al. 2004. The need for evidence-based conservation. *Trends in Ecology & Evolution*, 19, 305–308.

Taylor, C. M. & Hastings, A. 2004. Finding optimal control strategies for invasive species: a density-structured model for Spartina alterniflora. *Journal of Applied Ecology*, 41, 1049–1057.

Thomas, C. D., Cameron, A., Green, R. E., et al. 2004. Extinction risk from climate change. *Nature*, 427, 145.

Tilman, D. & Karieva, P. M. 1997. *Spatial Ecology: The Role of Space in Population Dynamics and Interspecific Interactions*. Princeton, NJ: Princeton University Press.

Turner, W., Spector, S., Gardiner, N., et al. 2003. Remote sensing for biodiversity science and conservation. *Trends in Ecology & Evolution*, 18, 306–314.

Watson, J. E. M., Evans, T., Venter, O., et al. 2018. The exceptional value of intact forest ecosystems. *Nature Ecology & Evolution*, 2, 599–610.

Yokota, F. & Thompson, K. M. 2004. Value of information literature analysis: a review of applications in health risk management. *Medical Decision Making*, 24, 287–298.

# CHAPTER TWELVE

# The natural capital approach to integrating science, economics and policy into decisions affecting the natural environment

IAN BATEMAN, AMY BINNER, BRETT DAY, MICHELA
FACCIOLI
*Land, Environment, Economics and Policy Institute*
CARLO FEZZI
*University of Trento*
ALEX RUSBY
*Harrow School*

and

GREG SMITH
*CSIRO*

## 12.1 The natural capital approach

The term natural capital refers to stocks of assets, provided for free by nature which, either directly or indirectly, deliver well-being for humans. Natural capital stocks in turn deliver flows of services, often called ecosystem services, which produce the benefits upon which humans depend. Natural capital assets include stocks of fresh water, fertile soils, clean air and biodiversity. These stocks may be either renewable (e.g. fish populations) or non-renewable (e.g. oil stocks). Both stock types are vital contributors to economic activity and well-being, but can be driven to exhaustion through human action. Economic activity therefore draws and depends upon natural capital, while also affecting the stock of those assets. This intimate relationship between the environment, economy and human well-being has caught the attention of governments internationally. In this chapter, we set out how governments should incorporate the notion of natural capital into policy- and decision-making. We also consider the means by which changes can be best directed to reflect the underlying science of the environment, the incentives of the economy and the preferences of society.

### 12.1.1  Mainstreaming natural capital: the drivers of change

Mainstreaming natural capital involves bringing nature's stock and flows of goods and services into decision-making. A key element of this is to provide decision-makers with an understanding of the factors that drive change in natural capital resource use. While analyses generally examine the advantage of moving from current to alternative resource use, they commonly fail to investigate how the move between these two states is to be effected. For example, it is relatively easy to demonstrate that a move from current intensive agricultural production practices to lower-input systems will deliver improvements in water quality, greenhouse gas emissions, wildlife habitat and greenspace access. These advantages are often rigorously demonstrated without guidance as to how such change should be delivered, leaving the decision-maker facing uncertainty regarding how best to act. Such natural capital analyses alone are of little practical value as they do not acknowledge that land-use change is driven by a wide array of socio-economic/market, policy and environmental forces. Understanding the drivers of change, and the consequences brought about by policy decisions, is one of the major reasons for bringing economists into decision-making.

### 12.1.2  Natural capital, ecosystem services, goods and values

When making policy decisions regarding the natural environment it is important to understand the linkages between the various forms of natural capital, the ecosystem services they provide and their transformation into valued goods and services (Figure 12.1). In the upper left of Figure 12.1 we have the raw inputs to this system: energy (from the sun) and matter (from the earth). Together these yield stocks of physical natural capital and natural processes. Combining these stocks and processes provides the myriad ecosystem service flows provided by the natural environment. However, as shown in the third column, goods are more typically obtained by combining ecosystem service flows with other human-derived forms of capital, such as labour, machinery and technology. Here the term 'goods' refers to anything which alters human well-being, ranging from tangible products like timber or food to non-tangibles, such as the positive emotions associated with knowing that biodiversity is being conserved. Similarly, while some of these goods are provided through markets and consequently have prices, others are provided outside markets and lack prices. Nonetheless, all are, by definition, of value.

Because natural capital and ecosystem services can be used to generate a wide variety of goods, it is useful to understand whether those resources could be used in better ways. In effect, we need some measure of the value of a set of goods (Figure 12.1). Many of the goods that contribute to human well-being can be assessed in economic values, and changes in these can be analysed in terms of the resultant benefits and costs. However, a few well-being–bearing goods

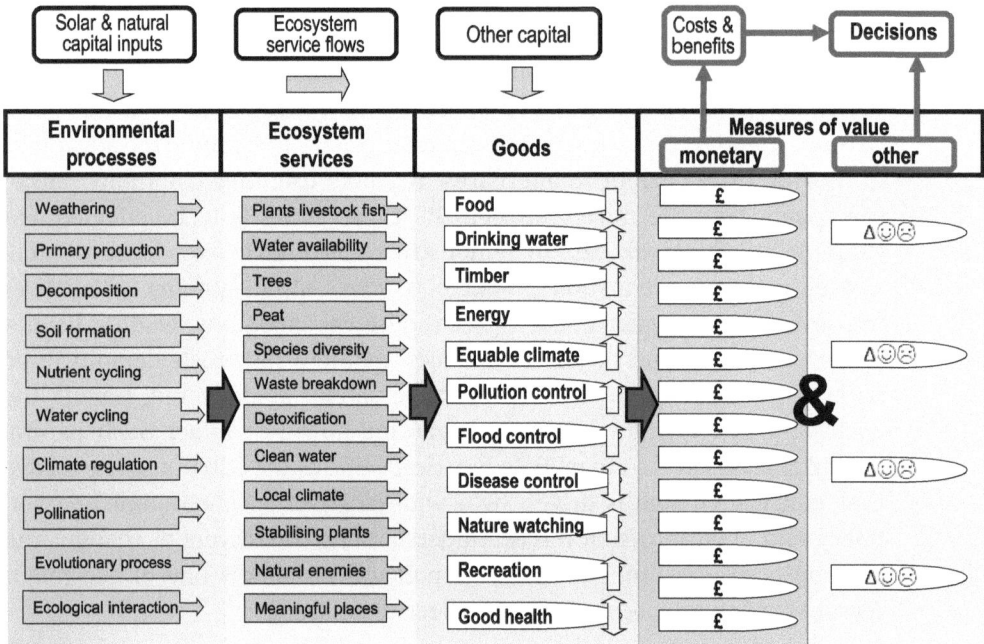

**Figure 12.1** Decision-making and the environment: from natural capital to decisions. The yellow arrows illustrate the multiple effects typical of a change in natural capital, in this case those arising from an investment to establish woodland on a currently farmed area. (A black and white version of this figure will appear in some formats. For the colour version, please refer to the plate section.)

cannot be robustly assessed in terms of economic value and therefore other, ideally quantitative, measures have to be incorporated into decisions.

In their raw, unused state, natural capital resources have high usefulness and can be employed to generate a wide range of goods, often simultaneously. However, this means that changes to the use of natural capital often generate multiple consequences. The environment is an interconnected system; changing its use in one way can have multiple effects, many of which might not have been anticipated by the decision-maker who prompted that original change (Figure 12.1). To illustrate, afforestation of farmland will typically reduce the amount of food produced. If the analysis is curtailed there, then an investment to convert farmland to woodland might often appear to yield poor value; timber values are long delayed and may well be less than the food value that can be generated over that period. Such restricted analysis is common, especially if food and timber are the only marketed, and hence priced, goods produced by such a change. However, afforestation can affect the production of a wide range of other goods. A shift from agriculture to woodland can often result in an improvement in water quality as forests require much lower inputs of fertiliser than farmland, reducing the run-off of nutrients into waterways, resulting in less-polluted rivers and higher water quality. In very many cases woodlands also reduce emissions of air pollution and store carbon, helping reduce climate change. Similarly, woodlands typically provide much greater recreational benefits than many forms of agriculture. To improve decisions regarding natural capital we need to assess all the major trade-offs arising from a proposed change and ensure that they are valued on a level playing field.

### 12.1.3  Decisions, trade-offs and valuation
#### 12.1.3.1  *Two inescapable facts*
The central challenge facing all decision-making can be encapsulated within two inescapable facts.

1.  Human wants (including those with the highest possible motivations such as improving society) exceed the resources available to satisfy them all.
2.  Because of these resource constraints, every time we decide to do one thing, we in effect decide not to do another; our decisions implicitly place values on each option.

This means that trade-offs are inevitable and valuations are unavoidable, as they are the essence of decision-making. The only real question is whether we leave those trade-offs and valuations implicit and hidden within a decision, or instead make them explicit and open to scrutiny. Economic analyses of environment-related investments are frequently the focus of criticism precisely because they make their valuations clear. However, failing to reveal valuations does not mean that decisions are being made without values. It merely means those values are being determined in an indistinct way, and are often not obvious even to those involved in the decision process.

#### 12.1.3.2  *The challenge of decision-making across integrated systems*
Low-entropy (i.e. previously unused or raw) natural capital resources have an amazing diversity of potential uses. The more that capital is used the greater its entropy and the less available it becomes for alternative uses. In some cases this is a simple binary choice (e.g. using a soil resource to grow food often means that it cannot be simultaneously used to produce timber). Nevertheless, the relationship is frequently more complex (e.g. using water for intensive food production does not necessarily mean that it is not subsequently available for drinking, but can mean that it has to be treated before consumption). Any decision that ignores this interconnection and its consequences is clearly flawed, whether it understates or overestimates the net effects, or results in decisions that are wholly deleterious for society.

Unfortunately, such incomplete analyses are commonplace. Some decision-makers may have preconceived notions of what is important and focus upon those consequences rather than the bigger picture. Often this is because the remit of the decision is constrained. So a government department charged with increasing food security may fail to adequately consider the wider environmental and societal impacts of its actions. A classic example is the EU Common Agricultural Policy (CAP) designed to promote food security. While the CAP has been substantially revised and improved in recent years, its early operation focused almost exclusively on boosting the production of food without consideration of the environmental consequences. Indeed, an argument that one objective supersedes all others is a common hallmark of many poor policy decisions. These poor policies impose unjustified and avoidable costs upon society and natural capital, which always have to be addressed in the long term and are better avoided from the outset. The catalogue of policy reversals that characterise the history of the CAP illustrate the unsustainable nature of policies with limited focus (e.g. subsidies for hedgerow removal being superseded by subsidies for their replacement).

Within the private sector, businesses typically focus upon those consequences of investment decisions that improve profits for its owners and shareholders; this, in turn, can result in a focus upon the output of goods that have market-priced values, often at the expense of other non-market, unpriced goods. In our opinion this is not morally reprehensible as, in many legal contexts, the management of a firm is legally obliged to operate in ways that benefit its owners. However, it means that public regulators need to consider policy frameworks that align the profit incentives of businesses with the interests of wider society, including environmental sustainability.

### 12.1.3.3 *The challenge of decision-making across non-commensurate metrics*

If decision-makers are interested in the overall impact that changes will have upon society then appraisals need to be comprehensive and consider all of the impacts of an investment; not only the policy focus (e.g. boosting agricultural production) but also all consequent trade-offs ('externalities' such as water pollution), be they negative or positive. A substantial challenge is that impacts are often measured using an array of different metrics. For instance, flood control is most obviously assessed in terms of risk per household, drinking water quality in mg/litre of pollutants, greenhouse gases in tonnes of carbon equivalent, recreation as the number of visits, and so on. These measures are typically non-commensurate (how many recreational visits should be given up to sequester an additional tonne of a given greenhouse gas?). Given that the overall objective of natural capital investments is to improve sustainable well-being, then the logical approach is to assess the extent to which each trade-off contributes to well-being (either positively or negatively). But what is the best unit with which to assess changes in well-being? Ideally we would want a pure unit of well-being, or, as economists term it, utility. Unfortunately, this does not exist. Therefore, an alternative is to use a unit that people commonly use to express the well-being they obtain from the gain or loss of a good. This, of course, is not a challenge that is confined to natural capital, and throughout history society has solved the problem of how to exchange different goods through the medium of money.

Using money as a unit of well-being for making commensurate the multiple trade-offs associated with natural capital change has important benefits. A commonly claimed advantage is that decision-makers are familiar with money, yet this general assertion hides a more important truth. If investments are being considered by the public sector, then the government needs to ensure that the limited tax funds at its disposal are allocated wisely, in the way that will maximise well-being. Society needs a robust natural capital base and high-quality environment. However, it also needs a health service, education, transport infrastructure, employment, security, etc., all of which draw upon the finite financial resources available to the government.

This is not to claim that money is the perfect common unit with which to express diverse benefits. Conversion problems abound, but these are even more challenging when other units are used. Indeed, it would be more accurate to argue that money is simply the least-worst common unit available. The long-term failure to assess the benefits of investing in the natural environment in monetary terms has coincided with long-term over-use and degradation of natural capital, as it is seen as a net cost yielding little obvious benefit. Certainly the case for increasing spending on the environment is difficult to make when expressed in diverse and unfamiliar units. Given this, it is hardly surprising that public spending on the environment typically represents a tiny fraction of GDP.

While marketed goods are often valued with reference to their prices, a range of methods have been developed for valuing non-market goods (Freeman et al., 2014; Champ et al., 2017). These methods can be broadly divided into three categories:

- production function methods, which examine how changes in the environment and ecosystem services affect economic output (e.g. how changes in the climate affect agricultural production; Fezzi & Bateman, 2015);
- revealed preference methods, which infer individuals' preferences and hence values through observing behaviour (e.g. looking at the time/expenditure which visitors spend to reach preferred recreational sites; Herriges & Kling, 2008);
- stated preference methods, which use experiments or surveys to ask respondents to either directly state their willingness to pay for changes,

or to choose between alternative outcomes with differing costs (e.g. examining choices between different levels of water bill according to the quality of river water they offer; Metcalfe et al., 2012).

Non-market valuation methods are important tools in the estimation of the multiple values that can arise from changes to natural capital. For example, impacts on recreation can be valued by looking at choices made by visitors across sites and relating these to the costs they incur to visit those sites (Herriges & Kling, 2008). If changes in recreational access can be shown to affect visitors' health or life expectancy, then this can be valued by examining people's willingness to pay for changes in health risk (Krupnick et al., 2002). Alternatively, estimates of health costs can be obtained either by looking at impacts on production (Murphy & Topel, 2006), or the avoided costs of illness (Tarricone, 2006). It is worth noting that these are social values, as reflected in individual behaviour, not the values postulated by economic experts.

### 12.1.3.4 Assessing impacts on biodiversity

While the majority of environmental costs and benefits can be robustly assessed using economic values, the valuation of biodiversity impacts is challenging. Certain aspects of biodiversity value can defensibly be estimated in economic terms (Hanley et al., 2015; Pascual et al., 2017). For example, provided that we have a clear understanding of the relationships between wild species, plant pollination and crop production, the monetisation of changes in output via crop market prices is relatively trivial (Losey & Vaughan, 2006; Melathopoulos et al., 2015; Breeze et al., 2016). Similarly, we can look at the increase in recreation values generated by biodiversity by examining how much further, or how often, people are prepared to travel for experiences such as viewing rare birds or hunting (USNCR, 1999; Kolstoe & Cameron, 2017). Nonetheless, it is also well established that biodiversity generates non-use value (e.g. from the knowledge that wild species continue to exist and will be bequeathed to future generations) (Kotchen & Reiling, 2000; Diafas et al., 2017). The lack of output effects or observable human behaviour in such cases means that production function and revealed preference methods are not applicable. Arguably they may be inferred by examining direct payments for conserving wild species through donations, memberships of conservation groups and legacies (Pearce, 2007; Simpson, 2007; Atkinson et al., 2012). However, such approaches will at best provide poor underestimates of true value (an expectation confirmed by the low values reported by such analyses), well out of synch with other measures of biodiversity conservation concern.

In theory, the non-use values associated with biodiversity can be directly estimated using stated preference methods, such as contingent valuation or choice experiments (Hanley et al., 2003; Christie et al., 2004; Morse-Jones et al.,

2012). In practice, these exercises face a number of challenges. One problem is that many studies have found the general public to have 'low awareness and poor understanding' of what biodiversity means (Christie et al., 2006, p. 305). Communicating such information to survey respondents is difficult as it can alter preferences and values, making them no longer representative of the social values researchers are seeking to estimate (Samples et al., 1986). Furthermore, studies seeking to estimate conservation values often cannot use scenarios in which the respondents are forced to make payments (unlike water bills as 'payment vehicles' for delivering changes in water quality).

So, how do we ensure that preferences regarding non-monetised values are not ignored? Fortunately, in the case of biodiversity we have plenty of other evidence regarding preferences that we can bring into play. For example, the most recent UK *Public attitudes and behaviours towards the environment* survey (National Statistics, 2009) revealed that 91% of respondents agreed that 'there are many natural places that I may never visit but I am glad they exist', while 85% agreed that 'I do worry about the loss of species of animals and plants in the world'. This provides us with a simple yet effective way of incorporating this preference information into decision analyses, by simply requiring that any potential change to natural capital should avoid the loss of, or enhance, biodiversity. Furthermore, alongside its direct use and non-use value, biodiversity supports a variety of ecosystem service–related benefits, most of which may be too complex and poorly understood to be adequately captured in an assessment (Turner & Daily, 2008; Mace, 2014; Mace et al., 2015; Bolt et al., 2016). A precautionary, standards-based approach should therefore be taken (Bateman et al., 2011a; Harper, 2017). Indeed, legislative support for stricter requirements being placed upon investments is evidenced in the UK Government's 25 Year Environment Plan, which sets out the principle of net environmental gain associated with new development of land (HM Government, 2018). For simplicity, however, we adopt a no-loss constraint in this chapter, confining ourselves to proving the point that biodiversity can be defensibly integrated into a natural capital decision-making approach without having to resort to dubious estimates of the economic value of the non-use benefits it provides.

### 12.1.4 Payment mechanisms: uniting payers and providers of ecosystem services

As part of any investment analysis, consideration needs to be given to who will provide and fund a given natural capital change, with the 'payment mechanism' being an important element of the appraisal process (Table 12.1). The provision of non-market environmental goods is most commonly funded by the public sector, while the private sector provides the goods (e.g. farmers subsidised to provide conservation services). A common challenge for public funding schemes is that subsidies are often allocated as untargeted flat-rate

Table 12.1 *The payer–provider matrix of payment mechanisms for environmental goods*

|  |  | Provider (of goods) | |
| --- | --- | --- | --- |
|  |  | Private sector | Public sector |
| Payer (for goods) | Private sector | Payments for ecosystem services; profitable environmental improvements | Corporate social responsibility projects |
|  | Public sector | Payments for ecosystem services; subsidies to businesses | Taxation-funded public provision |

payments across all locations, whereas the provision of biodiversity and eco-system services varies spatially. While such an approach is easy to administer, it is highly inefficient. By combining environmental modelling and economic valuation, interventions can be targeted to where they will yield greater benefits. This ensures that funders, ultimately tax payers, receive better value for money. It also means that the same level of resource generates enhanced environmental outcomes. Further improvements in the efficiency and impact of funding can be delivered through the use of 'natural capital markets' to allocate support payments. By creating competitive market struc-tures (so-called 'reverse auction' markets; Elliott et al., 2015; Fooks et al., 2015) which induce competition between ecosystem service providers, the incen-tive for private firms to over-charge for their actions is reduced.

Of course, from a public-sector perspective, these mechanisms are further enhanced if the private sector finances these initiatives. Corporate social respon-sibility investments now represent a substantial source of private-sector funding for environment projects involving major multinational corporates. For exam-ple, since 2012 Microsoft's global operations have been completely carbon-neutral (Microsoft Corp., 2017), an initiative recently taken up by Google (Google, 2016; Hölzle, 2016). While such investments clearly represent short-term costs to such companies, the social and reputational benefits generated by environmental improvements may well raise sales, generate price premiums and hence improve profits (e.g. Bateman et al., 2015). Moving more in the direction of conventional profit-bearing activities, many companies invest in areas that overtly yield a mix of both private and public benefits. For example, Häagen-Dazs (2017) has invested substantially in approaches to sustain honeybee populations, recognising that they are of considerable non-use value to society, as well as being vital to the ingredients supply chain of the ice cream manufacturer. Combining these activities with competitive Payments for Ecosystem Service markets allow companies to achieve cost reductions or revenue increases at minimum cost, thereby maximising the profitability of such actions (Day et al., 2013; Bateman et al., 2018).

### 12.1.5  Spatial scaling and targeting

From a pure natural science perspective it can be argued that there is no single perfect scale for decision-making involving an ecological system. This situation is further complicated by intersecting administrative jurisdictions and boundaries defined by the geographical extent of the economic benefits generated by ecosystem services (Bateman et al., 2006). We have to recognise these boundaries, overlaps and conflicts when making decisions to delineate the spatial scale that is most suitable for the investment. As highlighted above, a further spatial issue concerns the degree to which policies are untargeted, effectively ignoring the natural variation in the environment. These challenges have to be acknowledged and incorporated within decision-making systems if we are to achieve the levels of value for money that limited public funding requires. In particular, the tendency towards simplistic administrative methods has to be resisted. What appears to be financially cheap can often be economically very expensive in terms of the high opportunity costs and poor value for money delivered.

## 12.2  Analysis for natural capital decision-making: a national-level case study

### 12.2.1  Background

The Millennium Ecosystem Assessment (2005) highlighted global ecosystem service degradation and urged action at all governmental levels to address this problem. The first major national level response to this challenge was provided by the UK through its National Ecosystem Assessment (NEA). The NEA sought to assess the consequences of natural capital use and land-use change, and showed that over 30% of the services provided by the UK's natural environment are in decline.

   The data provided by the NEA (UK NEA, 2011) formed the basis of the models used in the assessment outlined in this case study (Bateman et al., 2011b, 2013, 2016). A wide range of highly detailed, spatially referenced, environmental data covering all of Great Britain were collected, ranging from soil characteristics (e.g. susceptibility to water logging), climate variables (e.g. temperature, rainfall) and land use (e.g. agricultural output) (Figure 12.2). This was complimented by similar spatially and temporally referenced data on market variables (e.g. prices, costs) and

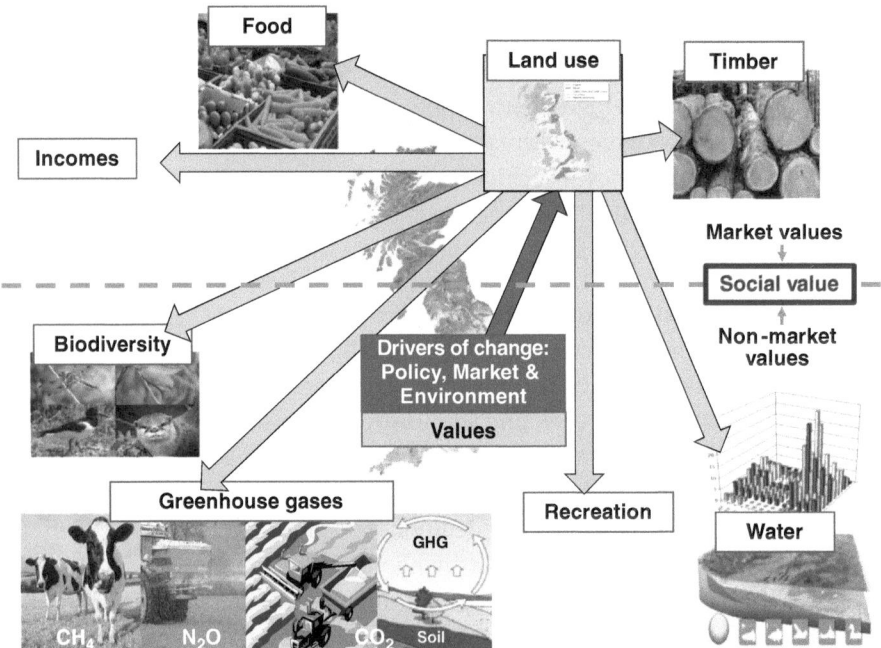

**Figure 12.2** The drivers, consequences and values of land-use change, associated with agricultural land use in Great Britain and incorporated within the conceptual framework of the National Ecosystem Assessment (Mace et al., 2011). (A black and white version of this figure will appear in some formats. For the colour version, please refer to the plate section.)

policy (e.g. subsidies, regulations such as land-use constraints). The analysis linked environmental, economic and policy factors to examine both the market and non-market consequences and values generated by land use and changes thereto. The spatial nature of these analyses also demonstrated how future policy can be targeted to most efficiently allocate available resources to maximise their net benefits.

Each analysis began from an econometric model of the environmental, economic and policy drivers of land-use (Fezzi & Bateman, 2011). This model drew upon long-term (~50 year) and high-resolution (2 × 2 km grid square or finer) national-scale data sets. The NEA set out to consider six policy scenarios (UK NEA, 2011; Bateman et al., 2013), each of which integrated both high and low future greenhouse gas (GHG) emission trends (Fezzi et al., 2014). Each predicted land use served as the base data, inputting to a series of interlinked ecosystem service impact and economic valuation models detailing the delivery of food production, emission and sequestration of greenhouse gases (including $CO_2$, $CH_4$ and $N_2O$), expected numbers of open-access recreational visits, levels of urban greenspace amenity and biodiversity metrics (Abson et al., 2014; Bateman et al., 2014; Fezzi et al., 2014; Perino et al., 2014; Sen et al., 2014).

## 12.2.2  Land-use–derived ecosystem services and their economic valuation

The major ecosystem services in the analyses were valued using a mix of market and non-market valuation techniques, with biodiversity set as a no-loss constraint, as follows.

- Food output provided the key, market-valued ecosystem service, determining approximately 75% of land use in the UK, including cropland, grassland, mountain, moor and heathland environments (Bateman et al., 2013).
- GHG sequestration had a non-market value. The quantity of GHG emission/storage associated with land was determined by the use and management of that land (e.g. cattle stocking density of cattle, other major methane producers, machinery emissions), annual flows of soil carbon due and accumulation/emission of carbon dioxide via terrestrial vegetative biomass. GHG values can be obtained through various routes, including estimates of the expected damage of climate change, the cost of abating emissions and the values of carbon traded in emission markets (Abson et al., 2014).
- Open-access recreational visits had a non-market value that varied across environments (e.g. mountains, coasts, forests, urban greenspaces) and location (Sen et al., 2014).
- Urban greenspace had a non-market value reflecting aesthetic, physical and mental health, neighbourhood, noise regulation and air pollution reduction benefits (Perino et al., 2014).
- Wild bird species diversity was used to represent biodiversity, because these species are high in the food chain and are often considered to be good indicators of wider ecosystem health (Gregory et al., 2005). As discussed previously, current estimates of biodiversity values and, in particular, pure non-use existence values are insufficiently robust. Following the reasoning set out above, we imposed a 'no-loss' constraint on biodiversity as a consequence of land-use change (Bateman et al., 2013).

## 12.2.3  Identification of the beneficiaries

The same change can yield very differing consequences to different groups of people. So we considered both the market and non-market net benefits to farmers, foresters, recreationalists, wildlife enthusiasts, etc. This allows the decision-maker to comparatively assess the scenarios and understand which provides the best value for money to society (both nationally and globally). Here, we ignore these distributional issues (but see Bateman et al., 2011b; Perino et al., 2014) and focus upon the overall benefits to society. The major beneficiaries of alternative land-uses included the following.

- Farmers: the latitude and generally colder climate of the UK means that temperature rises are likely to result in farmers increasing their profits and intensive arable production in areas that are not liable to drought (Fezzi et al., 2014; Fezzi & Bateman, 2015). However, in turn, this will probably negatively impact upon water quality due to nutrient pollution (Fezzi et al., 2015). Lower river water quality will also impact negatively upon freshwater biodiversity and river-related recreational values (Bateman et al., 2016).
- Recreationalists: open-access recreational sites benefit individuals who visit them, with the net benefit declining as distance from an individual's home or outset point grows.
- Urban residents: urban greenspace value is reflected in local property and rental value, with the value generally decaying as distance increases (Day et al., 2007; Andrews et al., 2017). Increasing access to urban greenspace typically generates significant aggregate social benefits. However, the distribution of benefits can be uneven and result in gentrification, which has the potential to push poorer families out to less-advantaged areas. Recently developed techniques such as Equilibrium Sorting Analyses seek to capture this effect and bring it into decision-making (Binner & Day, 2015).
- Biodiversity beneficiaries: improvements in species diversity not only benefit the species being directly or indirectly (e.g. through food chains) conserved, but people who value such improvements through use (e.g. hunter, fisherman, wildlife watchers) or non-use (existence values). Biodiversity also indirectly delivers value through roles in ecosystem functioning and service provision.

### 12.2.4 Analysing trade-offs across alternative land-use scenarios

For simplicity, we considered the two most extreme policy scenarios in this chapter. The World Markets scenario prioritises economic growth by completely liberalising trade, removing tariffs and trade barriers and ending agricultural subsidies; as a result, farming moved towards large-scale, intensive production methods. By contrast, the Nature@Work scenario priority is to adapt to climate change and enhance ecosystem service provision.

While considering market goods alone and ignoring non-market impacts captures only a single dimension of impact, the World Markets scenario indicated values which are frequently given primacy in policy decisions. This scenario saw agricultural value increase £1.03 billion per annum because of a shift towards more intensive production (Table 12.2). Conversely, the Nature@Work scenario led to agricultural values declining by £0.13 billion per annum as farmland was converted to urban-fringe and recreational greenspace. So, if we restricted our analysis to market-priced goods alone, then the

**Table 12.2** *Policy scenario effects on ecosystem service values in Great Britain (£ millions per annum), adapted from Bateman et al. (2014). All values are given in real (inflation-adjusted) 2010 values. Positive values indicate net gains, negative values show net losses. The two scenarios use high GHG emissions*

| Scenario | Market agricultural output values | Non-market GHG emissions | Non-market recreation | Non-market urban greenspace | Total mon-etised values | Biodiversity |
|---|---|---|---|---|---|---|
| World Markets | 1030 | −440 | −1180 | −18,400 | −18,990 | − |
| Nature@Work | −130 | 230 | 13,060 | 4760 | 17,920 | + |

World Markets scenario almost always appeared justified. This conclusion was unaffected by varying the degree of climate change across our analysis (Bateman et al., 2011a, p. 1268).

However, when we extended our assessment to consider the impacts of land-use change upon non-market goods, we find that the Nature@Work scenario consistently yielded preferable outcomes (Table 12.2). GHG emission values in the World Markets scenario were negative in nearly all areas. In contrast, under the Nature@Work scenario, most areas saw benefits in terms of increased carbon storage; the exceptions were upland areas dominated by fragile peatlands which were vulnerable to both agricultural intensification in the World Markets scenario and increasing forestry in the Nature@Work scenario. The World Markets scenario saw losses in visitor values in almost all areas across the country, while the Nature@Work scenario led to recreational benefits over the large majority of the country. Similar results were seen for urban greenspace values. Our biodiversity metric clearly shows that the World Markets scenario resulted in major declines across large swathes of the country. In comparison, the Nature@Work scenario generated improvements across the lowlands (and, therefore, much of the UK), although the picture in the uplands was more mixed, with insignificant or weakly negative effects. This suggests that an optimal solution would combine elements of multiple policies.

In summary, the World Markets scenario increased the production of marketed agricultural output at the cost of significant declines in all other ecosystem services, which strongly outweighed the value of agricultural gains. It therefore lowered overall social value very substantially. In contrast, the Nature@Work scenario reversed this pattern, causing a relatively modest reduction in agricultural production in return for very substantial increases in all other non-market ecosystem service–related goods, and a correspondingly major increase in overall social value. This disparity was further reinforced when we considered the non-monetised biodiversity measures. If we applied our constraint that any decision that would lower biodiversity in an area is ruled ineligible then, at a national level, the World Markets scenario was unacceptable. A spatially targeted optimisation approach could avoid biodiversity losses in local areas and further enhance decision-making.

### 12.2.5 Policy implications

The UK Government responded quickly and positively to the challenge of the National Ecosystem Assessment, adopting an overarching policy goal to be 'the first generation to leave the natural environment in a better state than it inherited' (HM Government, 2011, 2018; House of Commons, 2012). As part of this ambition, the UK has invested in research seeking to develop a 'natural capital approach' to decision-making, which explicitly recognises the dependence of economic value and well-being on the natural capital stocks provided by the environment and the ecosystem service flows which those assets provide. To help guide this process, the 2011 Natural Environment White Paper (HM Government, 2011) set up the world's first independent Natural Capital Committee (NCC) to advise on the restoration and improvement of natural capital as a means of sustaining and enhancing economic growth in the UK (Defra, 2012; NCC, 2013). Importantly, while it has a close relationship with the UK's environmental department, the NCC actually reports to the country's finance ministry. Indeed, the UK's Chief Finance Minister, the Chancellor of the Exchequer, chairs the Economic Affairs Committee (EAC, 2017), which the NCC formally advises (NCC, 2017a).

The NCC has reported extensively on methods to 'mainstream' natural capital considerations into both policy and business decision-making (NCC, 2017a, 2017b). Furthermore, it has also provided extensive advice on the valuation, accounting and financing of natural capital enhancement (NCC, 2017a, 2017c). Additionally, the NCC proposed and advised on a 25-year plan for the natural environment, focusing upon the need to ensure sustainable flows of ecosystem services from the UK's natural capital (NCC, 2015, 2017d), a recommendation which was then adopted by all of the major UK political parties and government (HM Government, 2018). This places the natural capital approach at the heart of decision- and policy-making over both the short and long term.

### 12.3 Acknowledgements

The authors are grateful to the Editors for their superb input to this chapter. Support for the work was provided by the consortium of funders underpinning the UK-NEA and the NERC SWEEP programme

# References

Abson, D., Termansen, M., Pascual, U., et al. 2014. Valuing climate change feedback effects upon UK agricultural GHG emissions: spatial analysis of a regulating ecosystem service. *Environmental and Resource Economics*, 57, 215–231. DOI:10.1007/s10640-013-9661-z

Andrews, B., Ferrini, S. & Bateman, I. J. 2017. Good parks – bad parks: the influence of location on WTP and preference motives for urban parks. *Journal of Environmental Economics and Policy*, 6, 204–224. DOI:10.1080/21606544.2016.1268543

Atkinson, G., Bateman, I. J. & Mourato, S. 2012. Recent advances in the valuation of ecosystem services and biodiversity. *Oxford Review of Economic Policy*, 28, 22–47.

Bateman, I. J., Abson, D., Andrews, B., et al. 2011b. Valuing changes in ecosystem services: scenario analyses. In *The UK National Ecosystem Assessment Technical Report*, UK National Ecosystem Assessment, UNEP-WCMC, Cambridge. Also available from http://uknea.unep-wcmc.org/.

Bateman, I. J., Agarwala, M., Binner, A., et al. 2016. Spatially explicit integrated modeling and economic valuation of climate change induced land-use change and its indirect effects. *Journal of Environmental Management*, 181, 172–184, http://dx.doi.org/10.1016/j.jenvman.2016.06.020

Bateman, I. J., Binner, A., Day, B. H., et al. 2018. Blended mechanisms: public and private sector payments for ecosystem services: integration, valuation, targeting and efficient delivery in the UK. In Daily, G. C., Mandle, L. & Salzman, J., editors, *International Experience in Green Growth: Mainstreaming the Values of Natural Capital into Policy and Finance Worldwide* (pp. 139–160). Stanford, CA: Natural Capital Project, and the Paulson Institute, Stanford University.

Bateman, I. J., Coombes, E., Fitzherbert, E., et al. 2015. Conserving tropical biodiversity via market forces and spatial targeting. *Proceedings of the National Academy of Sciences*, 112, 7408–7413. DOI:10.1073/pnas.1406484112

Bateman, I. J., Day, B. H., Georgiou, S., et al. 2006. The aggregation of environmental benefit values: welfare measures, distance decay and total WTP. *Ecological Economics*, 60, 450–460. DOI:10.1016/j.ecolecon.2006.04.003

Bateman, I.J., Harwood, A., Abson, D., et al. 2014. Economic analysis for the UK National Ecosystem Assessment: synthesis and scenario valuation of changes in ecosystem services. *Environmental and Resource Economics*, 57, 273–297. DOI 10.1007/s10640-013-9662-y

Bateman, I. J., Harwood, A. R., Mace, G. M., et al. 2013. Bringing ecosystem services into economic decision-making: land-use in the United Kingdom. *Science*, 341(6141), 45–50.

Bateman, I. J., Mace, G. M., Fezzi, C., et al. 2011a. Economic analysis for ecosystem service assessments. *Environmental and Resource Economics*, 48, 177–218. DOI 10.1007/s10640-010-9418-x.

Binner, A. & Day, B. H. 2015. Exploring mortgage interest deduction reforms: an equilibrium sorting model with endogenous tenure choice. *Journal of Public Economics*, 122, 40–54.

Bolt, K., Cranston, G., Maddox, T., et al. 2016. *Biodiversity at the Heart of Accounting for Natural Capital: The Key to Credibility*. Cambridge: Cambridge Conservation Initiative, www.conservation.cam.ac.uk/sites/default/files/file-attachments/CCI%20Natural%20Capital%20Paper%20July%202016_web%20version.pdf

Breeze, T. D., Gallai, N., Garibaldi, L. A., et al. 2016. Economic measures of pollination services: shortcomings and future directions. *Trends in Ecology and Evolution*, 31, 927–939. http://dx.doi.org/10.1016/j.tree.2016.09.002

Champ, P. A., Boyle, K. & Brown, T. C. (editors). 2017. *A Primer on Non-market Valuation (Second Edition), The Economics of Non-Market Goods and Services*, Volume 13. Dordrecht: Springer.

Christie, M., Hanley, N., Warren, J., et al. 2006. Valuing the diversity of biodiversity. *Ecological Economics*, 58, 304–317.

Christie, M., Warren, J., Hanley, N., et al. 2004. *Developing Measures for Valuing Changes in Biodiversity*. London: Defra.

Day, B. H., Bateman, I. J. & Lake, I. 2007. Beyond implicit prices: recovering theoretically consistent and transferable values for noise avoidance from a hedonic property price model. *Environmental and Resource Economics*, 37, 211–232.

Day, B. H., Couldrick, L., Welters, R., et al. 2013. The Fowey River Improvement Auction, PES Pilot Research Project – NE0131. Final Report to Defra, University of East Anglia. Available from http://randd.defra.gov.uk/Default.aspx?Menu=Menu%26Module=More%26Location=None%26ProjectID=18245

Defra. 2012. Defra announces five members of the Natural Capital Committee. Press release, Department for Environment, Food and Rural Affairs. Available from www.gov.uk/government/news/defra-announces-five-members-of-the-natural-capital-committee

Diafas, I., Barkmann, J. & Mburu, J. 2017. Measurement of bequest value using a non-monetary payment in a choice experiment – the case of improving forest ecosystem services for the benefit of local communities in rural Kenya. *Ecological Economics*, 140, 157–165. http://dx.doi.org/10.1016/j.ecolecon.2017.05.006.

EAC. 2017. Economic Affairs Committee website, available at: www.parliament.uk/business/committees/committees-a-z/lords-select/economic-affairs-committee/

Elliott, J., Day, B., Jones, G., et al. 2015. Scoping the strengths and weaknesses of different auction and PES mechanisms for Countryside Stewardship. Defra project LM0105. Final report, ADAS.

Fezzi, C. & Bateman, I. J. 2011. Structural agricultural land-use modeling for spatial agro-environmental policy analysis. *American Journal of Agricultural Economics*, 93, 1168–1188. doi:10.1093/ajae/aar037

Fezzi, C. & Bateman, I. J. 2015. The impact of climate change on agriculture: nonlinear effects and aggregation bias in Ricardian models of farmland values. *Journal of the Association of Environmental and Resource Economists*, 2(1), 57–92.

Fezzi, C., Bateman, I. J., Askew, T., et al. 2014. Valuing provisioning ecosystem services in agriculture: a climate scenario analysis for the United Kingdom, *Environmental and Resource Economics*, 57, 197–214. Doi:10.1007/s10640-013-9663-x

Fezzi, C., Harwood, A. R., Lovett, A. A. & Bateman, I. J. 2015. The environmental impact of climate change adaptation: land-use and water quality. *Nature Climate Change*, 5, 255–260. doi:10.1038/nclimate2525.

Fooks, J. R., Messer, K. D. & Duke, J. M. 2015. Dynamic entry, reverse auctions, and the purchase of environmental services. *Land Economics*, 91, 57–75.

Freeman, A. M. III, Herriges, J. A. & Kling, C. L. 2014. *The Measurement of Environmental and Resource Values: Theory and Methods*, 3rd edition. Oxford: RFF Press.

Google. 2016. Achieving our 100% renewable energy purchasing goal and going beyond. Available from: https://static.googleusercontent.com/media/www.google.com/en//green/pdf/achieving-100-renewable-energy-purchasing-goal.pdf

Gregory, R. D., van Strien, A., Voříšek, P., et al. 2005. Developing indicators for European birds. *Philosophical Transactions of the Royal Society B*, 360, 269–288.

Häagen-Dazs. 2017. Häagen-Dazs loves honey bees. Available from: www.haagendazs.us/about/news/haagendazsloveshoneybees/

Hanley, N., Breeze, T. D., Ellis, C., et al. 2015. Measuring the economic value of pollination services: principles, evidence and knowledge gaps. *Ecosystem Services*, 14, 124–132. http://dx.doi.org/10.1016/j.ecoser.2014.09.013.

Hanley, N., MacMillan, D., Patterson, I., et al. 2003. Economics and the design of nature conservation policy: a case study of wild goose conservation in Scotland using choice experiments. *Animal Conservation*, 6, 123–129. doi:10.1017/S1367943003003160

Harper, M. 2017. A message to those working on the 25 year environment plan: why biodiversity targets are essential to make natural capital approaches work. Available from: www.rspb.org.uk/community/our work/b/martinharper/archive/2017/09/13/a-message-to-all-those-working-on-the-25-year-environment-plan-why-biodiversity-targets-are-essential-to-make-natural-capital-approaches-work.aspx #3pRbbbOQeCFD3zRg.99

Herriges, J. A. & Kling, C. L. 2008. *Revealed Preference Approaches to Environmental Valuation, Volumes I and II.* Aldershot: Ashgate Publishing Limited.

HM Government. 2011. *The Natural Choice: Securing the Value of Nature.* Available from: www.gov.uk/government/uploads/system/uploads/attachment_data/file/228842/8082.pdf

HM Government. 2018. *A Green Future: Our 25 Year Plan to Improve the Environment.* Available from: www.gov.uk/government/publications

House of Commons. 2012. *Natural Environment White Paper, HC 492.* House of Commons Environment, Food and Rural Affairs Committee. London: The Stationery Office Limited.

Hölzle, U. 2016. We're set to reach 100% renewable energy – and it's just the beginning. Statement from Urs Hölzle, Google Senior Vice President, Technical Infrastructure. Available from: www.blog.google/topics/environment/100-percent-renewable-energy/

Kolstoe, S. & Cameron, T. A. 2017. The non-market value of birding sites and the marginal value of additional species: biodiversity in a random utility model of site choice by eBird members. *Ecological Economics*, 137, 1–12. http://dx.doi.org/10.1016/j.ecolecon.2017.02.013.

Kotchen, M. J. & Reiling, S. D. 2000. Environmental attitudes, motivations, and contingent valuation of nonuse values: a case study involving endangered species. *Ecological Economics*, 32, 93–107. http://dx.doi.org/10.1016/S0921-8009(99)00069-5.

Krupnick, A., Alberini, A., Cropper, M., et al. 2002. Age, health and the willingness to pay for mortality risk reductions: a contingent valuation survey of Ontario residents. *Journal of Risk and Uncertainty*, 24, 161–186.

Losey, J. E. & Vaughan, M. 2006. The economic value of ecological services provided by insects. *BioScience*, 56, 311–323. https://doi.org/10.1641/0006-3568(2006)56[311:TEVOES]2.0.CO;2

Mace, G. 2014. Whose conservation? *Science*, 345, 1558–1560. doi:10.1126/science.1254704

Mace, G. M., Hails, R. S., Cryle, P., et al. 2015. Review: Towards a risk register for natural capital. *The Journal of Applied Ecology*, 52, 641–653. http://doi.org/10.1111/1365-2664.12431

Melathopoulos, A. P., Cutler, G. C. & Tyedmers, P. 2015. Where is the value in valuing pollination ecosystem services to agriculture? *Ecological Economics*, 109, 59–70. http://dx.doi.org/10.1016/j.ecolecon.2014.11.007.

Metcalfe, P., Baker, B., Andrews, K., et al. 2012. An assessment of the non-market benefits of the Water Framework Directive for households in England and Wales. *Water Resources Research*, 48, W03526. doi:10.1029/2010WR009592

Microsoft Corp. 2017. Neutralizing our carbon emissions. Available from www .microsoft.com/about/csr/environment/

Millennium Ecosystem Assessment. 2005. *Ecosystems and Human Well-Being: Synthesis.* Washington, DC: Island Press.

Morse-Jones, S., Bateman, I. J., Kontoleon, A., et al. 2012. Stated preferences for tropical wildlife conservation amongst distant beneficiaries: charisma, endemism, scope and substitution effects. *Ecological Economics*, 78, 9–18. doi:10.1016/j. ecolecon.2011.11.002.

Murphy, K. M. & Topel, R. H. 2006. The value of health and longevity. *Journal of Political Economy*, 114, 871–904.

National Statistics. 2009. *2009 Public Attitudes and Behaviours Towards the Environment.* London: Defra. Available from www.defra.gov.uk /evidence/statistics/environment/pubatt

NCC. 2013. *The State of Natural Capital: Towards a Framework for Measurement and Valuation.* London: Natural Capital Committee, Department for Environment, Food and Rural Affairs (Defra). Available from www .gov.uk/government/publications/natural-capital-committees-first-state-of-natural-capital-report

NCC. 2015. *The State of Natural Capital: Protecting and Improving Natural Capital for Prosperity and Wellbeing.* Third report to the Economic Affairs Committee, Natural Capital Committee, Department for Environment, Food and Rural Affairs (Defra), London.

NCC. 2017a. *How To Do It: A Natural Capital Workbook.* London: Natural Capital Committee, Department for Environment, Food and Rural Affairs (Defra). Available from www.gov.uk/government/uploads/ system/uploads/attachment_data/file/ 608852/ncc-natural-capital-workbook.pdf

NCC. 2017b. *Improving Natural Capital: An Assessment of Progress.* London: Natural Capital Committee, Department for Environment, Food and Rural Affairs (Defra). Available from www.gov.uk/gov ernment/uploads/system/uploads/attach

ment_data/file/585429/ncc-annual-report -2017.pdf

NCC. 2017c. *Natural Capital Valuation.* London: Natural Capital Committee, Department for Environment, Food and Rural Affairs (Defra). Available from www.gov.uk/gov ernment/uploads/system/uploads/attach ment_data/file/608850/ncc-natural-capital-valuation.pdf

NCC. 2017d. *Advice to Government on the 25 Year Environment Plan.* London: National Capital Committee, Department for the Environment Food and Rural Affairs (Defra). Available from www.gov.uk/gov ernment/publications/natural-capital-committee-advice-on-governments-25-year -environment-plan

Pascual, U., Balvanera, P., Díaz, S., et al. 2017. Valuing nature's contributions to people: the IPBES approach. *Current Opinion in Environmental Sustainability*, 26, 7–16. http://dx.doi.org/10.1016/j .cosust.2016.12.006.

Pearce, D. W. 2007. Do we really care about Biodiversity? *Environmental and Resource Economics*, 37, 313–333.

Perino, G., Andrews, B., Kontoleon, A., et al. 2014. The value of urban green space in Britain: a methodological framework for spatially referenced benefit transfer. *Environmental and Resource Economics*, 57, 251–272. Doi:10.1007/s10640-013-9665-8

Samples, K. C., Dixon, J. A. & Gowen, M. M. 1986. Information disclosure and endangered species valuation. *Land Economics*, 62, 306–312.

Sen, A., Harwood, A., Bateman I. J., et al. 2014. Economic assessment of the recreational value of ecosystems in Great Britain. *Environmental and Resource Economics*, 57, 233–249. Doi:10.1007/ s10640-013-9666-7

Simpson, R. D. 2007. David Pearce and the economic valuation of biodiversity. *Environmental and Resource Economics*, 37, 91–109.

Tarricone, R. 2006. Cost-of-illness analysis: what room in health economics? *Health Policy*, 77, 51–63.

Turner, R. K. & Daily, G. C. 2008. The Ecosystem Services Framework and Natural Capital Conservation, *Environmental and Resource Economics*, 39, 25–35.

UK NEA. 2011. *UK National Ecosystem Assessment: Technical Report*. Cambridge: United Nations Environmental Programme–World Conservation Monitoring Centre (UNEP-WCMC).

USNCR. 1999. *Perspectives on Biodiversity: Valuing Its Role in an Everchanging World*. Washington, DC: National Research Council (US) Committee on Noneconomic and Economic Value of Biodiversity National Academies Press. Available from: www.ncbi.nlm.nih.gov/books/NBK224412/

# Working with government – innovative approaches to evidence-based policy-making

EDITH ARNDT
*University of Melbourne*
MARK BURGMAN
*Imperial College London*
KAREN SCHNEIDER AND ANDREW ROBINSON
*University of Melbourne*

## 13.1 Introduction

Governments internationally have long aspired to ground policy in rigorous evidence. Without evidence, policy-makers must rely on intuition, ideology, conventional wisdom or, at best, theory (Banks, 2009). Their evidence requirements span the physical, natural and social sciences. Policy issues in environment, natural resource management and biosecurity, in which risk and uncertainty are inherent, are prime examples. The UK government's White Paper on Modernising Government (1999) pledged to improve the use of evidence and research to better understand policy problems (Blair & Cunningham, 1999). Over the past three decades, the UK government has promoted evidence-based over ideologically driven policy (Banks, 2009). Likewise, the Australian government's 2012 Blueprint for Reform recommended strengthening relationships with academia to enhance strategic policy capabilities and drive innovation (Department of the Prime Minister and Cabinet, 2010). Such relationships help ensure that the government's significant investment in science, research and innovation is harnessed to engage with contemporary policy challenges (DIISRTE, 2012).

There has been much consideration of how scientists and government policy-makers interact and of the impediments to effective communication between science and policy. Organisational structures and social norms may impede the incorporation of science into policy development, as may the different timeframes over which science and policy are developed (Burgman, 2015a). Governments and researchers use different approaches to improve the delivery of policy-relevant science and to enhance the likelihood that science will contribute directly to policy decisions. The working model that is used depends on different factors, such as the degree of willingness to incorporate science into policy-making, the strength of existing relationships and available funding. This chapter first outlines the factors influencing science–policy relationships and then presents possible ways for scientists and policy-makers to work together. We introduce an innovative model of research collaboration that has had practical impacts on policy in Australia. In conclusion, we reflect on the implications of these innovations for interactions between science and government elsewhere.

## 13.2  The science–policy interface – how well does it work?

Government policy-makers and applied scientists frequently share the aspiration that science should contribute directly to policy decisions. Despite this, significant gaps can remain between the kinds of information that scientists provide and the kinds of inputs that government policy-makers find useful. The reasons for this can depend on culture, context and values, or on the relationships between individual scientists and policy-makers.

Different workplace cultures can impede the adoption of science in policy. Scientists are not always policy-literate and can fail to understand the complexity of the policy environment. This may include the wide range of inputs required, the interactions with other policies, the intensive scrutiny to which new policy proposals are exposed and the fact that policies are not made in isolation but are typically built on existing policy positions (Tyler, 2013; see also Chapter 2). The context in which policy-makers propose solutions to challenging problems is complex and characterised typically by competing, and at times conflicting, objectives among diverse stakeholders. The task of the policy-maker is to balance these objectives while being guided by political mandates and the public good. In these circumstances, policy-makers may appear to disregard scientific advice for reasons that scientists might support if they were privy to the full context of the decisions. For example, a solution that is suboptimal from a single scientific perspective may be the only tenable outcome in the short term and may contribute to a more ambitious policy objective in the longer term (Burgman, 2015a). Similarly, policy-makers often lack the skills to interpret science effectively and rigorously for their purpose, including understanding the quality, limitations and biases of evidence (Sutherland et al., 2013). These impediments are compounded when there is insufficient incentive for scientists and policy-makers to collaborate.

Policy-making is rarely an entirely objective process that leads to a single rational outcome. Decisions in complex situations involve both facts and values. Facts are not always certain and can be influenced by values, perceptions and emotions (Slovic, 1999; Burgman, 2015b). There is no single right way of assessing values (Gregory et al., 2012). Nor are scientists entirely objective and independent (Krinitzsky, 1993; O'Brien, 2000). Lack of objectivity can sometimes lead to situations in which scientific expertise is used deliberately and strategically to support a particular policy outcome. This can be especially strident where issues are emotionally or politically charged – the science of global climate change is a contemporary example (Burgman, 2015a). In most practical situations, the pool of scientific experts on which policy-makers can call is small and composed of people with differing values and partially overlapping experiences (French, 2012). In these circumstances, conventional science can help to clarify what might be lost or gained as a consequence of a policy decision, but can offer little to evaluate differences of opinion and the trade-offs that are often necessary to make a decision. Decision theory (French, 2012; Gregory et al., 2012) can provide a platform for structuring problems, engaging stakeholders, assessing alternatives and finding a solution that best achieves the aspirations of government.

The rewards systems in governments and academia are also frequently incompatible. The determinants of academic advancement are commonly skewed towards publication records, although there is a growing emphasis on the importance of practical research impact. Indeed, all major international university ranking systems now include a measure of research impact. Unfettered academic publication can be impeded by the policy-making process, in which control over the flow of information may be necessary to manage policy change among diverse stakeholders (Burgman, 2015a). Conversely, most government institutions do not readily reward involvement of their staff in what may be considered speculative scientific research.

The timeframes over which science and policy are developed can also be a barrier for the effective use of science in policy-making. Policy-makers can be unaware of and unable to absorb scientific evidence or emerging scientific methods in the short time horizons that are often imposed on policy development. Conversely, the development of good science can be a lengthy process that lags behind the response times required by new policy challenges. In other circumstances, where relevant science already exists, scientists can underestimate the time that it takes to implement policy change, including the time taken to evaluate the social, economic and political implications of potential change.

Limited access to data and research outputs may impede policy-makers' use of scientific evidence. This can be a simple communication issue, because it is not straightforward to write and disseminate research findings in a way that can be readily interpreted and applied by the policy community. More problematically, policy-makers may look to scientists to provide certainty. Scientists may be motivated not to disclose the full weight of uncertainty in their assumptions and results, or may be unaware of it, or not know how to communicate it to policy-makers (Sutherland et al., 2013). This low accessibility creates an imperative for policy-makers to understand the limitations and the context of the scientists themselves, and to cross-examine their evidence.

Useful and 'usable' science most often arises when researchers and policy-makers work closely together to iterate through problem formulation and solutions (Dilling & Lemos, 2011; Burgman 2015a; Chapter 10). In many cases, science contributes to public policy effectively because researchers and government policy-makers have developed personal relationships (Gibbons et al., 2008). In these instances, the 'literacy' barrier on both sides is reduced. However, roles and responsibilities can change frequently, especially in government, and can undermine the time taken to establish effective personal relationships (Burgman, 2015a). It is rare that informed personal relationships will consistently overcome all of the substantial barriers to the effective use of scientific evidence in policy-making.

## 13.3 Ways of working with government

Issues related to context, values, culture, timeframes, communication and relationships can thwart the effective use of science in policy. Participants attempt to bridge the gap between science and policy, using a range of ways of working together (Table 13.1). Here, we discuss models for science–policy interactions along a spectrum of time investment and complexity. This is not a complete list, and concepts and strategies for improving the effectiveness of partnerships evolve over time. Corroborating the dynamic nature of these elements, a recent survey indicated that Canadian scientists' and policy-makers' ideal way of working in the future would involve collaborative study design and analysis, indicating a shift of focus from knowledge dissemination to knowledge generation (Choi et al., 2016).

### 13.3.1 Policy briefs

At one end of the spectrum, strategies include one-off events or communication products. For example, policy briefs are succinct documents that address a single policy issue of high interest to policy-makers. The analysis of a priority policy problem is context-specific, incorporates solutions and implementation considerations and is usually completed within days (Lavis et al., 2009a). Policy briefs are an acknowledged method for disseminating knowledge to policy-makers and are often used in the health and social sciences sectors (Lavis et al., 2009b; Rajic et al., 2013; Balian et al., 2016). The Food and Agriculture Organization of the United Nations adopted policy briefs to disseminate information about agricultural development issues to the general public. However, the impact of policy briefs depends on the reader. Experts are less likely to change their opinion after reading a brief than non-experts (Masset et al., 2013).

**Table 13.1** *Examples of working strategies between scientists and policy-makers to improve the effective use of science in policy, including a brief description and relevant references*

| Working strategy | Description | References |
|---|---|---|
| Policy briefs | A short but comprehensive analysis and discussion of a high-priority issue including solution statements and implementation considerations | Balian et al. (2016)<br>Lavis et al. (2009a)<br>Masset et al. (2013)<br>Rajic et al. (2013) |
| Science–policy forums | A networking event allowing policy dialogue. Researchers and policy-makers present research findings and policy requirements in an interactive knowledge-sharing setting | Lavis et al. (2009b)<br>Boydell et al. (2017)<br>Gregory et al. (2008) |
| Training courses, exchange programmes and job-shadowing | Theoretical or practical learning settings that aim to convey to scientists and policy-makers a better understanding of the content and the circumstances in which science and policy operate | DIISRTE (2012)<br>Gibbons et al. (2008)<br>Young et al. (2014) |
| Knowledge brokers | Intermediaries who facilitate interactions between scientists and end users but remain impartial to the decision-making process | Rajic et al. (2013)<br>Ward et al. (2009)<br>Meagher and Lyall (2013) |
| Informal working groups | Ad-hoc arrangements where scientists and policy-makers collaboratively address a policy problem | Burgman (2015a)<br>Gibbons et al. (2008)<br>Nichols et al. (2015) |
| National funding schemes | Funding schemes that explicitly support research with strong links to the objectives of other organisations such as government, industry and business | Australian Research Council (2018)<br>Cooperative Research Centres (2018) |
| Shared governance model (coproduction) | Government-funded research centres where the development of research priorities and achievement of outcomes is shared between policy-makers and scientists | Van Kerkhoff and Lebel (2015)<br>Burgman (2015a) |

### 13.3.2  Science–policy forums

A science–policy forum, or policy dialogue, brings stakeholders and scientists together. In contrast to policy briefs, policy dialogues may concentrate on actions in response to research evidence. The main aim of this tool is to facilitate discussion (Lavis et al., 2009b). Policy dialogues can be time-intensive to plan and organise but provide an opportunity to hear about experiences from a diversity of stakeholders. They may establish and cultivate ongoing personal relationships between decision-makers and researchers (Boydell et al., 2017). Deliberate engagement techniques, such as policy dialogues, can generate confidence among participants that their inputs will guide policy development (Gregory et al., 2008).

### 13.3.3  Training courses, exchange programmes and job-shadowing

Training courses for researchers and policy-makers may support translation skills, communication and networking skills or understanding of subject matter or of government processes, so individuals can communicate more effectively with their counterparts (Young et al., 2014). Exchange programmes such as secondments are a useful way for scientists to learn how to translate their knowledge to generate benefits in the specific decision-making contexts in which policy-makers work. They can also catalyse new relationships (Gibbons et al., 2008). The National Environmental Research Program in Australia 2010–2015 aimed, in part, to enhance mutual understanding by offering short-term secondments for researchers into policy settings (DIISRTE, 2012). Job-shadowing, in which individuals accompany high-level policy-makers in their daily professional interactions, is also valuable for improving understanding of the realities of decision-making (Young et al., 2014).

### 13.3.4  Knowledge brokers

One outcome of theoretical or practical learning may be the emergence of so-called knowledge brokers, individuals or groups that facilitate interactions and knowledge transfer between researchers and end users (Rajic et al., 2013) by understanding and serving the needs of both. However, the effectiveness of such arrangements is not often evaluated (Ward et al., 2009; Meagher & Lyall, 2013).

### 13.3.5  Informal working groups

When scientists and policy-makers have established relationships, they may create ad-hoc working groups to address public policy issues (Burgman, 2015a). If participants define problems and outputs well, and consider incentives for both parties, then working groups offer shared responsibility for objectives and the prospect of effective outcomes for policy needs (Gibbons et al., 2008). Working groups have the potential to grow into longer term arrangements. For example, in the USA, the ad-hoc formation of a working group of waterfowl managers and biologists from federal and state agencies led to the development of a now long-running programme based on adaptive resource management principles (Nichols et al., 2015).

### 13.3.6 National funding schemes

National funding schemes can aim to bring scientists and policy-makers closer together by creating policy-relevant incentives for research institutions. The Australian Research Council (ARC) linkage funding scheme, for example, encourages the development of partnerships between science and government, business, industry and community organisations. ARC has also created Centres of Excellence, consisting of long-term collaborations between eligible higher education organisations and partner businesses and agencies. They focus on priority research that is identified by the Australian Government, and operate within clearly articulated governance structures (ARC, 2018). The Australian Cooperative Research Centres Association programme was established in 1990 to bring large groups of researchers in the public and private sectors together with end users (CRCA, 2018). The role of the end users is to help plan the direction of the research and monitor its progress (Burgman, 2015a; CRCA, 2018).

In the UK, from the early 1900s, the Haldane Principle guided government investment in research based on the philosophy that decisions about research priorities should be made by researchers. In 1972, this was replaced by the Customer Contractor Principle, which introduced a market-orientated approach to government support for research (Kogan et al., 2006; Daniels et al., 2014). The 2014 UK Research Excellence Framework (HEFCE, 2018) guided national research investment in universities and used impact to assess the benefits of research beyond academia (Greenhalgh & Fahy, 2015). Similarly, in the USA, the Office of Productivity, Technology & Innovation was created in the Department of Commerce in 1981 to advocate Research and Development Limited Partnerships at universities to accelerate the transfer and private appropriation (through patents) of federally funded technology. The US National Science Foundation now considers the benefits for society of scientist's discoveries when allocating funding (Wiley, 2014; N. Voulvoulis & M. Burgman, unpublished data).

### 13.3.7 Shared governance

Long-term arrangements, such as Centres of Excellence and Research and Development Limited Partnerships, focus on joint research priorities. However, research centres operating under a model of shared governance go a step further. In the shared governance model, scientists and policy-makers co-develop and co-manage research priorities, business cases and project plans, and the delivery of research outcomes. Shared governance, also referred to as 'co-production', between scientists and policy-makers is possible when partners 'have sufficient trust, willingness and institutional room to man-oeuvre to share information and decision-making power' (Van Kerkhoff & Lebel, 2015). This model encourages the formation of research–policy partnerships built on strong personal relationships (Gibbons et al., 2008) and has the potential to overcome many of the issues limiting the effective use of science in policy. The Centre of Excellence for Biosecurity Risk Analysis (CEBRA) is one example (Burgman, 2015a).

## 13.4 The Centre of Excellence for Biosecurity Risk Analysis – a collaborative approach to bring science to policy

In the biosecurity domain, CEBRA and its predecessor, the Australian Centre of Excellence for Risk Analysis (ACERA), are examples of governance arrangements that encourage close science–government interaction. ACERA was established in 2006 to develop state-of-the-art methods (tools, guidelines, procedures) to enhance risk analysis in the Australian Government. It was a collaborative agreement between the Australian Government Department of Agriculture and Water Resources and the University of Melbourne. In 2014 the partnership expanded to include New Zealand's Ministry for Primary Industries and sharpened its focus on biosecurity risk, continuing under the new name of CEBRA. The two governments provide the majority of the financial resources to operate the centre and have signed a research agreement with the university provider.

CEBRA's governance arrangements and operational practices include a number of features that have evolved to avoid or overcome some of the most pervasive impediments to effective communication between scientists and policy-makers. They aim to maximise the likelihood that CEBRA's research outputs will generate pragmatic policy outcomes. A key characteristic of the governance model is shared responsibility for the development of research themes, priorities and the delivery of outcomes.

In CEBRA, policy-makers identify research themes, ideas and priorities on an annual basis, under the guidance of a steering committee that comprises senior executives of both the Australian and New Zealand governments, and considering other biosecurity research efforts in which the governments participate. CEBRA researchers and their government counterparts then collaborate to develop the prioritised research ideas into detailed project descriptions and budgets, including implementation plans. The final set of projects to be undertaken depends on the priority list and the available budget. Both the Australian and New Zealand governments have prioritised some multi-year projects that contribute to important strategic objectives. The balance between applied and more speculative research is achieved by earmarking 20% of the budget for 'blue-sky' research, focusing on topics that are relevant to CEBRA's mission but that may not solve the most immediately pressing policy questions.

Shared responsibility between researchers and policy-makers extends to meeting milestones and generating deliverables. On each project, a research leader from CEBRA is teamed with a project manager from government who provides research and administrative support. In addition, a senior government executive sponsors each project and champions its delivery through government, including, where necessary, facilitating acquisition of relevant data and allocating staff time and other resources. CEBRA is responsible for finding experts to deliver the research projects, either from its own staff or in collaboration with researchers from other institutions. A science advisory committee provides assurance of the scientific integrity of project proposals and the scientific quality of research outputs, overseeing peer review and encouraging publication of results. It comprises independent and appropriately experienced scientists, who assess scientific integrity and quality using a process comparable to the peer-review process of international journals. The centre's strategic direction and governance arrangements are overseen by an independent advisory board, comprising university, government and independent members, under an independent chair.

CEBRA's experience has been that the close working relationships fostered between researchers and policy-makers under this model benefit the delivery of pragmatic research outcomes and increase the likelihood that research findings will be implemented. Somewhat unexpectedly, the policy demands of government led to the development of research agendas in entirely new areas. For example, CEBRA's early investment in research on expert judgement led to a suite of experiments, tests and empirical results that have wide applications outside biosecurity (Burgman, 2015b), including in geopolitical forecasting for security and intelligence (Wintle et al., 2012), and conservation biology (Martin et al., 2012). Increasing levels of trust over time have enhanced researchers' understanding of the context in which biosecurity decisions are made and the constraints inherent in the policy-making process. This includes the timeframes for providing usable science outputs. Conversely, policy-makers teamed with researchers have the opportunity to participate in science to achieve policy-relevant outcomes, better understanding the limitations and uncertainties of the scientific results. This has proven effective even where policy-makers have minimal previous scientific experience.

A further advantage of the model is that scientists maintain their independence and are perceived to be independent by other stakeholders in industry and the wider community (Burgman, 2015a). The agreement between government and the university stipulates that the Centre's work should be in the public domain. This is important for government, because biosecurity decisions can be highly contestable, including at the international level. Part of this independence is that scientists are free to publish their work or comment with the usual academic freedom. Policy-makers may or may not decide to endorse the products of the research and can dissociate themselves from advice or commentaries that they consider to be inaccurate, inappropriate or in conflict with public policy (Burgman, 2015a). Under this model, university researchers are able to undertake work that is directly relevant to public policy, where it can have immediate and significant impact, while maintaining their traditional academic freedoms.

Creating policy impact has been a key objective of CEBRA since its establishment and a number of projects have achieved this. For example, CEBRA designed a monitoring system for aircans (containers for aeroplane baggage) that significantly reduced the burden of intervention for the then Australian Quarantine and Inspection Service in the wake of the 2001 foot-and-mouth disease outbreak in the United Kingdom. CEBRA developed a monitoring regime for aircans based on applied statistics and the operational experience of stakeholders, but also considered the constraints of different regional offices. Under the current system (Robinson et al., 2011), the Australian Government inspects a maximum of 15,000 aircans a year, out of the almost 400,000 that arrive, while assuring the government that the pathway continues to present a very low risk.

In the area of biosecurity intelligence, CEBRA and its government collaborators found a way to monitor publicly available information on the global spread of pests and diseases systematically and cost-effectively. The department now uses innovative software, the International Biosecurity Intelligence System (IBIS), to search open-source information for emerging pest and disease threats, providing early warning. It generates daily reports that effectively monitor the disease status of Australia's trading partners. Government staff convert the information IBIS generates into usable intelligence that informs risk identification, assessment and prioritisation (see Chapter 3 for more details of this process).

A third CEBRA research programme has led to a shift in thinking about biosecurity inspection rules and their implementation. A suite of subprojects developed and applied economic experiments and drew on principles from behavioural economics and micro-economic theory to better understand how importers react to incentives within a new compliance-based inspection scheme for a range of plant–product import pathways (Robinson et al., 2012; Rossiter et al., 2015; Rossiter & Hester, 2017; Leibbrandt et al., 2018). The government uses this scheme to reward consistently compliant importers by imposing reduced inspections. While this work is ongoing it has had some significant practical impacts on compliance-based inspection schemes.

## 13.5 Lessons learnt

There are many ways in which governments work with scientists to maximise the opportunity to apply sound evidence in the policy-making process. Since its establishment in 2006, CEBRA and its predecessor ACERA have developed a model based on shared responsibility for the development of a research agenda, priorities and the delivery of outcomes. This close relationship between research objectives and policy needs has contributed to the strong uptake of research outcomes. The relationship between policy-makers and scientists has evolved since 2006 to one of mutual respect for the complementary roles and skills that each brings. This has been key to the success of the organisation.

CEBRA's shared governance arrangement respects the conventional academic reward system. It encourages peer-reviewed publication of articles. Staff present papers at international conferences and CEBRA hosts scientists from other institutions for working groups, workshops, research projects and sabbaticals. This supports traditional pathways to advancement through the university system. Less traditionally, but just as importantly, the collaborative nature of working on public policy issues with government staff can contribute to overall job satisfaction, especially when applied research outcomes positively influence biosecurity policy or operations.

Some CEBRA projects started as one-year projects and expanded into multi-year projects. CEBRA's longer-term funding model allows more in-depth scientific discourse on research questions related to specific policy needs. Continuation of work leads to greater development of expertise and is more likely to result in satisfactory practical outcomes for biosecurity policy. If a research project team has a productive partnership with their policy counterparts, then long-term (multi-year) projects benefit.

While the shared governance model delivers many positive outcomes for scientists and policy-makers, some challenges persist. Working in close proximity to the machinery of government, researchers may be subject to novel administrative obligations. For example, there can be a requirement for frequent verbal or written progress reports. Further, the collaborative development of a detailed business case can be time-consuming because it is an iterative process involving a number of contributors, and proposals for new projects require formal approval by senior government officials. Government internal quality assurance and contract management processes in general might have an impact on researchers' workloads and project timeframes, although these are generally no more onerous than writing and managing conventional grants.

A close relationship between project sponsor and research provider may also lead to pressure on researchers to expand the scope of a project when new insights emerge during its progress. In contrast, researchers working under a shared governance arrangement may not put enough effort into achieving project milestones because of the long-term nature of the research centre contract. It is an issue that can be resolved, however, through a responsive, structured and transparent process of change management where all involved parties are informed of and agree to changes in project deliverables or timeframes.

One challenge for research scientists in the shared governance model is shared by all other modes of interaction. That is, the researchers have to at least partially subordinate their interests to those of their research partner. It is not enough to have an idea or a skill and to look for opportunities to apply it. Rather, the researchers have to listen carefully and understand the context of their colleague's operational environment. Only then can they draw on the suite of skills and experience they have acquired to solve problems. They also have to be patient and persistent in searching for ways of presenting the solutions they discover in an accessible and useable form. Not all researchers are capable of such adjustments.

In conclusion, biosecurity in an Australian context has provided an example in which government regulation has been enhanced by the application of good science. The CEBRA model of collaborative governance arrangements underpinning pragmatic policy outcomes could be applied to other areas of government policy-making in which scientific considerations are important. Potential examples include public health, natural resource management and environmental issues, including conservation policy.

## References

ARC. 2018. Australian Research Council Linkage Program. Australian Government, Canberra. Available from www.arc.gov.au/linkage-program (accessed 30 July 2018).

Balian, E. V., Drius, L., Eggermont, H., et al. 2016. Supporting evidence-based policy on biodiversity and ecosystem services: recommendations for effective policy briefs. *Evidence & Policy*, 12, 431–451. doi:10.1332/174426416X14700777371551

Banks, G. 2009. Evidence-based policy-making: What is it? How do we get it? ANU Public Lecture Series, presented by ANZSOG, 4 February., Productivity Commission, Canberra.

Blair, T., & Cunningham, J. 1999. *Modernising Government*. London: Prime Minister and Minister for the Cabinet Office.

Boydell, K. M., Dew, A., Hodgins, M., et al. 2017. Deliberative dialogues between policy-makers and researchers in Canada and Australia. *Journal of Disability Policy Studies*, 28, 13–22. doi:10.1177/1044207317694840

Burgman, M. A. 2015a. Governance for effective policy-relevant scientific research: the shared governance model. *Asia and the Pacific Policy Studies*, 2, 441–451. doi:10.1002/app5.104

Burgman, M. A. 2015b. *Trusting Judgements: How to Get the Best Out of Experts*. Cambridge: Cambridge University Press.

Choi, B. C. K., Liping, L., Yaogui, L., et al. 2016. Bridging the gap between science and policy: an international survey of scientists and policy-makers in China and Canada. *Implementation Science*, 11, 16. doi:10.1186/s13012-016-0377-7

CRCA. 2018. Cooperative Research Centres Association. Available from http://crca.asn.au/about-the-crc-association/about-crcs/ (accessed 9 February 2018).

Daniels, R. J., Spector, P. M. & Goetz, R. 2014. Fault lines in the compact: higher education and the public interest in the United States. In Weber, L. E. & Duderstadt, J. J., editors, *Preparing the World's Research Universities to Respond to an Era of Challenges and Change* (pp. 127–140). Glion Colloquium Series No 8. Geneva: ECONOMICA.

Department of the Prime Minister and Cabinet. 2010. *Ahead of the Game: Blueprint for the Reform of Australian Government Administration*. Canberra: Author.

Dilling, L. & Lemos, M. C. 2011. Creating usable science: opportunities and constraints for climate knowledge use and their implications for science policy. *Global Environmental Change*, 21, 680–689. doi.org/10.1016/j.gloenvcha.2010.11.006

DIISRTE. 2012. APS200 Project: The place of science in policy development in the public service. Australian Government Department of Industry, Innovation, Science, Research and Tertiary Education, Canberra. Available from www.industry.gov.au/science/Pages/APS200ProjectScienceinPolicy.aspx

French, S. 2012. Expert judgment, meta-analysis and participatory risk analysis. *Decision Analysis*, 9, 119–127.

Gibbons, P., Zammit, C., Youngentob, K., et al. 2008. Some practical suggestions for improving engagement between researchers and policy-makers in natural resource management. *Ecological Management & Restoration*, 9, 182–186. doi:10.1111/j.1442-8903.2008.00416.x

Greenhalgh, T. & Fahy, N. 2015. Research impact in the community-based health sciences: an analysis of 162 case studies from the 2014 UK Research Excellence Framework. *BMC Medicine*, 13, 232. doi:10.1186/s12916-015-0467-4

Gregory, R., Failing, L., Harstone, M., et al. 2012. *Structured Decision Making. A Practical Guide to Environmental Management Choices*. Chichester: Wiley-Blackwell.

Gregory, J., Hartz-Karp, J. & Watson, R. 2008. Using deliberative techniques to engage the community in policy development. *Australia and New Zealand Health Policy*, 5, Art. 16. doi:10.1186/1743-8462-5-16

HEFCE. 2018. REF Impact. Higher Education Funding Council for England. Available from www.hefce.ac.uk/rsrch/REFimpact/ (accessed 16 August 2018).

Krinitzsky, E. L. 1993. Earthquake probability in engineering – Part 1: the use and misuse of expert opinion. The Third Richard H. Jahns Distinguished Lecture in Engineering Geology. *Engineering Geology*, 33, 257–288. doi:10.1016/0013-7952(93)90030-G

Kogan, M., Henkel, M. & Hanney, S. 2006. *Higher Education Dynamics, Volume 11. Government and Research. Thirty Years of Evolution* (2nd ed.). Dordrecht: Springer.

Lavis, J. N., Boyko, J. A., Oxman, A. D., et al. 2009b. SUPPORT tools for evidence-informed health policymaking (STP) 14: organising and using policy dialogues to support evidence-informed policymaking. *Health Research Policy and Systems*, 7(Suppl 1), S14. doi:10.1186/1478-4505-7-S1-S14

Lavis, J. N., Permanand, G., Oxman, A. D., et al. 2009a. SUPPORT tools for evidence-informed health policymaking (STP) 13: preparing and using policy briefs to support evidence-informed policymaking. *Health Research Policy and Systems*, 7(Suppl 1), S13. doi:10.1186/1478-4505-7-S1-S13

Leibbrandt, A., Rossiter, A., Hester, S., et al. 2018. Testing compliance-based inspection protocols. CEBRA final report for project 1404C. The Centre of Excellence for Biosecurity Risk Analysis, the University of Melbourne.

Martin, T. G., Burgman, M. A., Fidler, F., et al. 2012. Eliciting expert knowledge in conservation science. *Conservation Biology*, 26, 29–38. doi:10.1111/j.1523-1739.2011.01806.x

Masset, E., Gaarder, M., Beynon, P., et al. 2013. What is the impact of a policy brief? Results of an experiment in research dissemination. *Journal of Development Effectiveness*, 5, 50–63. doi:10.1080/19439342.2012.759257

Meagher, L. & Lyall, C. 2013. The invisible made visible: using impact evaluations to illuminate and inform the role of knowledge intermediaries. *Evidence & Policy*, 9, 409–418. doi:10.1332/174426413X14818994998468

Nichols, J. D., Johnson, F. A., Williams, B. K., et al. 2015. On formally integrating science and policy: walking the walk. *Journal of Applied Ecology*, 52, 539–543. doi:10.1111/1365-2664.12406

O'Brien, M. 2000. *Making Better Environmental Decisions: An Alternative to Risk Assessment*. Cambridge: MIT Press.

Rajic, A., Young, I. & McEwen, S. A. 2013. Improving the utilization of research knowledge in agri-food public health: a mixed-method review of knowledge translation and transfer. *Foodborne Pathogens and Disease*, 10, 397–412. doi:10.1089/fpd.2012.1349

Robinson, A., Burgman, M. A. & Cannon, R. 2011. Allocating surveillance resources to reduce ecological invasions: maximizing detections and information about the threat. *Ecological Applications*, 21, 1410–1417. doi:10.1890/10-0195.1

Robinson, A., Bell, J., Woolcott, B., et al. 2012. AQIS Quarantine Operations Risk Return ACERA 1001 Study J: Imported Plant-Product Pathways. Australian Centre of Excellence for Risk Analysis, University of Melbourne, Project 1001 J.

Rossiter, A. & Hester, S. 2017. Designing biosecurity inspection regimes to account for stakeholder incentives: an inspection game approach. *Economic Record*, 93(301), 277–301. doi:10.1111/1475-4932.12315

Rossiter, A., Hester, S., Aston, C., et al. 2015. Incentives for Importer Choices. Centre of Excellence for Biosecurity Risk Analysis, University of Melbourne, Project 1304C, Final Report 1: Overview.

Slovic, P. 1999. Trust, emotion, sex, politics, and science: surveying the risk-assessment battlefield. *Risk Analysis*, 19, 689–701. doi:10.1111/j.1539-6924.1999.tb00439.x

Sutherland, W. J., Spiegelhalter, D. & Burgman, M. A. 2013 Policy: twenty tips for interpreting scientific claims. *Nature*, 503, 335–337. doi:10.1038/503335a

Tyler, C. 2013. Top 20 things scientists need to know about policy-making. *The Guardian*. Available from www.theguardian.com/science/2013/dec/02/scientists-policy-governments-science (accessed 31 January 2018).

Van Kerkhoff, L. E. & Lebel, L. 2015. Coproductive capacities: rethinking science–governance relations in a diverse world. *Ecology and Society*, 20, 14. doi.org/10.5751/ES-07188-200114

Ward, V, House, A. & Hamer, S. 2009. Knowledge brokering: the missing link in the evidence to action chain? *Evidence & Policy*, 5, 267–279. doi:10.1332/174426409X463811

Wiley, S. L. 2014. Doing broader impacts? The National Science Foundation (NSF) broader impacts criterion and communication-based activities. Graduate Theses and Dissertations. 13734. Available from https://lib.dr.iastate.edu/etd/13734 (accessed 16 August 2018).

Wintle, B., Mascaro, M., Fidler, F., et al. 2012. The intelligence game: assessing Delphi groups and structured question formats. In Corkill, J., Coole, M. & Valli, C., editors, *Proceedings of the 5th Australian Security and Intelligence Conference* (pp. 14–26). Perth: Security Research Institute, Edith Cowan University.

Young, J. C., Waylen, K. A., Sarkki, S., et al. 2014. Improving the science–policy dialogue to meet the challenges of biodiversity conservation: having conversations rather than talking at one another. *Biodiversity Conservation*, 23, 387–404. doi:10.1007/s10531-013-0607-0

# CHAPTER FOURTEEN

# Approaches to conflict management and brokering between groups

JULIETTE YOUNG
*NERC Centre for Ecology and Hydrology*
CLIVE MITCHELL
*Scottish Natural Heritage*

and

STEPHEN MARK REDPATH
*University of Aberdeen*

## 14.1 What do we mean by conservation conflicts and their management?

Conflicts in conservation arise between individuals or groups of stakeholders whose strongly held opinions clash over conservation objectives and when one party is perceived to assert its interests at the expense of another (Redpath et al., 2013). Such conflicts can take many forms. For example, conflicts may occur between those wanting to conserve large carnivores and those wanting to control them due to their impacts on livestock, or between those wanting to conserve habitats in protected areas and the communities being moved out of those areas. In light of the potential negative impacts on conservation, livelihoods and well-being, managing such conflicts is key to enabling effective conservation.

Conflicts around conservation derive from the fact that the state of nature is socially constructed and has different meanings to different people. Conflicts arise from issues of identity and choices about how the land and sea are used, as well as the uneven distribution of the associated costs and benefits associated with the conservation of biodiversity and ecosystems. These issues reflect the power relations acting across societies over time (Radkau, 2008). The state of nature, which ties into ideas of what is 'natural' and 'acceptable', is therefore inherently mainly a political matter. As such, conflict, defined as 'the pursuit of incompatible goals by different groups' (Ramsbotham et al., 2011, p. 30), is intrinsic to its conservation (Adams, 2015).

Redpath et al. (2013, 2015a) discuss several types of conflict in the field of nature conservation: conflicts of interest, conflicts over beliefs and values, over process or over information, structural conflicts (often involving power relations) and interpersonal conflicts. Often the characteristics of a conflict between people over nature are unclear and it may take considerable expertise to unpick them, but unless we do this, significant time and resources may be invested into one aspect of a problem (e.g. gathering information and evidence), when the conflict is really about something else (e.g. beliefs and values). Another key aspect of defining a conflict is understanding that the people involved will have different and varied values, worldviews and perspectives on the situation and how it should be managed, depending on their roles and agendas. Exploring the different perspectives and goals of people involved in conflicts, and being clear about the problem, its character and various dimensions, are the first steps towards finding a solution.

Finding 'solutions' to these problems is, however, almost as contentious as the conflicts themselves. In certain situations some stakeholders may see the solution as maintaining the status quo, if this fits with their agenda. In others, stakeholders may seek to 'win' the battle by imposing their own approach or views at the expense of the other party. Nevertheless, many stakeholders seek an improvement on the current situation through conflict resolution, transformation or management. In the field of peace studies, the paradigm is shifting from conflict resolution, where the emphasis is on reaching jointly agreed long-term outcomes to conflicts, to the more challenging transformation of conflicts, involving profound change in terms of outcome and process (Mitchell, 2002). This implies fundamental shifts in the ways in which the people involved in the conflict reflect on the real point of conflict and the paradigms and approaches used to mitigate it, leading to the transformation of the institutions and discourses, as well as in the relationships within and between the conflict parties (Ramsbotham et al., 2011). Such shifts have yet to occur in the conservation world.

## 14.2  General approaches to conflict in practice

There are several challenges to understanding and managing conflict. Conflict management usually refers to the containment of conflict, but can also be used generically, to refer to all handling of conflict. We use management here to refer to any positive approach to handling a conflict (Ramsbotham et al., 2011).

Many of the challenges revolve around issues related to knowledge, communication, representation, trust and leadership (Sjölander-Lindqvist et al., 2015). However, problems can arise at the outset from the way these issues are framed. For instance, in the field of human–wildlife conflicts they are often presented as a struggle between animals and people, and the conflict between different human interest groups is ignored (Peterson et al., 2010; Redpath et al., 2013). In reality, most of these conflicts are between conservation interests and other human interests, such as farming, hunting or fishing (Redpath et al., 2015b). Representing these issues as conflicts between farmers and predators is misleading and limits the opportunities for management. To help delineate these two dimensions, Young et al. (2010) distinguished between human–wildlife impacts and human–human conflicts.

The problem of framing is further compounded by the fact that it is often the conservationists who, although not neutral in such settings, are the ones driving the development of management strategies. Clearly, they are likely to be biased in seeking outcomes that benefit conservation, and may not be trusted by the other party or parties. For example, a government conservation organisation may decide to tackle a conflict around a protected species. Because of the background of that organisation, other stakeholders, such as hunters, may assume that the goals are biased towards conservation interests and opposed to hunting interests, and may decide either not to engage in the process or to actively fight against it. A critical step, then, is to be aware of the framing of conflicts around the state of nature and the position different parties take. Having neutral, trusted facilitators, mediators or negotiators can help in the search for potential solutions.

Traditionally, approaches to dealing with human–wildlife conflict have largely been driven by the knowledge created by ecological research and technical fixes. Consequently, efforts to understand and manage conflicts over predators have tended to focus on monitoring, collecting genetic material, estimating predation rates and mitigation methods (such as chilli fences to discourage elephants from destroying crops, diversionary feeding of hen harriers to minimise their impacts on grouse, adapted fishing gear to reduce accidental by-catch). While ecological and technical factors are important aspects of conflict management, social aspects must also be considered. Without insight into the needs, values and positions of the people involved, it is likely that time and money will be wasted and frustration at the continuing conflict will build. This human dimension needs to be understood at both the individual and the collective scale. How do individuals perceive the conflict and react to the species, the other stakeholders involved and the different types of mitigation proposed (Johansson et al., 2012)? At a collective scale it is important to address how the institutions and governance structures are set up. What roles do government and stakeholders play? Who has a say in the decisions?

Knowledge is not simply a product of research by academics from the natural and social sciences and humanities. Substantial knowledge is held by farmers, fishermen and foresters, arising from their experiences, and is often called 'local knowledge'. Typically, ecological scientific knowledge drives conflict management, while the perceptions and understanding held by local knowledge-holders is ignored or dismissed as anecdote. This is compounded by the fact that many of the administration or policy advisors also come from an ecological tradition, and may treat local knowledge in a similar way. This can create major problems for conflict management and contribute to perverse outcomes, such as the illegal killing of wolves in Finland (Pohja-Mykrä & Kurki, 2014). One way around this issue is for researchers to collaborate with other stakeholders in transdisciplinary teams (Butler et al., 2015). The essential value of these co-management approaches is that they are likely to broaden the scope and trust in science, and provide stakeholders with some psychological ownership of the results (Matilainen et al., 2017).

Two other barriers to effective management of conflicts can arise at the policy interface. First, the response to conflicts tends to be reactive (Young et al., 2016a). This has been seen clearly in conflicts over geese, where populations of several species have been increasing rapidly in different regions (Fox & Madsen, 2017), with impacts on crops and farmers' livelihoods. Discussions about conflict management only generally begin once the conflicts have become serious. Conflict management will inevitably be more effective if the process starts earlier and invests in building relationships between stakeholder groups, as well as committing to an improved understanding of the conflict, the people involved and their views, perceptions and values (Young et al., 2016a, 2016b). Second, policy-makers often want quick fixes and rapid conflict *resolution*. Yet, these conflicts are ubiquitous and persistent. We know of no example where a wildlife conflict is considered to have been resolved. Indeed, there are very few instances where they have been effectively *managed* in the long term to reduce conflict, although there have been some short-term, local successes. For example, the Moray Firth Seal Management Plan was developed by fishermen and other key stakeholders from conservation, government agencies, science and tourism in the north-east of Scotland striving to reach a balance between seal conservation and salmon fishing (e.g. Young et al., 2012; Butler et al., 2015). One possible approach to overcome these hurdles would be to horizon scan for emerging conflicts and build relationships, understanding and trust between groups before they escalate.

A further problem is that we currently do not have an informed understanding of which approach to conflict management is most effective under various circumstances. Treves et al. (2017) argue for more top-down approaches, with expert panels, strong policy and enforcement. Conversely, Redpath et al. (2017) argue for more bottom-up governance processes, built on engagement and trust.

To help overcome many of the challenges associated with wildlife conflict management, Young et al. (2016a) developed a decision-support tool with a government agency using a transdisciplinary approach. The tool uses a systematic stepwise approach when faced with management decisions, with six distinct stages: (i) establishing whether there is a conflict or an impact; (ii) understanding the context of the conflict, including the stakeholders affected; (iii) developing shared understanding of the conflict and goals; (iv) building a consensus on how to reach the goals; (v) implementing measures; and (vi) monitoring the outcomes. The authors argue that this new tool has wide applicability and democratic legitimacy, and offers an exciting and practical approach to improve the management of conservation conflicts (see Figure 14.1).

## 14.3  The limitations and challenges of conflict management

Policies seek to resolve disputes by establishing practices and standards with which relevant actors must comply. A naïve view, held by many natural scientists, is that as long as they have a working knowledge of how policy-making and

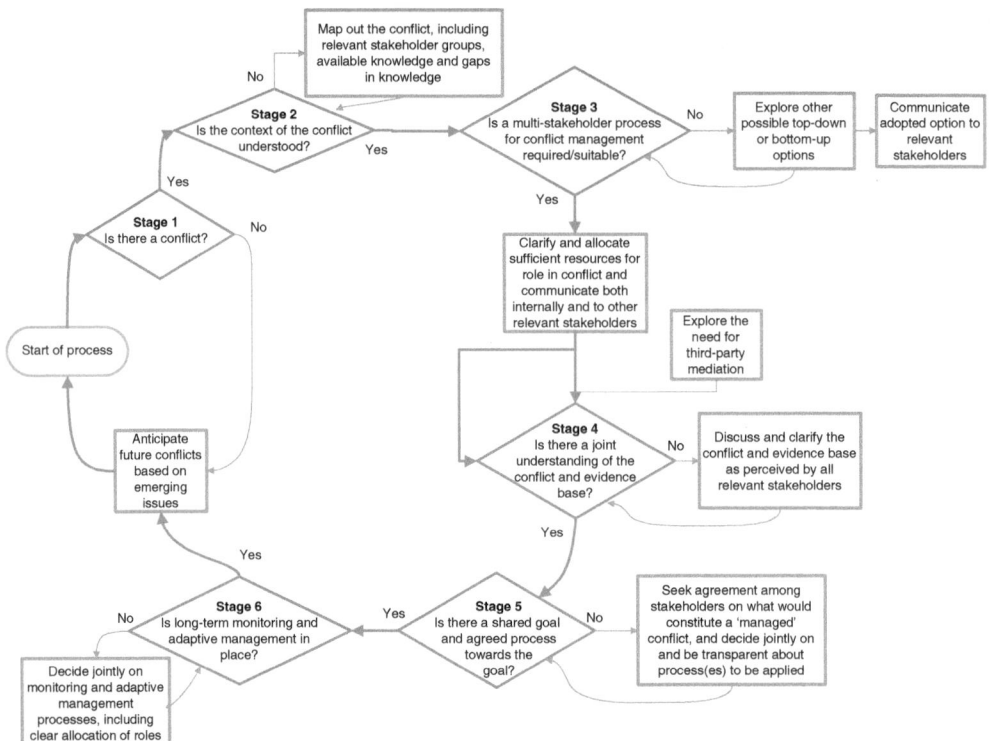

**Figure 14.1** Stepwise approach aimed at enabling decision-makers to identify, manage and monitor conservation conflicts. Diamond shapes indicate the six key decision stages. Squares state what needs to happen to go from one decision stage to the next. Adapted from Young et al. (2016a). (A black and white version of this figure will appear in some formats. For the colour version, please refer to the plate section.)

conflict management function and relate to each, they can make timely contributions that will inform and improve decision-making. However, the decision of whether to conserve or exploit nature is a political and value-based choice. While the focus might appear to be on nature, conservation is also about identity, resource allocation and making choices between people. Therefore, it is intimately bound up in the political economy and granularity of governance. This is a messy business and there are many examples where policy has failed to respond to credible early evidence of problems arising across a range of environmental issues, from lead in petrol to climate change to pesticide use.

Despite the existence of more sophisticated frameworks describing the reality of policy-making, such as the Advocacy Coalition Framework (e.g. Jenkins-Smith & Sabatier, 1994) and Multi-systems Approach (e.g. Cairney & Jones, 2016), much of the policy training in the public sector uses the 'policy wheel'. In general, the process is assumed to start with a problem, which provides a rationale for a policy intervention. Objectives are then set, options appraised and a decision made. The policy is implemented and its effectiveness monitored. The outcomes are evaluated, and the lessons learned contribute to refinements of the policy or inform the definition of the next problem and new policy cycle. This schema works well for problems that are well-defined, tightly bounded and relatively uncontroversial, but there are few such examples in conservation. For more complex issues, which typify conflicts over nature, there are potential difficulties at every step in the cycle.

Many disciplines, including ecological science, history, political science, economics, anthropology, law, psychology, ethics, sociology and peace studies, can be drawn upon to understand conflicts in conservation, as well as practice in areas such as farming, forestry, fisheries and infrastructure development (Redpath et al., 2015a). Nevertheless, the natural sciences still tend to dominate in shaping policy and practice (e.g. Stirling, 2015), with many practitioners believing that 'science speaks truth to power' (e.g. Collingridge & Reeve, 1986). There are a number of fundamental problems linked to this belief.

First, the belief that science trades in facts and that these are unambiguous. This is a realist ontological view that there is 'a' truth to reveal to those in power (e.g. Moses & Knutsen, 2012). If there is doubt, further research will fill in the blanks to reveal the true picture. While this may apply in some cases, it does not hold for much of the field of scientific endeavour, which seeks to deepen our understanding of the world and how it works based on theoretical frameworks (e.g. Moon & Blackman, 2014). The natural sciences typically reveal multiple 'truths' supported by evidence, and the most successful of these can be judged based on their explanatory power and degree of consilience. Knowledge is therefore always shifting (Gee et al., 2013), meaning that conflict can arise from policy and practice that is out of step with current knowledge or specific contexts.

Second, the belief that science and 'facts' are independent of social context. Again, this may be true for some observations, but not for the meanings associated with them (Funtowitz & Ravetz, 1991), and it is often the distinction between observation and meaning that is critical. Many scientists hold that 'matters of fact' lead directly to 'matters of concern', but in practice facts are filtered through individual 'narratives' or worldviews to determine matters of concern (Latour, 2004). These worldviews, which we all have, often remain unspoken, but fuel conflicts of interest. They significantly constrain the scope and relevance of 'expert views' (Sutherland & Burgman, 2015), which are often brought forward to support one position or another in conflicts.

Third, even when science provides a more compelling account of natural phenomena than the alternatives, it requires belief or faith in the scientific method. Many people may struggle to accept a scientific view of an issue over another narrative that reinforces their sense of identity and worldview. Well-reasoned scepticism (Stirling, 2015) is essential to guard against a potential progression to populism, 'fake news' and lobbying for policy that flies in the face of evidence (Corner, 2017).

Marquand (2004) observed the paradox of the requirement for both a strong citizenry, needed for an inclusive public domain, as well as the availability of expert professional viewpoints, which are by definition exclusive, to achieve evidence-based and accountable decision-making. The paradox is how professional views, where knowledge is held by the few rather than the many, contribute to the public domain. This is not necessarily a problem if professional views are in alignment with the public interest, but various checks and balances are required to control for professional interests/institutions and associated power relations. This paradox can be resolved if professionals, including ecologists and conservationists, earn and retain the trust of citizens. Funtowicz and Ravetz (1991), Marquand (2004), Radkau (2008) and Stirling (2015) are among many who advocate for a more participatory approach by which science can act in the public interest on complex issues, in which the evidence from science (including social science) *and* local knowledge is co-created and co-produced (e.g. Fazey et al., 2018) or co-assessed (Sutherland et al., 2017). This potentially allows stakeholders in conflicts to give legitimacy to the authority of professionals (Fazey et al., 2018), thereby addressing issues of trust, bias and power.

## 14.4  Trust, bias and power in conflicts

Power is the uneven distribution of agency (Stirling, 2016), and is a defining and unavoidable characteristic of all social interactions. It is not necessarily bad, as it can get things done. However, whether power is 'good' or 'bad' depends on your viewpoint and, hence, power and politics are intimately linked. Criticism is valid when power is neglected or denied. Similarly, every-one is biased to some extent. This is as true in science as any other field. Like power, bias is problematic when it is neglected or denied.

Decisions about natural resource use and the state of nature involve issues of trust, bias and power, which are inevitable in any set of social interactions (e.g. Young et al., 2016b). How well they are resolved depends on the governance contexts in which decisions are taken. These bring together the personal relationships of the private domain, access to wealth and power in the market domain, and the public interest of the public domain (Marquand, 2004). The more diverse, plural and different the views from stakeholders that are expressed and integrated into decisions about the natural environment the better (e.g. Young et al., 2016b), with power relations and biases acknowledged to keep incumbent hegemonies and vested interests in check (Stirling, 2015). This is not to argue that the process is easy or that everyone can always agree, but that people can agree to differ through a well-structured process and move on from conflict: a 'solution' that involves winners and losers will always resurface as a conflict (Young et al., 2016a). This argues directly against cen-tralisation, often a dominant force in 'command-and-control' politics (e.g. Cooke & Muir, 2012).

The extent to which administrative and institutional arrangements are able to respond flexibly, in a scale-appropriate manner, and quickly to reflect the character of real-world problems, is a critical factor in successfully translating evidence into effective policy and practice (e.g. Sparrow, 2011). However, there is a great deal of inertia in institutions, often as a result of their struc-tures, processes and associated habits and ways of working. Internal arrange-ments designed for one set of problems may be ill-suited to others. An important distinction is whether organisations (including government) exist to 'deliver' or 'enable'. The latter is essential when creating the conditions that facilitate participative approaches and the development of trusting relations.

## 14.5 An outlook on conflict management: focusing on worldviews around the state of nature

Identity, and specifically the worldviews on the state of nature, are of critical importance in conflict management, including the question of whether peo-ple are seen or see themselves as a part of, or apart from, nature (Fischer & Young, 2007). This can influence the understanding and mental constructs around terms such as biodiversity, nature, ecosystem health, native, natural-ness, integrity, sustainability, resilience, stability, balance, wild, land-sparing and land-sharing. In short, all of the language, concepts and ideas of conserva-tion are open to different interpretations, which perhaps testifies to the idea that the state of 'nature' and 'conservation' are social constructs. In turn, this has implications for the institutional arrangements and approaches to conservation (e.g. what we measure, performance management frameworks). The idea that nature is unambiguous and categorical sits comfortably with more rigid measurement frameworks informed by authoritative science and used to 'deliver' conservation objectives. In contrast, a more fluid relationship between people and nature, based on a broad range of knowledge and possible truths, is better aligned to situational, participative and co-produced approaches.

This is not to suggest that worldviews (whether people are part of, or apart from, nature) and their consequences can be readily polarised. Indeed, these worldviews are not necessarily mutually exclusive: some people may gravitate more to one than the other, while others may hold both simultaneously. Similarly, while debates between utilitarian and intrinsic values greatly exercise many conservationists, many people hold both together without conflict. However, it appears that utilitarian values are often associated with general and replicable issues and intrinsic values are often more situational and associated with personal experience and knowledge. This serves only to illustrate that worldviews can and do shape evidence, institutional arrangements and approaches to conservation, including the way in which conflicts are managed.

## References

Adams, W. 2015. The political ecology of conservation conflicts. In Redpath, S. M., Gutierrez, R. J., Wood, K. A., et al., editors. *Conflicts in Conservation: Navigating Towards Solutions*. Ecological Reviews (pp. 64–75). Cambridge: Cambridge University Press.

Butler, J. R. A., Young, J. C., McMyn, I. A. G., et al. 2015. Evaluating adaptive co-management as conservation conflict resolution: learning from seals and salmon. *Journal of Environmental Management*, 160, 212–225.

Cairney, P. & Jones, M. D. 2016. Kingdon's Multiple Streams Approach: what is the empirical impact of this universal theory? *Policy Studies Journal*, 44, 37–58.

Collingridge, D. & Reeve, C. 1986. *Science Speaks Truth to Power: The Role of Experts in Policy Making*. London: Frances Pinter.

Cooke, G. & Muir, R., editors. 2012. *The Relational State: How Recognising the Importance of Human Relationships Could Revolutionise the Role of the State*. London: IPPR.

Corner, J. 2017. Fake news, post-truth and media – political change. *Media, Culture & Society*, 39, 1100–1107.

Fazey, I., Moug, P., Allen, S., et al. 2018. Transformation in a changing climate: a research agenda. *Climate and Development*, 10, 197–217.

Fischer, A. & Young, J. C. 2007. Understanding mental constructs of biodiversity: implications for biodiversity management and conservation. *Biological Conservation*, 136, 271–282.

Fox, A. D. & Madsen, J. 2017. Threatened species to super-abundance: the unexpected international implications of successful goose conservation. *Ambio*, 46, 179–187.

Funtowicz, S. O. & Ravetz, J. R. 1991. A new scientific methodology for global environmental issues. In Costanza, R., editor, *Ecological Economics: The Science and Management of Sustainability* (pp. 137–152). New York, NY: Columbia University Press.

Gee, D., Grandjean, P., Hansen, S. F., et al. 2013. Late Lessons from Early Warnings: Science, Precaution, Innovation. Copenhagen: European Environment Agency.

Jenkins-Smith, H. C. & Sabatier, P. A. 1994. Evaluating the advocacy coalition framework. *Journal of Public Policy*, 14, 175–203.

Johansson, M., Karlsson, J., Pedersen, E., et al. 2012. Factors governing human fear of brown bear and wolf. *Human Dimensions in Wildlife*, 17, 58–74.

Latour, B. 2004. Why has critique run out of steam? From matters of fact to matters of concern. *Critical Inquiry*, 30, 225–248.

Marquand, D. 2004. False friend: the state and the public domain. *The Political Quarterly*, 75 (S1), 51–62.

Matilainen, A., Pohja-Mykrä, M., Lähdesmäki, M., et al. 2017. "I feel it is mine!" – Psychological ownership in relation to natural resources. *Journal of Environmental Psychology*, 51, 31–45.

Mitchell, C. 2002. Beyond resolution: what does conflict transformation actually transform? *Peace and Conflict Studies*, 9(1), 1–23.

Moon, K. & Blackman, D. 2014. A guide to understanding social science research for natural scientists. *Conservation Biology*, 28, 1167–1177.

Moses, J. W. & Knutsen, T. 2012. *Ways of Knowing: Competing Methodologies in Social and Political Research*. Bsingstoke: Palgrave Macmillan.

Peterson, M. N., Birckhead, J. L., Leong, K., et al. 2010. Rearticulating the myth of human–wildlife conflict. *Conservation Letters*, 3, 74–82.

Pohja-Mykrä, M. & Kurki, S. 2014. Strong community support for illegal killing challenges wolf management. *European Journal of Wildlife Research*, 60, 759–770.

Radkau, J. 2008. World history and environmental history. *Encyclopedia of Life Support Systems (EOLSS)*, 976–86.

Ramsbotham, O., Miall, H. & Woodhouse, T. 2011. *Contemporary Conflict Resolution*. Cambridge: Polity.

Redpath, S. M., Bhatia, S. & Young, J. 2015b. Tilting at wildlife: reconsidering human–wildlife conflict. *Oryx*, 49, 222–225.

Redpath, S. M., Gutiérrez, R. J., Wood, K. A., et al., editors. 2015a. *Conflicts in Conservation: Navigating Towards Solutions*. Camabridge: Cambridge University Press.

Redpath, S., Linnell, J., Festa-Bianchet, M., et al. 2017. Don't forget to look down – collaborative approaches to predator conservation. *Biological Reviews*, 92, 2157–2163.

Redpath, S., Young, J., Evely, A., et al. 2013. Understanding and managing conflicts in biodiversity conservation. *Trends in Ecology and Evolution*, 28, 100–109.

Sjölander-Lindqvist, A., Johansson, M. & Sandström, C. 2015. Individual and collective responses to large carnivore management: the roles of trust, representation, knowledge spheres, communication and leadership. *Wildlife Biology*, 21, 175–185.

Sparrow, M. K. 2011. *The Regulatory Craft: Controlling Risks, Solving Problems, and Managing Compliance*. Washington, DC: Brookings Institution Press.

Stirling, A. 2015. Emancipating transformations: from controlling 'the transition' to culturing plural radical progress. In Scoones, I., Leach, M. & Newell, P., editors, *The Politics of Green Transformations* (pp. 54–67). Abingdon: Routledge.

Stirling, A. 2016. Knowing doing governing: realizing heterodyne democracies. In Voß, J.-P. & Freeman, R., editors, *Knowing Governance* (pp. 259–289). Basingstoke: Palgrave Macmillan.

Sutherland, W. J. & Burgman, M. A. 2015. Use experts wisely. *Nature*, 526(7573), 317–318.

Sutherland, W. J., Shackleford, G. & Rose, D. C. 2017. Collaborating with communities: co-production or co-assessment? *Oryx*, 51, 569–570. doi:10.1017/S0030605317001296

Treves, A., Chapron, G., López-Bao, J. V., et al. 2017. Predators and the public trust. *Biological Reviews*, 92, 248–270.

Young, J., Butler, J. R. A., Jordan, A., et al. 2012. Less government intervention in biodiversity management: risks and opportunities. *Biodiversity and Conservation*, 21, 1095–1100.

Young, J., Marzano, M., White, R. M., et al. 2010. The emergence of biodiversity conflicts from biodiversity impacts: characteristics and management strategies. *Biodiversity & Conservation*, 19, 3973–3990.

Young, J. C., Searle, K. R., Butler, A., et al. 2016b. The role of trust in the resolution of conservation conflicts. *Biological Conservation* 195, 196–202.

Young, J. C., Thompson, D., Moore, P., et al. 2016a. A conflict management tool for conservation agencies. *Journal of Applied Ecology*, 53, 705–711.

# CHAPTER FIFTEEN

# Conservation goals in international policies

ALETTA BONN

*Helmholtz-Centre for Environmental Research – UFZ*
*Friedrich Schiller University Jena*
*German Centre for Integrative Biodiversity Research (iDiv)*

MARIANNE DARBI

*Helmholtz-Centre for Environmental Research – UFZ*

HYEJIN KIM

*Martin Luther University Halle-Wittenberg*
*German Centre for Integrative Biodiversity Research (iDiv)*

and

ELISABETH MARQUARD

*Helmholtz-Centre for Environmental Research – UFZ*

## 15.1 Introduction

Biodiversity and its importance has long been recognised and enshrined in national and international policies. While the earliest conservation policies were framed around 150 years ago and mainly consisted of national policies to protect biodiversity, over the last century conservation policies have undergone a significant shift in emphasis towards integration of, and alignment with, societal goals (Mace, 2014). Moving from a sole focus on species and habitat protection in the early twentieth century, or 'Nature for itself' as framed by Mace (2014), policies have gradually aligned with other societal aims. This started with a recognition of ecosystem services (Daily, 1997), as the benefits people derive from nature ('Nature for People'), which was brought into the mainstream by the Millennium Ecosystem Assessment (MA, 2005). There has since been a move away from utilitarian values to consider 'Nature and People' (Mace, 2014; Díaz et al., 2018) as a more inclusive concept to better support synergies and negotiate trade-offs of conservation and societal goals. In this chapter, we aim to demonstrate and discuss how this increasingly integrative view is reflected in the development of international conservation policies and related institutions. After briefly sketching the historical origins of current international conservation policies, we focus on the Convention on Biological Diversity (CBD), which couples its core objective of nature conservation with human well-being. Next, we show how an integrative view on nature conservation has shaped the Intergovernmental Science–Policy Platform on Biodiversity and Ecosystem Services (IPBES). Finally, we explore the

Sustainable Development Goals (SDGs) as a third global enterprise that closely links the conservation of nature to other societal aspirations. Using these three examples, we address the following questions.

1. How do these three agreements function and how are decisions made?
2. What is the role of science and evidence in the CBD, IPBES and the SDGs?
3. What are the achievements so far, and how can scientists engage to foster progress?

## 15.2  A short history of conservation policies

To understand current conservation policies, it is useful to reflect briefly on their development. Historically, conservation policies were created in response to a realisation of loss of natural habitat, and led to national conservation designations, notably the first big national parks. In the USA, Yellowstone was established as the first National Park worldwide by the Yellowstone National Park Act in 1872, withdrawing almost one million hectares from further land use development to be 'dedicated and set apart as a public park ... for the benefit and enjoyment of the people'. In Europe, the UK was the first country to establish national parks under the 1949 National Parks and Access to the Countryside Act, also born out of a strong demand for open public access to private land. The Peak District National Park, designated in 1951, remains one of the most-visited national parks worldwide. Many more national parks followed in the 1970s and 1980s in Africa, Europe and across all continents. Often, however, these designations showed little consideration of local communities and their livelihoods ('Nature despite people'; Mace, 2014), leading at times to violations of rights of indigenous people and severe conflicts (Colchester, 2004). Protected areas continue to provide crucial cornerstones of local, regional and international strategies for biodiversity conservation. They have significantly contributed to halting losses of species and habitats, although their performance is at times mixed and often not known (Gaston et al., 2008; Mora & Sale, 2011).

International conservation policy development started with a series of global conventions in the 1970s and 1980s focusing on species and habitat protection (Table 15.1). Once countries ratified these multi-lateral environmental agreements, they proved to be drivers for national law development. For example, the US Endangered Species Act of 1973 was developed as a response to the Convention on International Trade in Endangered Species of Wild Fauna and Flora (CITES) that had entered into force the same year. As another example, the European Union met its obligations for bird species under the Bern Convention (1979) and Bonn Convention (1979) through the *Council Directive 79/409/EEC on the conservation of wild birds* (*Birds Directive*) adopted in 1979. This has since been substantially amended several times to the Directive 2009/147/EC adopted in

**Table 15.1** *Important multi-lateral environmental agreements in the nature conservation context. Information retrieved from the treaty's websites or from www.informea.org (accessed 9 December 2018)*

| Treaty name | Abbreviation | Adoption | Entry into force | Parties* | Main target |
|---|---|---|---|---|---|
| Convention on Wetlands of International Importance | Ramsar Convention | 1971 | 1975 | 170 | Conservation and sustainable use of wetlands |
| Convention Concerning the Protection of the World Cultural and Natural Heritage | WHC/World Heritage Convention | 1972 | 175 | 193 | Protection of the world cultural and natural heritage |
| Convention on International Trade in Endangered Species of Wild Fauna and Flora | CITES | 1973 | 1975 | 183 | Regulation of trade of wild plants and animals |
| Convention on the Conservation of European Wildlife and Natural Habitats | Bern Convention | 1979 | 1982 | 51 | Conservation of wild flora and fauna and their natural habitats, and promotion of European cooperation |
| Convention on the Conservation of Migratory Species of Wild Animals | CMS/Bonn Convention | 1979 | 1983 | 126 | Conservation and sustainable use of migratory animals and their habitats |
| United Nations Framework Convention on Climate Change | UNFCCC | 1992 | 1994 | 197 | Prevention of dangerous anthropogenic interference with the climate system, slowing global warming and mitigating its impact |

**Table 15.1** (*cont.*)

| Treaty name | Abbreviation | Adoption | Entry into force | Parties* | Main target |
|---|---|---|---|---|---|
| Convention on Biological Diversity | CBD | 1992 | 1993 | 196 | Conservation of biological diversity, the sustainable use of its components, and the fair and equitable sharing of benefits arising from the use of genetic resources |
| United Nations Convention to Combat Desertification | UNCCD | 1994 | 1996 | 197 | Prevention of desertification and land degradation |

* Number of member states as of December 2018.

2009 and sits alongside the *Council Directive 92/43/EEC on the conservation of natural habitats and of wild fauna and flora* (Habitats Directive) adopted in 1992. Legal mechanisms for the achievement of international conventions at national scales are at the discretion of each member state.

During the 1980s, environmental pollution, the over-use of resources and the resulting loss of species and natural habitats gained increasing attention from the public and political representatives. This led to the 'Rio World Summit' in 1992 (United Nations Conference on Environment and Development, UNCED), at which three new conventions were opened for signature: the United Nations Framework Convention on Climate Change (UNFCCC), the United Nations Convention to Combat Desertification (UNCCD) and the Convention on Biological Diversity (CBD). Further details of the set up, operation and achievements of these three conventions are described in the sections below.

## 15.3  General set up and mode of operation

### 15.3.1  The Convention on Biological Diversity (CBD)

The CBD is, with regards to goals addressed, the most comprehensive global treaty dealing with nature conservation. Its three overarching objectives are (Article 1 of the Convention):

(a)  the conservation of biological diversity,
(b)  the sustainable use of its components and
(c)  the fair and equitable sharing of the benefits arising out of the utilisation of genetic resources.

Thus, the CBD's objectives refer to both intrinsic and instrumental values of biodiversity. It does so by including an unconditional call for the conservation of biodiversity in combination with the acknowledgement that people depend on nature and need to make use of it, as well as a call for dividing the benefits that are derived from nature equitably.

In total, the Convention's text contains 42 Articles that further define aims and assign duties to the bodies of the Convention. The CBD's clear recognition of the interaction between nature-related and societal goals is also codified in its principles. For example, the first CBD principle states that the 'objectives of management of land, water and living resources are a matter of societal choices', while the twelfth acknowledges that 'the ecosystem approach should involve all relevant sectors of society and scientific disciplines'. The CBD is a legally binding treaty. Thus, a state that has signed and ratified the Convention is obliged to implement the Convention on its territory through national policies and practical management. Every two years, representatives of the member states meet at the Conference of the Parties (COP). The COP is the highest decision-making body of the CBD and it operates according to the consensus principle. This means that the text of a decision is negotiated until a compromise is reached among all parties present. If no consensus is reached, parties do not vote. Instead, only text to which no party objects is agreed upon and a decision on unresolved questions is postponed. A CBD COP decision therefore almost always represents a compromise between states with differing views. This 'consensus principle' has been criticised for preventing progress and watering down any suggestion to the lowest common denominator, often resulting in general, vague or ambiguous text (Kanie, 2014; Kemp, 2016). However, a shift from the consensus principle to a voting system faces many obstacles, e.g. the fear that parties could perceive this as a loss of sovereignty and could therefore drop out of the Convention, or that such a reform would open a 'Pandora's box' and encourage open disputes on, and possibly change in, other principles or rules of procedure (Kemp, 2016).

To facilitate negotiations under the consensus principle, the CBD parties are divided into groups of states that discuss and align their positions; one of their members is then responsible for representing them in the plenary of the COP. Important associations of states are the European Union and the official United Nations Regional Groups (African Group, Asia–Pacific Group, Eastern European Group, Latin America and Caribbean Group, Western European and Others Group), alongside some informal groups, such as an alliance of industrialised non-EU countries called JUSCANNZ (i.e. Japan, United States, Switzerland, Canada, Australia, Norway, New Zealand).

Meetings of the CBD COP and of many other CBD bodies (e.g. of the Subsidiary Body of Technical and Technological Advice – SBSTTA, see 15.5.1) are open to so-called 'observers'. The observer status can be obtained by, for example, non-governmental organisations, business associations or scientific institutions and it gives the right to speak in plenary but not to veto a decision.

One way in which the CBD fosters progress towards its objectives is by setting up particular Programmes of Work, each with a vision and suggested actions that CBD parties are encouraged to support. These are concerned with topics related to Agricultural Biodiversity, Dry and Sub-humid Lands Biodiversity, Forest Biodiversity, Inland Waters Biodiversity, Island Biodiversity, Marine and Coastal Biodiversity and Mountain Biodiversity. The CBD also dedicates work to cross-cutting issues, such as Climate Change and Biodiversity; Communication, Education and Public Awareness, Economics, Trade and Incentives Measures or Identification, Monitoring, Indicators and Assessments. It aims to link work on these themes closely with other UN Conventions by collaborating with, for example, UNFCCC and UNCCD secretariats (www.unccd.int/convention/about-convention/unccd-cbd-and-unfccc-joint-liaison-group).

Approximately every five years, parties must report the steps taken to implement the CBD provisions and their effectiveness to the CBD Secretariat. These 'National Reports' are used by the CBD Secretariat to gain an overview of global trends in the implementation process. However, as the parties are sovereign entities, they decide individually about their national implementation approaches, and are free to set own priorities (with the exception of EU member states who coordinate their efforts and are committed to EU regulations). There are no established CBD non-compliance procedures. The degree of compliance therefore varies widely and, overall, has proven to be generally insufficient, as the CBD's goals and targets, formulated in the Convention's Strategic Plans, have been repeatedly missed. For the period 2002–2010, the core element of the CBD's Strategic Plan was the '2010 Target': a 'significant reduction of the current rate of biodiversity loss at the global, regional and national level as a contribution to poverty alleviation and to the benefit of all life on Earth' (COP-Decision VI/26). However, this 2010 Target was widely missed (Butchart et al., 2010; Dirzo et al., 2014).

For the following decade, the level of ambition was raised further: 'to halt the loss of biodiversity' by 2020. To better address the underlying causes of biodiversity loss and be more explicit about what needed to be done to make progress towards the CBD objectives, the Strategic Plan for 2011–2020 was underpinned with five strategic goals and 20 'Aichi Biodiversity Targets' that formed the backbone of the Plan (see Figure 15.1). Setting up such a comprehensive framework that addressed the direct and indirect drivers of

**Figure 15.1** The 20 Aichi Biodiversity Targets. Image: Copyright BIP/SCBD. (A black and white version of this figure will appear in some formats. For the colour version, please refer to the plate section.)

the ongoing biodiversity crises was seen as a major achievement. Furthermore, the Strategic Plan 2011–2020 has been highly relevant, beyond the global biodiversity agenda; it was endorsed by the UN General Assembly and other multi-lateral environmental agreements and therefore formed the principle global roadmap for the conservation of nature. The 20 Aichi Biodiversity Targets that formed the core of the Strategic Plan 2011–2020 were also incorporated into the global development agendas and fed into the Millennium Goals (until 2015) and subsequently the Sustainable Development Goals (until 2030).

However, despite this high political recognition, the Aichi Targets were not on track in 2018 and most will be widely missed by 2020, as indicated by the fourth Global Biodiversity Outlook report (Leadley et al., 2014) and the IPBES Global Assessment (IPBES/7/10/Add.1). Despite progress towards some Targets, the overall picture leaves no doubt: efforts need to be increased dramatically to halt and reverse the current situation, in which the drivers of biodiversity loss worldwide strongly override conservation efforts. There have been accelerated policy and management responses to the biodiversity crisis, but these are unlikely to significantly reverse trends in the state of biodiversity by 2020 (Tittensor et al., 2014).

For the post-2020 period, it is therefore crucial to focus on the implementation of the new CBD strategic framework that will then be in place. This needs to be achieved, in the first place, by the parties at the national level. Therefore, besides increased globally concerted efforts, place-based and context-specific approaches are essential for monitoring, conserving and sustainably using biodiversity.

## 15.3.2 Intergovernmental Science–Policy Platform on Biodiversity and Ecosystem Services (IPBES)

As a response to knowledge needs that became evident in the context of the CBD and other multi-lateral environmental agreements, the Millennium

Ecosystem Assessment (MA, 2005) was conducted in 2005, followed by several national ecosystem assessments (Schröter et al., 2016). Building on this experience (Carpenter et al., 2009) and modelled on the Intergovernmental Panel on Climate Change (IPCC), the Intergovernmental Science–Policy Platform on Biodiversity and Ecosystem Services (IPBES) was established in 2012 to generate an integrative knowledge foundation on biodiversity, ecosystems, ecosystem services and their impact on human and societal well-being (UNEP, 2012). IPBES is not a convention but a science–policy interface that supports governments and stakeholders in decision-making at multiple scales by providing policy-relevant and scientifically credible information on the status and trends of nature and its contributions to people (Brooks et al., 2014). IPBES does not enforce decisions on conventions or countries, but aspires to develop an expert-based platform that provides an accessible, useful and scientifically rigorous evidence base to support biodiversity-related decision-making by national governments and international conventions (e.g. CBD, RAMSAR, CITES, UNCCD).

To achieve this, IPBES operates via four main functions – assessment, knowledge generation, policy support and capacity-building – that are implemented through voluntary participation of experts chosen by governments and organisations globally, with balanced representation across regions, gender and disciplines (IPBES, 2014). Over the coming years, IPBES aims to continue bringing together the best knowledge-holders and institutions on biodiversity around the globe, synthesising the complex dynamics of nature and their impact on human societies and the planet, providing the most credible information available through research and practice, and catalysing the generation of new knowledge to fill critical gaps in order to better conserve nature and ensure human and societal well-being (Figure 15.2).

The IPBES Plenary, where 130 member states form a governing body, meets annually to track the progress of the work programme and to make decisions on the way forward. A Multidisciplinary Expert Panel (MEP) advises on scientific and technical aspects of the programme. The expert groups, taskforces and assessment authors are the scientists and knowledge-holders. Stakeholders and observers also play significant roles in IPBES by providing diverse perspectives and forms of knowledge and acting as catalysts for conservation in their respective communities of practice. In particular, IPBES is developing a mechanism to better integrate holders of indigenous and local knowledge into the process for a more comprehensive understanding and outlook on nature's values and futures (IPBES, 2014).

The decision-making process of IPBES is lengthy but transparent, due to the nature of the intergovernmental plenary system (Figure 15.3 shows the participants).

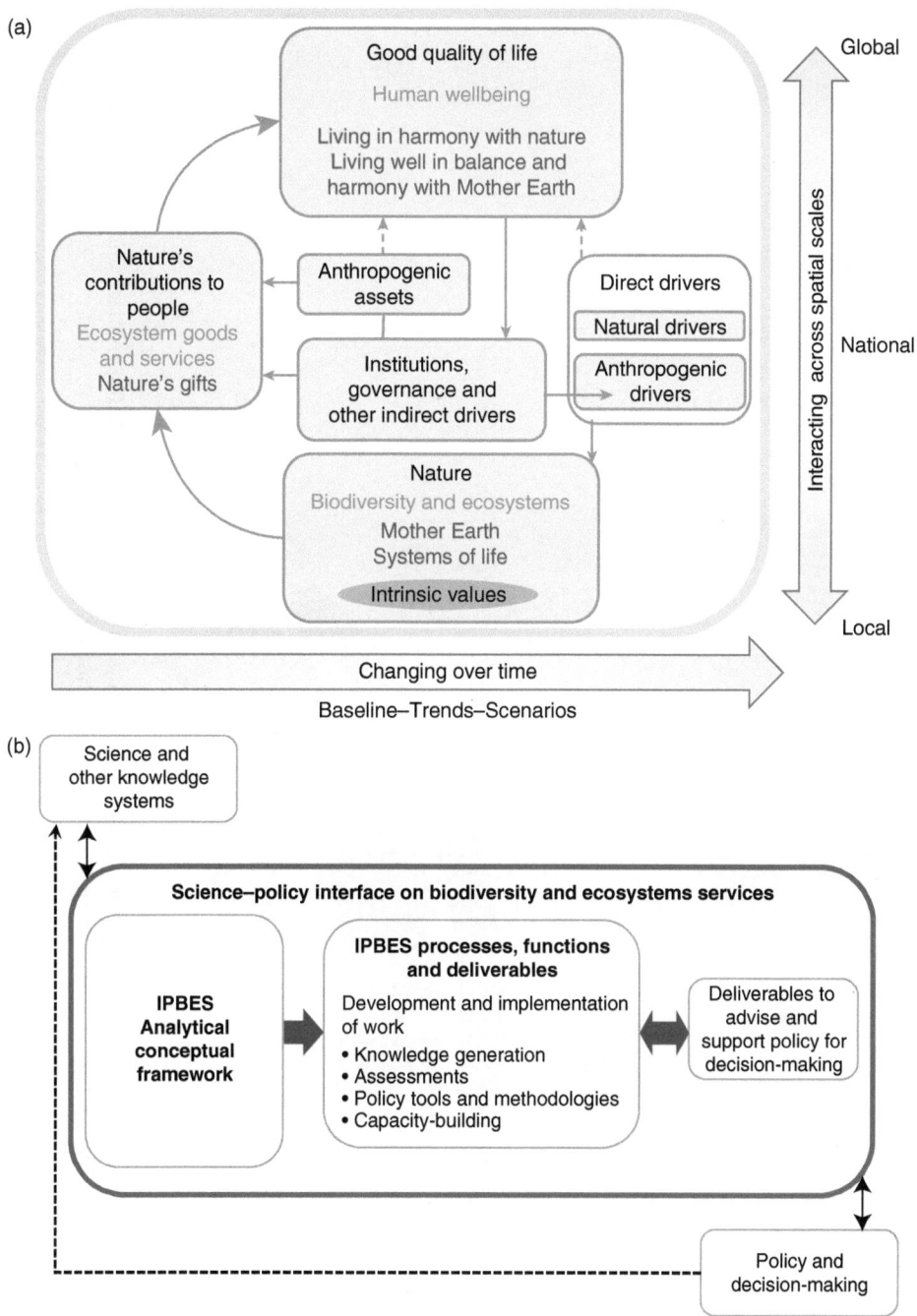

**Figure 15.2** (a) IPBES operational model of the Platform (adapted from IPBES, 2014), (b) analytical conceptual framework of assessments (adapted from Díaz et al., 2015). (A black and white version of this figure will appear in some formats. For the colour version, please refer to the plate section.)

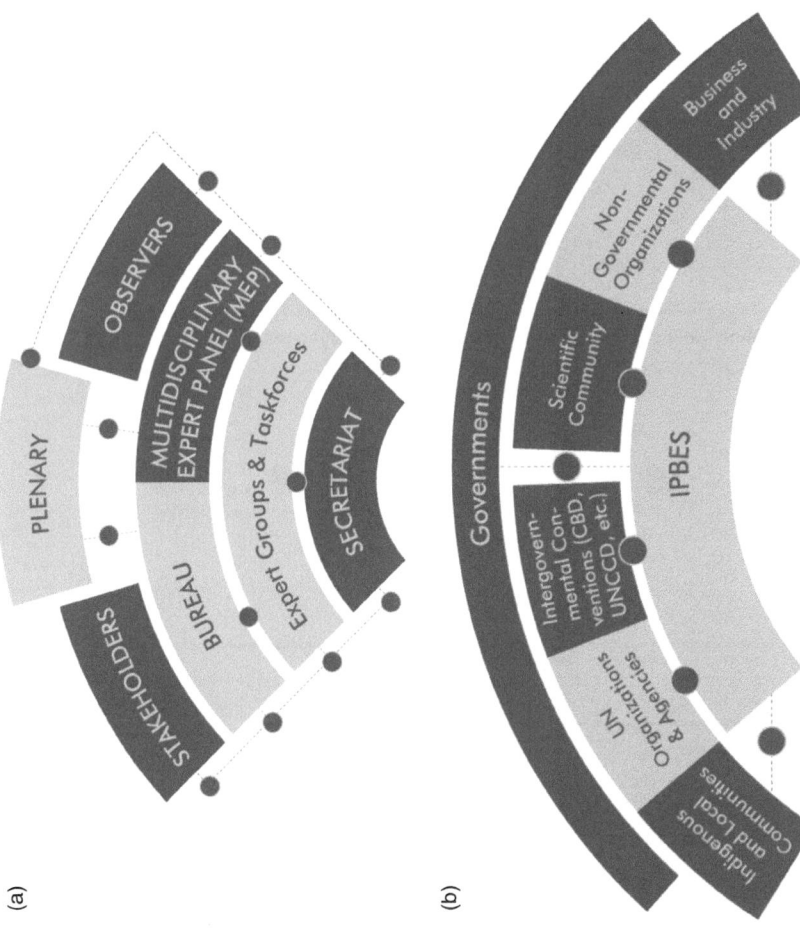

**Figure 15.3** Structures of IPBES (a) science–policy platform, (b) intergovernmental plenary (IPBES, 2018b). (A black and white version of this figure will appear in some formats. For the colour version, please refer to the plate section.)

IPBES is an independent intergovernmental platform that works in partnership with the large United Nations Programmes such as the UN Environment Programme (UNEP), the UN Educational, Scientific and Cultural Organization (UNESCO), the Food and Agriculture Organization of the UN (FAO) and the UN Development Programme (UNDP). Its work is aligned to the CBD and other international Conventions (e.g. Ramsar, CITES, as well as the UNCCD). Its unique role is to mobilise scientific communities from multiple disciplines to harmonise research agendas on biodiversity and its impact on societies among key organisations, such as the International Union for the Conservation of Nature (IUCN), Future Earth and the Group On Earth Observations Biodiversity Observation Network (GEO BON) (IPBES, 2018a). While the social sciences and humanities are still underrepresented in the process (Vadrot et al., 2018), IPBES aims to attract more social scientists.

### 15.3.3 The Sustainable Development Goals

The establishment of IPBES was well timed to coincide with the inception of United Nation's new global agenda, the Sustainable Development Goals (SDGs) (UN, 2015). Historically, the concept of sustainability builds on more than 30 years of intense political discourse, following the Brundtland Commission (1987), the Rio Declaration on Environment and Development (UN, 1992) and the eight Millennium Development Goals (MDGs) (McArthur, 2014). These included a goal to 'ensure environmental sustainability', but did not relate to biodiversity specifically. Based on the MDGs, the SDGs were developed as a more holistic and integrated approach to development following the United Nations Conference on Sustainable Development in 2012. In January 2016, the *2030 Agenda for Sustainable Development*, comprising 17 SDGs with 169 targets and a declaration, were officially approved during a UN Summit attended by 193 member states (UN, 2015). The 2030 Agenda aimed to stimulate action in areas of critical importance for humanity and the planet with a set of approved goals (Figure 15.4). It provides a holistic strategy that combines economic development, social inclusion and environmental sustainability and applies to all countries – poor, rich and middle-income alike – and to all segments of society (ICSU, 2017); this is the major novelty and strength of this framework, in which biodiversity conservation is no longer isolated.

Its main decision body, the High-level Political Forum, provides a central platform for all member states to review progress towards the 2030 Agenda for Sustainable Development and the SDGs. To foster the implementation of the SDGs, the United Nations partnered with several governmental and non-governmental organisations worldwide to ensure commitment to this cause and also enhance synergies across global conventions. Several international coalitions, including the G20 and G8, have incorporated the 2030 Agenda

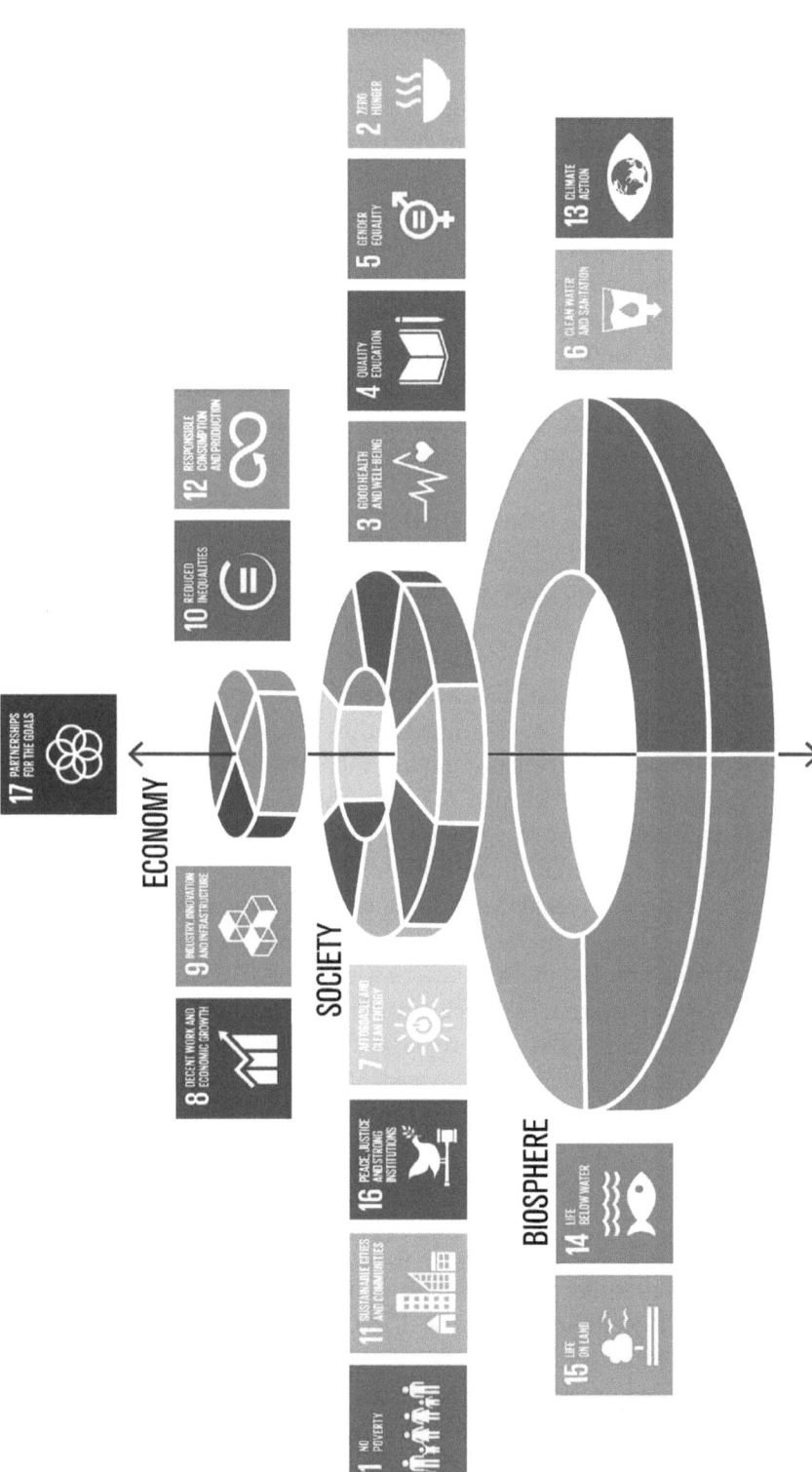

**Figure 15.4** The Sustainable Development Goals 'wedding cake' (source/credit: Azote Images for Stockholm Resilience Centre, Stockholm University). (A black and white version of this figure will appear in some formats. For the colour version, please refer to the plate section.)

into their policy frameworks, although reviews have indicated that the implementation of SDGs in general and the biodiversity goals in particular (SDG 14 life below water and SDG 15 life on land) are not yet sufficiently incorporated into national policies of either OECD or non-OECD countries (O'Connor et al., 2016; Schmidt-Traub et al., 2017). Achieving the SDGs requires a willingness to cooperate at the international level and sustainable development to be anchored as a guiding principle in all policy fields at national, European and international levels (Schmidt-Traub et al., 2017). However, the achievement of many SDGs depends largely on action taken in member states and above all requires the development and implementation of strong operative concepts at national and regional levels (Schmidt-Traub et al., 2017). Governments and other stakeholders are expected to mobilise efforts to establish national and regional plans towards implementation of the SDGs (ICSU, 2017). This requires a balance between addressing the scope and systemic nature of the 2030 Agenda with budgetary, political and resource constraints that inevitably mean countries prioritise certain targets (ICSU, 2017) and the associated risk of negative effects for 'non-prioritised' ones, particularly if they are in a conflicting, even mutually exclusive, relationship (Schmalzbauer & Visbeck, 2016). Furthermore, the goals are rarely independent and consequently failures in one area can quickly undermine progress in other areas (Schmalzbauer & Visbeck, 2016). National policy-makers thus face the challenge of understanding the inter-dependencies across the SDGs and achieving coherent implementation to ensure that progress in some areas is not made at the expense of progress in others. In addition, national policies often have implications on neighbouring countries or across globalised value chains, i.e. we need to avoid pursuing objectives in one region that negatively affect other countries' pursuit of their objectives (ICSU, 2017).

## 15.4  Joint working of the CBD and SDG 2030 Agenda
According to the CBD, the Strategic Plan for Biodiversity and the 2030 Agenda are consistent with each other and mutually supportive (CBD et al., 2017). The central role of the biosphere is explicitly acknowledged in the new illustration of the SDGs, as layers in a 'wedding cake' that build on one another, developed by the Stockholm Resilience Centre (see Figure 15.4). It implies a transition away from sectoral approaches embedding economy and society as parts of the biosphere and recognises that the related goals of promoting human dignity and prosperity can only be achieved sustainably if the Earth's vital biophysical processes and ecosystem services are safeguarded (ICSU, 2017). However, working towards the implementation of the SDGs in UN member states requires a process of prioritisation. This poses a fundamental challenge and possibly a genuine risk to

biodiversity conservation, as biodiversity concerns may not always be adequately anchored in other non-environmental policy sectors and thus may be overridden by other interests, especially when trade-offs arise between short-term development achievements and long-term sustainability (Schmalzbauer & Visbeck, 2016). These trade-offs will often be at the expense of biodiversity (SDGs 14 and 15), with likely negative consequences for several other SDGs, such as those related to food security, water supply and climate change mitigation. There have been some attempts to analyse these links further (Scharlemann et al., 2016; SRC 2016; CBD et al., 2017), but the critical question of how to resolve potential trade-offs in practice remains to be negotiated at the local, national and regional scales.

## 15.5 Role of science and evidence
### 15.5.1 CBD
To conserve biodiversity, it is important to devise action on reliable, sound knowledge about its components. The CBD has incorporated this principle by obliging all contracting parties to identify and monitor particularly diverse ecosystems and habitats, threatened species and other biodiversity components of ecological, social, economic, cultural or scientific importance (Article 7 and Annex 1 of the Convention). To effectively conserve biodiversity, it is furthermore crucial to build action on sound evidence about the factors that lead to its loss and measures to reduce their impact, e.g. possible policy and management responses and their effectiveness.

The CBD collates, utilises and synthesises such knowledge in various ways. The CBD secretariat, for example, regularly publishes notifications that call for input with regard to particular questions. Approximately every five years, it publishes the 'Global Biodiversity Outlook', an assessment of global biodiversity states and trends and of the progress toward the CBD objectives (Leadley et al., 2014).

The CBD's Subsidiary Body on Scientific, Technical and Technological Advice (SBSTTA) is responsible for processing knowledge-related tasks and providing advice and guidance to the COP with respect to scientific (and technical and technological) questions. The SBSTTA plays a crucial role because it presents recommendations that are often later followed by the COP (sometimes with modifications). Therefore, its meetings are highly politicised and cannot provide a comprehensive and balanced evidence base with regard to upcoming COP negotiations. This has long been a major criticism of the SBSTTA and was one of the major motivations for creating the Intergovernmental Platform on Biodiversity and Ecosystem Services.

### 15.5.2 IPBES
As a platform of scientific communities and knowledge-holding networks, IPBES is expected to play a critical role in providing the best available, rigorous

and comprehensive scientific evidence to various biodiversity-related conventions and international initiatives. Since its establishment in 2012, IPBES has brought together more than a thousand scientists and knowledge-holders from around the globe to integrate knowledge systems from multiple disciplines. The main IPBES products and deliverables are assessments, which synthesise scientific findings and evidence on biodiversity change and its impact on human well-being to inform policy decisions.

One of the first IPBES assessments, the IPBES pollination assessment (IPBES, 2016) has made a significant global impact on policy development. For instance at the 13th Conference of the Parties to the Convention on Biological Diversity in Mexico in 2016 (CBD COP13), a COP decision recognised its relevance for the planned fifth edition of the Global Biodiversity Outlook and listed it among the best available scientific information. The COP also encouraged parties, other governments, relevant organisations, the scientific community and stakeholders, as well as indigenous peoples and local communities, to develop and use these tools and contribute to their further development (CBD, 2016a). The pollination assessment provides a best-practice 'toolkit' of the approaches that can be used to decide policies and actions by governments, the private sector and civil society. Different valuation methodologies are evaluated according to different visions, approaches and knowledge systems, as well as their policy relevance, based on the diverse conceptualisation of values of biodiversity and nature's benefits to people, including provisioning, regulating and cultural services. As such, this assessment has generated a wide range of follow-up products, actions and policy initiatives, including the following.

- A formal endorsement of the key messages of the assessment by the parties to the CBD at the 13th Conference of the Parties (COP13) in Mexico (CBD, 2016b).
- The formation of a 'Coalition of the Willing' by a growing number of governments around the world, inspired by the assessment to act nationally to protect pollinators and promote pollination (Promote pollinators, 2018).
- Publications in high-ranking scientific journals building on and reviewing the assessments (Potts et al., 2016; Díaz et al., 2018).
- An expanding list of national strategies and action plans on pollination in countries including, among others, Brazil, France, Germany, the Netherlands, the Republic of Korea and South Africa.

The IPBES scientific community also made significant contributions to the controversial discourse on the appropriateness of the ecosystem service concept and paved the way to reconciling differing views on conceptualisation of the human–nature relationship (Díaz et al., 2018; Stenseke & Larigauderie, 2018). It should be recognised, however, that the community will continue to use many different terms for ecosystem services or the contributions people receive from nature, depending on context, and this plurality should be welcomed (Peterson et al., 2018). Both the open-ended stakeholder network and the new concept of nature's contributions to people reflect the co-design and co-development aspects of IPBES as a learning organisation.

The challenges posed in IPBES are many, including a more balanced integration of scientists and experts from both natural and social sciences for a holistic understanding of biodiversity and its interactions with society and humanity (Jetzkowitz et al., 2018; Stenseke & Larigauderie, 2018). A more thorough consideration of, and improvement in, achieving the balance and quality of geographic, gender and disciplinary representations will be critical in filling the knowledge gaps and adding interdisciplinary value to the IPBES assessments (Obermeister, 2017; Heubach & Lambini, 2018). Moving forward, it will be important for IPBES to liaise with the private sector for greater impact on socially responsible and sustainable development, and with the public in disseminating scientific knowledge to promote changes in individual behaviour and decisions conscious of biodiversity conservation.

### 15.5.3 SDGs

It is crucial that progress in the implementation of the SDGs in national policy processes is adequately monitored (Hák et al., 2016; Reyers et al., 2017). To track the SDGs, the UN Statistics Commission has recommended over 230 official indicators, and countries are invited to submit voluntary national reviews of their progress to the High-Level Political Forum (Sachs et al., 2017). However, not all of the indicators have well-established definitions or data for all UN member states. A review of reports submitted so far (Bizikova & Pinter, 2017) found they were particularly weak on the environmental SDGs 12–15 (Sachs et al., 2017) and the assessment of interlinkages, synergies and trade-offs between targets (Allen et al., 2018). The evaluation of SDGs and tracking the progress to their achievement requires holistic scientific approaches to better understand the linkages between the SDGs and their underlying challenges, to understand thresholds, rebound effects and tipping points, and to explain the benefits and trade-offs of a range of development pathways that could lead to a more sustainable global society (Schmalzbauer & Visbeck, 2016).

The IPBES community of scientists can also provide best expert knowledge and scientific evidence for the sustainable development of the planet to inform the SDGs. For example, the recent IPBES assessment of land degradation and restoration (IPBES, 2018c) mapped the relevance of land degradation against the SDG goals. This may help to mainstream biodiversity across sectors and societies and bring forth synergies between global initiatives. A well-functioning knowledge generation mechanism connecting scientific and policy bodies of the platform will be particularly important if IPBES is to become an effective catalyst and orchestrator of harmonised science, policy and practice for better conservation.

## 15.6  Achievements of the CBD, IPBES and SGDs

There are several developments at the national level that can directly be traced to the CBD, such as the adoption of National Biodiversity Strategies and Action Plans in 185 countries of the world (as of December 2018, according to the CBD website). Other examples of direct influence of the CBD on its member states are the national regulations that parties have adopted to comply with the provisions of the two Protocols that have arisen from the CBD: the Cartagena Protocol on biosafety and the Nagoya Protocol on Access and Benefit Sharing. However, the CBD's influence on biodiversity governance at the national scale still appears limited. This is partly due to the power imbalances that exist among global institutions, and strong global forces that prioritise economic considerations over nature conservation, as well as power relations and societal preferences at the national scale. Furthermore, the fact that the CBD lacks a non-compliance mechanism may further weaken its influence.

Nonetheless, the CBD has provided inspiration to a great variety of state and non-state actors to initiate conservation actions. For example, the Aichi Biodiversity Targets (included in the Strategic Plan of the CBD for the period 2011–2020) have sparked debates and research on biodiversity-related questions and serve as important reference points in calls for greater efforts in nature conservation (e.g. they are often referred to by non-governmental organisations). These Targets, along with the UN Decade on Biodiversity with the same timeframe (2011–2020), have also inspired numerous actions on the ground, as documented on the CBD website (www.cbd.int/2011-2020/). Furthermore, the CBD mobilises resources and may provide finances to developing countries for the purpose of implementing the Convention (e.g. via the Global Environment Facility).

An important area where the CBD and SDGs exert influence is through fostering collaborations, between different biodiversity-related conventions and among relevant organisations and stakeholder groups at all subglobal scales. Alongside IPBES, they have also raised awareness of the values of biodiversity and their integration in other societal goals.

## 15.7  What next – how to engage?

As demonstrated, the past decades have seen an alignment of biodiversity-related agendas with different sectoral policies. Now the Aichi Biodiversity Targets and the SDGs need an increased implementation effort to deliver tangible results. In the national policy context this hinges on ensuring consistency within and between these two agendas and other political processes, effective governance systems, institutions and partnerships, and intellectual and financial resources

(ICSU, 2017). Scientists can – jointly with societal and policy actors – help to provide supporting evidence (see also Schmalzbauer & Visbeck, 2016):

- to build new partnerships across disciplines, to engage different knowledge domains and thereby foster innovation;
- to develop problem- and solution-oriented metrics, tools and indicators to aid the process of continuous learning and adaptive management;
- to provide open-source and open-access data and infrastructure to share knowledge and good practice;
- to conduct economic, social and health cost–benefit analyses to assess joint action versus silo approaches;
- to assist forecasting and informed decision-making through scenarios and models.

In order to maximise the impact of science in society through international conventions, national policies and local implementations, scientists can:

- address conservation questions in their own research and proactively enhance the transferability of research results as evidence for real-world application;
- actively engage with government agencies, NGOs and the public to learn about their knowledge needs, the ongoing political processes and the mode of operation, to enhance the societal relevance of their own research and better frame and communicate own research findings in a policy context (see Chapters 10 and 13);
- attend meetings of CBD, SDG, IPBES and other relevant conventions and initiatives as experts, observers, stakeholders or delegations through the channels of organisations and countries;
- proactively engage as authors or reviewers in IPBES assessments or other science–policy reports and contribute scientific evidence throughout the process, even if not a formal contributing author. IPBES has open calls and is open for engagement on many levels;
- develop transdisciplinary research collaborations and networks with experts from agencies, NGOs and other civic organisations.

This engagement at the science–policy interface requires time, openness and willingness for true collaboration between scientists, policy advisors and practitioners. While not always easy in short-term research funding circles, this can be very rewarding for everyone involved. Overall, conservation can only move forward when aligned with other policy goals and through integral support of all disciplines and all sectors to work for 'People and Nature'.

## 15.8 Acknowledgements
The authors wish to thank UFZ and iDiv colleagues for inspiring discussions and the German Network Forum for Biodiversity Research (NEFO, FKZ 01LC0831 A2) for support.

# References

Allen, C., Metternicht, G. & Wiedmann, T. 2018. Initial progress in implementing the Sustainable Development Goals (SDGs): a review of evidence from countries. *Sustainability Science*, 13, 1453–1467.

Bizikova, L. & Pinter, L. 2017. Indicator preferences in national reporting of progress towards the Sustainable Development Goals. International Institute for Sustainable Development Briefing Note.

Brooks, T. M., Lamoreux, J. F. & Soberón, J. 2014. IPBES ≠ IPCC. *Trends in Ecology & Evolution*, 29, 543–545.

Brundtland Commission. 1987. *World Commission on Environment and Development: Our Common Future*. New York, NY: Oxford University Press.

Butchart, S. H. M., Walpole, M., Collen, B., et al. 2010. Global biodiversity: indicators of recent declines. *Science*, 328, 1164–1168.

Carpenter, S., Mooney, H., Agard, J., et al. 2009. Science for managing ecosystem services: beyond the Millennium Ecosystem Assessment. *Proceedings of the National Academy of Sciences of the USA*, 106, 1305–1312.

CBD. 2016a. Global Biodiversity Outlook and Intergovernmental Science–Policy Platform on Biodiversity and Ecosystem Services; Document CBD/COP/DEC/XIII/29. Available from www.cbd.int/doc/decisions/cop-13/cop-13-dec-29-en.pdf.

CBD. 2016b. Implications of the IPBES Assessment on pollinators, pollination and food production for the work of the Convention CBD/COP/DEC/XIII/15. Available from www.cbd.int/doc/decisions/cop-13/cop-13-dec-15-en.pdf.

CBD, FAO, WB, UNEP & UNDP. 2017. Biodiversity and the 2030 Agenda for Sustainable Development, Technical Note. Convention on Biological Diversity; Food and Agriculture Organization of the United Nations; The World Bank; UN Environment; United Nations

Development Programme. Available from www.cbd.int/development/doc/biodiversity-2030-agenda-technical-note-en.pdf.

Colchester, M. 2004. Conservation policy and indigenous peoples. *Environmental Science & Policy*, 7, 145–153.

Daily, G. C. 1997. *Nature's Services: Societal Dependence on Natural Ecosystems*. Washinton, DC: Island Press.

Díaz, S., Demissew, S., Joly, C., et al. 2015. A Rosetta Stone for nature's benefits to people. *PLoS Biology*, 13, e1002040.

Díaz, S., Pascual, U., Stenseke, M., et al. 2018. Assessing nature's contributions to people. *Science*, 359, 270–272.

Dirzo, R., Young, H. S., Galetti, M., et al. 2014. Defaunation in the Anthropocene. *Science*, 345, 401–406.

Gaston, K. J., Jackson, S. F., Cantú-Salazar, L., et al. 2008. The ecological performance of protected areas. *Annual Review of Ecology, Evolution, and Systematics*, 39, 93–113.

Hák, T., Janoušková, S. & Moldan, B. 2016. Sustainable Development Goals: a need for relevant indicators. *Ecological Indicators*, 60, 565–573.

Heubach, K. & Lambini, C. K. 2018. Distribution and selection of experts in the intergovernmental science–policy platform on biodiversity and ecosystem services (IPBES): the case of the regional assessment for Africa. *Innovation: The European Journal of Social Science Research*, 31, S61–S77.

ICSU. 2017. Guide to SDG Interactions: From Science to Implementation. International Council for Science, Paris, France. Available from https://council.science/cms/2017/05/SDGs-Guide-to-Interactions.pdf.

IPBES. 2014. Report of the second session of the Plenary of the Intergovernmental Science-Policy Platform on Biodiversity and Ecosystem Services; Document IPBES/2/17. Available from www.ipbes.net/system/tdf/downloads/IPBES_2_17_en_0.pdf?file=1%26type=node%26id=14621.

IPBES. 2016. *The Assessment Report on Pollinators, Pollination and Food Production of the Intergovernmental Science–Policy Platform on Biodiversity and Ecosystem Services* (edited by S. G. Potts, V. Imperatriz-Fonseca & H. Ngo). Bonn: IPBES Secretariat.

IPBES. 2018a. Information on collaboration and partnerships; Document IPBES/6/INF/21. Available from www.ipbes.net/system/tdf/ipbes-6-inf-21_-_re-issued.pdf?file=1%26type=node%26id=16534.

IPBES. 2018b. IPBES Science and Policy for People and Nature. Available from www.ipbes.net/

IPBES. 2018c. *Thematic Assessment Report on Land Degradation and Restoration of the Intergovernmental Science–Policy Platform on Biodiversity and Ecosystem Services* (edited by R. Scholes, L. Montanarella, A. Brainich, et al.). Bonn: IPBES Secretariat. Available from www.ipbes.net/assessment-reports/ldr.

Jetzkowitz, J., van Koppen, C., Lidskog, R., et al. 2018. The significance of meaning. Why IPBES needs the social sciences and humanities. *Innovation: The European Journal of Social Science Research*, 31, S38–S60.

Kanie, N. 2014. Governance with multilateral environmental agreements: a healthy or ill-equipped fragmentation? In Conca, K. & Dabelko, G. D., editors, *Green Planet Blues: Critical Perspectives on Global Environmental Politics* (pp. 67–86). London: Hachette.

Kemp, L. 2016. Framework for the future? Exploring the possibility of majority voting in the climate negotiations. *International Environmental Agreements: Politics, Law and Economics*, 16, 757–779.

Leadley, P. W., Krug, C. B., Alkemade, R., et al. 2014. Progress towards the Aichi Biodiversity Targets: An assessment of biodiversity trends, policy scenarios and key actions. Global Biodiversity Outlook 4 (GBO-4) Technical Report. CBD Technical Series No. 78. Secretariat of the Convention on Biological Diversity, Montreal, Canada. Available from www.cbd.int/GBO4.

MA. 2005. *Millenium Ecosystem Assessment*. Washington, DC: Island Press.

Mace, G. M. 2014. Whose conservation? *Science*, 345, 1558–1560.

McArthur, J. W. 2014. The origins of the millennium development goals. *SAIS Review of International Affairs*, 34, 5–24.

Mora, C. & Sale, P. F. 2011. Ongoing global biodiversity loss and the need to move beyond protected areas: a review of the technical and practical shortcomings of protected areas on land and sea. *Marine Ecology Progress Series*, 434, 251–266.

O'Connor, D., Mackie, J., Van Esveld, D., et al. 2016. Universality, integration, and policy coherence for sustainable development: early SDG implementation in selected OECD countries. Working Paper. World Resources Institute, Washington, DC. Available from www.wri.org/publication/universality_integration_and_policy_coherence.

Obermeister, N. 2017. From dichotomy to duality: addressing interdisciplinary epistemological barriers to inclusive knowledge governance in global environmental assessments. *Environmental Science & Policy*, 68, 80–86.

Peterson, G. D., Harmáčková, Z. V., Meacham, M., et al. 2018. Welcoming different perspectives in IPBES: "Nature's contributions to people" and "Ecosystem services". *Ecology and Society*, 23(1), Art. 39.

Potts, S. G., Imperatriz-Fonseca, V., Ngo, H. T., et al. 2016. Safeguarding pollinators and their values to human well-being. *Nature*, 540, 220.

Promote pollinators. 2018. Promote pollinators: Coalition of the Willing on Pollinators. Available from https://promotepollinators.org.

Reyers, B., Stafford-Smith, M., Erb, K.-H., et al. 2017. Essential variables help to focus sustainable development goals monitoring. *Current Opinion in Environmental Sustainability*, 26, 97–105.

Sachs, J., Schmidt-Traub, G., Kroll, C., et al. 2017. *SDG Index and Dashboards Report 2017*. New York, NY: Bertelsmann Stiftung and Sustainable Development Solutions Network (SDSN).

Scharlemann, J. P., Mant, R. C., Balfour, N., et al. 2016. *Global Goals Mapping: The Environment–Human Landscape. A Contribution towards the NERC, The Rockefeller Foundation and ESRC initiative, Towards a Sustainable Earth: Environment–Human Systems and the UN Global Goals*. Brighton:Sussex Sustainability Research Programme, University of Sussex, andCambridge:UN Environment World Conservation Monitoring Centre.

Schmalzbauer, B. & Visbeck, M. 2016. *The Contribution of Science in Implementing the Sustainable Development Goals*. Biological Conservation. Stuttgart/Kiel: German Committee Future Earth,

Schmidt-Traub, G., Kroll, C., Teksoz, K., et al. 2017. National baselines for the Sustainable Development Goals assessed in the SDG Index and Dashboards. *Nature Geoscience*, 10, 547–555.

Schröter, M., Albert, C., Marques, A., et al. 2016. National Ecosystem Assessments in Europe: a review. *Bioscience*, 66, 813–828.

SRC. 2016. *The 2030 Agenda and Ecosystems: A Discussion Paper on the Links between the Aichi Biodiversity Targets and the Sustainable Development Goals*. Stockholm: Stockholm Resilience Centre.

Stenseke, M. & Larigauderie, A. 2018. The role, importance and challenges of social sciences and humanities in the work of the intergovernmental science–policy platform on biodiversity and ecosystem services (IPBES). *Innovation: The European Journal of Social Science Research*, 31, S10–S14.

Tittensor, D. P., Walpole, M., Hill, S. L., et al. 2014. A mid-term analysis of progress toward international biodiversity targets. *Science*, 346, 241–244.

UN. 1992. *Rio Declaration on Environment and Development*. Rio: United Nations.

UN. 2015. Transforming our world: The 2030 agenda for sustainable development. A/RES/70/1. United Nations. Availabel from https://sustainabledevelopment.un.org/post2015/transformingourworld.

UNEP. 2012. Report of the Second Session of the Plenary Meeting to Determine Modalities and Institutional Arrangements for an Intergovernmental Science–policy Platform on Biodiversity and Ecosystem Services; Document UNEP/IPBES.MI/2/9. Available from www.ipbes.net/document-library-catalogue/unepipbesmi29.

Vadrot, A. B., Rankovic, A., Lapeyre, R., et al. 2018. Why are social sciences and humanities needed in the works of IPBES? A systematic review of the literature. *Innovation: The European Journal of Social Science Research*, 31, S78–S100.

# PART III

# Communicating the message

# CHAPTER SIXTEEN

# Citizens and science: media, communication and conservation

LIBBY LESTER

*University of Tasmania*

and

KERRIE FOXWELL-NORTON

*Griffith University*

## 16.1 Introduction

In 2016 a full-page advertisement was placed by 56 Australian scientists in the Brisbane *Courier Mail*. The context of the advertisement was the continuing commitment of Australian governments, federal and state, to coal mining and coal-fired power stations despite overwhelming evidence connecting this activity to the severe damage being suffered by the Great Barrier Reef (Hoegh-Guldberg, 2015). As well as presenting their scientific credentials in the advertisement – together they had devoted more than 1200 years to studying climate change, marine ecosystems and the Great Barrier Reef – the scientists prioritised the Reef's economic value over its conservation values. The burning of fossil fuels, they wrote, is 'directly threatening a major economic resource. The World Heritage listed Great Barrier Reef earns multiple billions for the economy and provides jobs to tens of thousands of Australians' (*Courier Mail*, 2016). '[T]here can be no new coal mines ...', the scientists demanded, and 'No new coal-fired power stations'.

This attempt to influence public opinion and thus political outcomes through media appeared in the face of what is now recognised as one of the world's most notable failures in conservation: the continuing destruction of a global nature 'superstar'. We suggest in this chapter that such public acts are often rendered futile because of a poor understanding of the communicative processes underpinning the research-to-policy pathway. This is troubling given the risks some scientists – working within expectations of independence and measured professional response – take when entering public debate. But this is only part of the story. While many scientists do not have the necessary communication skills or knowledge to join controversial debates (Besley & Tanner, 2011) or have been burned by previous experience (Dunwoody, 2015), there is also evidence that others see themselves as remote from the public sphere, a messy space of negotiation and contest that has a clearly troubled relationship with fact (Besley & Nisbet, 2013; Dudo & Besley, 2016; Simis et al., 2016).

In this chapter, we highlight aspects of this disconnection between environmental science and public debate and policy outcomes from a media and communication perspective. We begin by briefly outlining recent approaches to mediated environmental communication. We then turn to the communication of science more specifically. We argue that models of science communication and public engagement with science need to more explicitly acknowledge issues of power, complexity and conflict within the context of the contemporary media landscape. To conclude, we offer suggestions for how science and communication can be better equipped to influence environmental debate and decision-making.

## 16.2 Mediated environmental communication

As a starting point, we need to recognise the inherently political nature of environmental and conservation sciences – that even at their least political, they seek to influence behaviours and outcomes, and at their most political they are resisting global pressures for intensified use of land and water and increasing demand for and movement of resources. The politics of the environment consistently test our capacity to civilly negotiate a shared future (Cox, 2012; Dryzek, 2013), whether that concerns the composition of our atmosphere or the fate of a small localised fishery (Murphy, 2017). That environmental activists and journalists are greater targets of violence than ever before in many parts of the world is evidence not only that resource management and conservation are areas of conflict, but that what is said, how and to whom clearly matters (Cottle et al., 2016; Lester, 2017). Media and communication are central to this flow or containment of environmental information and meanings. As such, here we briefly outline key ideas from communication and media studies as they relate to environmental debate and decision-making.

As others before them, media and communication scholars have turned to nature for useful metaphors to help describe some of the dynamism and complexity they now witness. 'Media ecology' is a popular term to capture the interconnection of various media systems, platforms, technologies, genres, formats, and producer and audience practices driving media production and distribution (Altheide, 1994; Singer, 2018). How, and to what extent, this metaphor should be applied remains contested (Maxwell & Miller, 2012; Lester, 2019). Nevertheless, a focus on interconnectivity within media and communication is useful in highlighting the interactions and dynamism of contemporary spheres for public and political negotiation (Habermas, 1989; Fraser, 2007).

An immediate outcome of applying this metaphor is the redundancy of the definite article in relation to 'media'. Once it may have made sense to refer to 'the media' as a bounded entity, in which media companies hired journalists, editors and camera operators to produce information in the form of news and entertainment that was circulated via newspapers and broadcast outlets to readers and viewers. Now, the use of 'the' in front of 'media' is as anomalous as it would be if used in front of 'nature'. Media are no longer separable from our social lives or indeed our environmental futures (Deuze, 2012). Media shape and frame our everyday life, including political decisions. They are the principal means through which we form a shared understanding of the world and come together to debate and negotiate common risks and concerns.

A second outcome of recognising ecological-type interconnectivity within a media and communication context is the acknowledgement of interaction. It is almost impossible to isolate environmental concerns and risks and the decisions they prompt to a defined locality. When residents in Mackay, Queensland, protested against the impacts of the proposed port expansion on the Great Barrier Reef, they entered a world that stretched communicatively from their local newspaper, to a series of NGO-established hashtags, to transnational corporations that sell ice cream, to European banks, to a US president and his daughters, to international governance bodies (Lester, 2016; Foxwell-Norton & Lester, 2017). And back again. Claims by industry of a 'social licence to operate' can be challenged when an 'affected public' is no longer defined as those living within a 20-km radius of a development site. We might all consider ourselves affected when the future of the Great Barrier Reef is concerned, and media and communication provide us with the means of engaging, and the sense that we have a right and duty to be involved openly in decisions about its future (Volkmer, 2014).

Dynamism is the third element to be considered. As the traditional business model for the production of news has collapsed, numerous other forms of information production and circulation have emerged. All are constantly adjusting and changing their practices in relation to one another. NGOs collate and publish information on illegal logging in places where it is now too dangerous or expensive for income-losing news organisations to send their journalists. Citizens establish community websites for local audiences or single-issue blogs for targeted business readers. News outlets campaign on climate change to attract subscribers, or do not cover climate change at all if it attracts too few site visits. Other media outlets closely guard a political and/or conservative readership, muscling out potential competitors with tactics sometimes bordering on bullying, in order to maintain a reputation for political influence (McKnight, 2012). Meanwhile, audiences have more choices than ever on what news they will receive and via what platform, self-selecting, re-selecting and screening sources, topics and subject matter via news feeds, hashtags and new sites selection.

Power plays a key role in structuring this interconnected, interactive and dynamic system. Within media and communication, power appears in diverse and often surprising forms, and even ownership of mega-media companies is no guarantee of uninterrupted influence, as both Rupert Murdoch and Mark Zuckerberg have experienced. Power is never certain, although it holds true that some conditions enhance the capacity to control information as it travels. Information emanating from institutional settings, such as universities, scientific organisations, courts, parliaments or international governance bodies, can often travel with authority for longer than NGO-sponsored communications. However, the long-running clash in the Southern Ocean between the NGO, the Sea Shepherd Conservation Society, and the Japanese government-backed whaling fleet provides an excellent example of how geography impacts this. Throughout much of the conflict, Sea Shepherd was able to capitalise on the remote location of the conflict, from which journalists were absent, by producing and distributing images and messages that circulated within media relatively unchallenged. Symbolic power is key here. No amount of Japanese government-sponsored public relations or 'scientific knowledge' was able to successfully counter the messages carried by the bloodied corpses of 'charismatic megafauna' (McHendry, 2012; Cox & Schwarze, 2015).

Environmental NGOs have pioneered the strategic management of symbolic power within media and communication, and here conflict is often a necessary component. Sophisticated multi-pronged campaigns with minimal financial resources have threatened and interrupted the multimillion-dollar flow of goods and capital. The campaign aimed at Japanese buyers of Tasmanian native timbers involved a young woman in a tree with a laptop and a daily blog (albeit for over a year); a string of social media-active international backpackers and celebrity visitors; a single campaigner in Japan translating various media texts; and access to the email addresses of key corporate and social responsibility personnel in relevant Japanese companies (Lester, 2014). The Sarawak-based forestry company at the centre of the trade quickly altered its business practices in Tasmania once the Japanese companies withdrew from contracts rather than be seen to be failing to meet their own environmental procurement principles.

This terrain is media saturated, and the role of media and communication is more than mere conduits for data or messages. Modern environmental conflict is hugely influenced by media, as the 'product of mutually constitutive inter-actions between activism, journalism, formal politics, and industry' (Hutchins & Lester, 2015, p. 339) enacted in the public sphere. Activists' strategies and campaigns, journalistic practices and news reporting, formal politics and decision-making processes, and industry activities and trade coalesce to enact moments of environmental conflict in public view. These moments of conflict largely centre on the legitimate dimensions of local, national and international policy and law, underpinned by the pursuit of environmentally sustainable development (Konkes, 2018; Foxwell-Norton & Konkes, 2019).

For example, state, NGO and industry responses to Japanese whaling conflicts in the southern oceans drew heavily upon the duties of signatories to the International Convention for the Regulation of Whaling, that for over 30 years has delivered a commercial whaling moratorium. Sea Shepherd undertook protest action, with international laws and policy aiming to deliver whale conservation underpinning its media-based efforts, holding nations and industries to institutional and public account. Science was used both to support conservation via the International Whaling Commission (IWC) and to challenge it via the research claims of Japanese whaling fleets. Meanwhile, the IWC's pursuit of conservation management plans, sanctuaries and marine parks has been underpinned by science that seeks to balance whale populations with the impacts of industry, even when not explicit. Science and scientific knowledge are thus very much a part of these conflicts, powerful, contested factors in contemporary social relations.

Media and communication form an interconnected, interactive and dynamic system, in which power, conflict and threat to established practices and order are always evident. As with any complex ecology, this is delicately balanced and easily interrupted, constantly adjusting and shifting as its component parts struggle for sustainability and/or dominance. They remain integral to the formation of public opinion and the political influence that follows, but contemporary flows and networks of information make the paths from source to policy more difficult to predict than ever. In the next section, we contrast this view of media and communication with that circulating around environmental sciences.

### 16.3 Communicating environmental sciences

If the view we have presented of media and communication is of a highly political, dynamic and complex system – one that is central to social life and environmental decision-making, but that does not easily lend itself to being understood or charted via neat models – the environmental sciences can present a near opposite view. Communication here is often an add-on activity, and 'the media' considered a relatively stable platform or tool to deploy as needed in order to change public opinion and produce policy outcomes. Indeed, a key premise in recent literature is the idea of 'protecting science communication' from the dynamism and noise characteristic of public debate and controversy, and of an active separation of science communication from political communication (Hall Jamieson, 2017; Kahan et al., 2017). Here, 'science and its communication' rather than 'communication and its implications for science' has underpinned scholarship, leaving science seemingly remote from, rather than a part of, the public.

In considering how this situation has developed, we turn to a subset of literature that is not so interested in public understanding of science as scientists' understanding of 'the public'. In a review of findings from surveys of scientists, Besley and Nisbet (2011) found that, when asked about the role of the public, 'scientists may opt for some type of co-decision-making but also suggest a desire by scientists to differentiate themselves from the public'. Their relevant findings include the following.

- Scientists say the main barrier to 'greater understanding of science' among the public is lack of education. Media are second.
- Scientists see the public as homogenous – although experience interacting with the public can bring a more nuanced view. Scientists perceive policy-makers as the most important group with which to engage, with the public in the mid-range of importance – somewhat more important than young people or NGOs, but less important than the private sector and educators.
- Scientists appear to rely on a simple sender–receiver model of media effects that fits poorly with contemporary media research, that is, they 'tend to favour one-way communication with the public via the media, viewing engagement as chiefly about dissemination rather than dialogue' (Besley & Nisbet, 2011, p. 653).

Overall, scientists are willing to engage directly with citizens but 'such engagement is usually still framed in terms of providing information' 'to increase citizen knowledge' (Besley & Nisbet, 2011), while addressing the knowledge deficit and/or 'scientific literacy' still dominates scientists' communication goals (Peters & Dunwoody, 2016).

This transmission model of communication (Shannon & Weaver, 1949) – underpinned by a desire for a clear channel of communication that protects the message on its route from sender to receiver – has serious implications for public understanding, awareness and/or engagement with conservation and other sciences. It epitomises frustrated attempts to eliminate 'noise' – that is, to control the 'message' on a path to the public or policy and decision-makers. In the case of science, and more specifically conservation and ecology, the greatest 'noise' is the sound that resonates in the public sphere when citizens and scientific expertise collide. Exploring this noise requires a thoughtful and critical examination of the structural characteristics of this collision, and how this may impact the passage of scientific knowledge to citizens. This is difficult work, occurring in a space where diverse publics and communities with a range of understandings about scientific expertise and/or the primacy of economic imperatives reside.

Instead, a range of contexts, influences and often conflict await the path of scientific knowledge to the public. Public understandings of science cannot be divorced from these social processes, and a 'pure and protected' science message, unsullied by politics, is unlikely to arrive untouched at its destination audience.

Citizens enter the public communication of science as social, political and cultural beings with a range of historical and contextual nuances. The underlying assumption of communication as mere transmission of data – as a controllable process – will often fail to register the impacts sought and may act to reinforce the communicative distance between scientific expertise and the citizens to whom their message is directed. While some effort has been made to abandon communication models that are based upon 'knowledge deficit', the model is still evident in many attempts to distribute scientific research and findings to the public. A carefully crafted tweet, a multimillion-dollar documentary or a full-page advertisement framed by 1200 years of expertise and experience of Great Barrier Reef scientists or equivalent is communication that often underestimates the conditions within which these citizens reside. What is heard by the public can be quite distant from the sender's intent.

## 16.4  Better conservation communication

We suggest some key strategies that might help in the communication of conservation. The starting point must be a consciousness of one's own role – a critical self-reflexivity – that positions science and its communication as only one of many domains of legitimacy and authority in conservation debates and efforts. There are other sources that carry legitimacy and authority in the public and private lives of individuals, institutions and their societies and these also command a place in public communication about conservation. This 'communication noise' cannot be bypassed and is indeed a distinctive characteristic of the current era. When conservation science enters this messy sphere of debate, it becomes enmeshed in the public realm of politics and political communication. Efforts to 'secure' a message to an audience, even via the expensive production of one's own media content, underestimate communication's complexity and unstable networks of connectivity. Seeking innovative collaborations with communication scholars, and inviting their meaningful participation in the constitution and design of research projects, is one way in which conservation scientists might better prepare their work for public deliberations.

Popular messages are not necessarily wedded to scientific rigour, expertise or fact. In the twenty-first century, scientists are encouraged to communicate their knowledge widely, making it increasingly susceptible to challenge and disrepute. An understanding of how science is embedded and implicated in processes of public debate and negotiation may reorient these communication strategies. For example, by prioritising the scientific and economic imperatives to protect the Reef, as evident in our opening example, the scientists could actually have affirmed the powerlessness of the public in relation to the destruction of the Reef, especially when even experts are compelled to take out full-page advertisements in a state newspaper. Conversely, communicating the Reef as a scientific fact and an economic resource may alienate already marginalised public sentiments that do not prioritise this message in their own experience of or relationship with the Reef.

Further, when scientific messages are framed with deliberate reference to the 'economy', including the tourism and mining industries, the impacts of mining and tourism on the Great Barrier Reef and the science are (again) diluted by a perhaps unwitting collusion with industry – as has been repeated in the history of Reef policy and protest moments (see Foxwell-Norton & Lester, 2017; Foxwell-Norton & Konkes, 2019). Conservation science may do better to elevate the impact on the Reef's ecology, and return to its messages of connectedness between human and natural systems. Is the Reef not worth protecting in itself? In the 1960s, the emergent discipline of ecology was evoked to argue that a mining lease on one part of the Reef would have dire consequences for the entire Reef ecosystem (McCalman, 2013). This ecological approach requires ongoing critical reflection on the concept of 'ecologically sustainable development' and the relationship of research to a system of industrial development that threatens ecologies everywhere (Redclift, 2005). Suffice to say, much public trust in science is at stake in these reflections.

In the longer term, better conservation communication can also be fostered in training and development. The distance between the 'two cultures' or, more specifically, the humanities, arts and social sciences and that of the science, technology, engineering and mathematics disciplines, is shrinking, but not fast enough. Clearly, neither 'culture' alone is sufficient to arrest the current trajectory of ecological decline. As researchers, we must continue to challenge false dichotomies that diminish scholarly contributions to conservation efforts – from global superstar ecologies like the Great Barrier Reef to the local ecologies of the places we live (Foxwell-Norton, 2018). This distance can also be lessened in the design of degree programmes and training courses, giving current and next-generation science communicators access to different ways of thinking about their role, their potential place in public sphere debate, and the public.

In the twenty-first century, where networks of communication link individuals and civic institutions through digital media and mobile communication, a sophisticated understanding of communication is power (Castells, 2013). Communication scholars are well-equipped to assist scientists, and their disciplinary communicators, to extend existing understanding of communication, media and journalism. This entails a re-examination of what is meant by 'science communication' and its current strategies to engage citizens in support for, and trust in, its work and expertise. Currently, such collaborations overwhelmingly favour scientific expertise, leaving communication expertise (beyond media industry experience or production expertise) underrepresented, despite its potential to add critical dimensions to scientific research and projects. Deeper collaborations could better explore the challenges and capitalise on the opportunities that emerge where communication is pervasive, ubiquitous and complex.

## 16.5 Real 'citizen science'?

In liberal democratic societies, science enters the public sphere of debate with a menagerie of mitigating concessions and qualifications. Conservation ecology and science communication that seek to engage the public cannot be protected from these complexities: they are *sine qua non* to human societies. Communication between science and citizens in the twenty-first century is further impacted by the complex, interconnected network of communication technologies, practices and transnational flows characteristic of the modern experience. The public sphere that scientific knowledge enters is not a level playing field for all participants. Even 'pure' science messages are exposed to the unevenness wrought by conflict involving power, wealth, industry and politics.

Our Reef scientists and the scientific community are clearly attuned to the power of media in addressing environmental conflict and the public, hence the advertisement. We have questioned, however, whether such a blunt tool underpinned by a transmission model of communication is likely to result in the protection of the Reef intended by these scientists. We assert that messages, even those that seemingly carry the credibility and authority of scientific expertise, are confused and contorted by 'communication noise'. This embeds science in the dirty politics of public sphere debate, rather than beyond the politics of knowledge, position and power. Early communication scholar John Dewey expressed these ideas at the turn of the twentieth century:

Society not only continues to exist *by* transmission, *by* communication, but it may be fairly said to exist *in* transmission, *in* communication. There is more than a verbal tie between the words common, community and communication. Men live in a community in virtue of the things they have in common; and communication is the way in which they come to possess things in common. What they must have in common in order to form a community or a society are aims, beliefs, aspirations, knowledge – a common understanding – like mindedness as the sociologists say. Such things cannot be passed physically from one thing to another like bricks; they cannot be shared as persons would share a pie by dividing it into physical pieces.

(Dewey, 1916)

Opportunities are repeatedly missed and frustration grows in part because communication is assumed, and the scientists' 'camera' faces out when what is needed is a science 'selfie' – a critical self-reflexivity capable of understanding not only the science but how science might be heard once it leaves the minds of experts and enters the community (Foxwell-Norton, 2018). Understanding this requires 'knowing thyself' as a product of a peculiar set of historical circumstances that have legitimised and given authority to scientific messages but also as part of the politics of the public sphere – where citizens (including scientists) reside and knowledges circulate. Citizens must be the target of science messages in order to shift voting behaviour for a politics that gives due reference and regard to best conservation practice. This is clearly, from a communication perspective, the terrain upon which the Reef scientists are operating, albeit unconsciously. The core problem is that science communication understands itself, and largely gathers its authority and legitimacy, by defining its terrain in terms of 'science' rather than communication.

Science communication is very clear about the merits of bringing science to society, but is found wanting in the reverse, of the importance of bringing society to science. This is a tragic flaw, especially relevant at the current juncture when communication networks mean science is everywhere, visible and not, elevated and undermined, in every moment in society. As a starting point, there are a few key strategies that can begin to mitigate against the repetition of the 'communication breakdowns'.

- Improve scientists' understanding of the ways in which their knowledges enter the public sphere of political debate and the politicised nature of their own knowledge.
- Acknowledge that conservation science is understood by the public in terms mostly not answerable to, or cognisant of, scientific rigour or research.
- Enter the arena of media-immersed environmental conflict willing to participate alongside and through other interests of politics and decision-making, including activist groups, industries and government.
- Accept there can be no divorce of any aspect of conservation science from these politics, as it hampers meaningful engagement between science and its publics.
- Take the 'scientific selfie in society' that shows the flaws, the unknowns and the occasional exhilaration.

A thorough and candid examination of the relations between citizens and scientists in a media-saturated society is, we suggest, extraordinarily hard science. It is, however, science that is critical to the development of new directions in the public communication of conservation science.

## 16.6 Acknowledgements

This chapter draws on research supported by the Australian Research Council's Discovery Program (DP150103454 'Transnational Environmental Campaigns in the Australia-Asian Region') and the Griffith Centre for Social and Cultural Research.

## References

Altheide, D. L. 1994. An ecology of communication: toward a mapping of the effective environment. *The Sociological Quarterly*, 35, 665–683.

Besley, J. C. & Nisbet, M. 2013. How scientists view the public, the media and the political process. *Public Understanding of Science*, 22, 644–659.

Besley, J. C. & Tanner, A. H. 2011. What science communication scholars think about training scientists to communicate. *Science Communication*, 33, 239–263.

Castells, M. 2013. *Communication Power*. Oxford: Oxford University Press.

Cottle, S., Sambrook, R. & Mosdell, N. 2016. *Reporting Dangerously: Journalist Killings, Intimidation and Security*. London: Palgrave Macmillan.

*Courier Mail*. 2016. Climate change is destroying our reefs. We must phase out coal. *Courier Mail*. 21 April. Brisbane. Available from www.climatecouncil.org.au/reefstatement

Cox, R. 2012. *Environmental Communication and the Public Sphere*. Los Angeles, CA: Sage.

Cox, R. & Schwarze, S. 2015. Strategies of environmental pressure groups and NGOs. In Hansen, A. & Cox, R., editors, *The Routledge Handbook of Environment and Communication* (pp. 73–85). Abingdon: Routledge.

Deuze, M. 2012. *Media Life*. Cambridge: Polity Press.

Dewey, J. 1916. *Democracy and Education*. New York, NY: Courier Corporation.

Dryzek, J. S. 2013. *The Politics of the Earth: Environmental Discourses*. Oxford: Oxford University Press.

Dudo, A. & Besley, J. C. 2016. Scientists' prioritization of communication objectives for public engagement. *PLoS ONE*, 11(2), e0148867.

Dunwoody, S. 2015. Environmental scientists and public communication. In Hansen, A. & Cox, R., editors, *The Routledge Handbook of Environment and Communication* (pp. 63–72). Abingdon: Routledge.

Foxwell-Norton, K. 2018. *Environmental Communication and Critical Coastal Policy: Communities, Culture and Nature*. London: Routledge.

Foxwell-Norton, K. & Konkes, C. 2019. The Great Barrier Reef: News media, policy and the politics of protection. *International Communication Gazette*, 81(3), 211–234.

Foxwell-Norton, K. & Lester, L. 2017. Saving the Great Barrier Reef from disaster, then and now. *Media, Culture & Society*, 39, 568–581.

Fraser, N. 2007. Transnationalizing the public sphere: on the legitimacy and efficacy of public opinion in a post-Westphalian world. *Theory, Culture & Society*, 24(4), 7–30.

Habermas, J. 1989. *The Structural Transformation of the Public Sphere*, trans. Thomas Burger. Cambridge, MA: MIT Press.

Hall Jamieson, K. 2017. The need for a science of science communication: communicating science's values and norms. In Hall-Jamieson, K., Kahan, D. & Scheufele, D. A., editors, *The Oxford Handbook of the Science of Science Communication*. Oxford: Oxford University Press.

Hoegh-Guldberg, O. 2015. Coal and climate change: a death sentence for the Great Barrier Reef. *The Conversation*, 20 May.

Hutchins, B. & Lester, L. 2015. Theorizing the enactment of mediatized environmental conflict. *International Communication Gazette*, 77, 337–358.

International Whaling Commission. 2018. https://iwc.int/history-and-purpose (accessed 14 December 2018).

Kahan, D. M., Scheufele, D. A. & Hall Jamieson, K. 2017. Introduction: why science communication? In Hall-Jamieson, K., Kahan, D. & Scheufele, D. A., editors, *The Oxford Handbook of the Science of Science Communication*. Oxford: Oxford University Press.

Konkes, C. 2018. Green lawfare: environmental public interest litigation and mediatized environmental conflict. *Environmental Communication*, 12, 191–203.

Lester, L. 2014. Transnational publics and environmental conflict in the Asian century. *Media International Australia*, 150(1), 167–178.

Lester, L. 2016. Containing spectacle in the transnational public sphere. *Environmental Communication*, 10, 791–802. doi:10.1080/17524032.2015.1127849.

Lester, L. 2017. Environment and human rights activism, journalism and 'The New War'. In Tumber, H. & Waisbord, S., editors, *The Routledge Companion to Media and Human Rights* (pp. 268–276). Abingdon: Routledge.

Lester, L. 2019. *Global Trade and Mediatised Environmental Conflict: The View from Here*. Switzerland: Palgrave Macmillan.

Maxwell, R. & Miller, T. 2012. *Greening the Media*. Oxford: Oxford University Press.

McCalman, I. 2013. *The Reef: A Passionate History from Cook to Climate Change*. Brunswick: Viking/Penguin Books.

McHendry Jr, G. F. 2012. Whale Wars and the axiomatization of image events on the public screen. *Environmental Communication: A Journal of Nature and Culture*, 6, 139–155.

McKnight, D. 2012. *Rupert Murdoch: An Investigation of Political Power*. Sydney: Allen & Unwin.

Murphy, P. 2017. *The Media Commons: Globalization and Environmental Discourses*. Champaign, IL: University of Illinois Press.

Peters, H. P. & Dunwoody, S. 2016. Scientific uncertainty in media content: introduction to this special issue. *Public Understanding of Science*, 25, 893–908.

Redclift, M. 2005. Sustainable development (1987–2005): an oxymoron comes of age. *Sustainable Development*, 13, 212–227.

Shannon, C. E. & Weaver, W. 1949. *A Mathematical Model of Communication*. Champaign, IL: University of Illinois Press.

Simis, M. J., Madden, H., Cacciatore, M. A., et al. 2016. The lure of rationality: why does the deficit model persist in science communication? *Public Understanding of Science*, 25, 400–414.

Singer, J. B. 2018. Transmission creep: media effects theories and journalism studies in a digital era. *Journalism Studies*, 19, 209–226.

Volkmer, I. 2014. *The Global Public Sphere: Public Communication in the Age of Reflective Interdependence*. Hoboken, NJ: John Wiley & Sons.

# CHAPTER SEVENTEEN

# Campaigning to bring about change

CATHY DEAN
*Save the Rhino International*

and

AMY HINSLEY
*University of Oxford*

## 17.1 Introduction

This chapter examines campaigning: what it is, when it is needed and who conducts campaigns. Drawing upon examples from the NGO conservation sector, we discuss how to plan and execute a campaign, and explore the different types of campaign: behaviour change, policy change and fundraising. Finally, we consider some of the potential pitfalls, including a lack of a strong evidence base, overstating claims of success, the introduction of bias, conflicting views of co-organising partners, the inappropriate use of emotion and the risk of unintended consequences.

## 17.2 What is campaigning?

Campaigning, also described as influencing or advocacy, is about creating a change. Whether the aim is to reduce trade in the horn of a threatened species of rhino, protect the habitat of a rare population of wild orchids, raise funds for a workshop or the ongoing costs of species monitoring, or change the law on the import of hunting trophies into a country, conservation NGOs campaign to create change. The desired change may be to address the root cause of a conservation problem, such as demand-reduction or behaviour-change campaigns, or the campaign may be focused only on mitigating the effects of a problem, as in the case of grants to improve law enforcement activities that prosecute wildlife traffickers. Some organisations may decide to focus on campaigning to tackle both the cause and the effect.

## 17.3 When is campaigning appropriate?

Campaigning can be appropriate in a diverse range of situations: from local to global issues, from high-profile to emerging conservation problems, from long-term to opportunist responses. While campaigning is often on high-profile and well-known conservation problems, it may also be used to mobilise or harness existing public support for less well-known or emerging issues, or to tackle issues with impacts at a global scale.

In a recent opportunistic, but highly effective example, several NGOs launched campaigns to urge the public and policy-makers to phase out single-use plastics after the high-profile BBC documentary *Blue Planet II*, screened in the UK in December 2017, highlighted the problem of plastic pollution in the world's oceans. The programme showed footage of a pilot whale cow carrying her dead calf for days, with the calf's death linked to the possibility of its mother's milk being poisoned with toxins accumulated through the food she had been eating. The combined messaging gained considerable public attention, and in April 2018 the UK Government launched a consultation to explore the possibilities of banning plastic straws and other single-use plastics. While this consultation follows on from other action to reduce plastic usage that took place before these campaigns, such as the introduction of charges for plastic bags in 2015, increased public pressure likely highlighted the issue as a priority at this time. Indeed, the then Environment Secretary Michael Gove reportedly stated that he had been moved by the BBC programme (Rawlinson, 2017). In addition, several large companies responded to pressure from consumers by pledging to reduce or phase out single-use plastics.

Campaigning can also be used to give a voice to those without one. NGOs focusing on humanitarian relief or disadvantaged groups of people will often tell the story of a single person as a microcosm of the wider issue. Conservation causes, whether endangered species or ecosystems, are not able to speak for themselves, and NGOs often use 'ambassador' animals, such as Sudan, the last male Northern white rhino (euthanised in March 2018 after experiencing an increasing number of age-related problems), which came to embody the long, sorry history of the doomed attempts to conserve the species. Sudan became the focus of numerous fundraising campaigns to generate income for assisted reproduction technologies to try to 'recreate' the subspecies.

Finally, campaigning is sometimes the only action possible, especially when the scale of the problem is large or cannot be addressed without state or international intervention (such as plastics in the ocean). One successful example took place in 2002, when campaigning by Project Seahorse played a central role in the listing of all seahorse species on Appendix II of the Convention on the International Trade in Endangered Species of Wild Flora and Fauna (CITES), meaning that international seahorse trade was regulated and monitored for the first time (Project Seahorse, 2018). Through policy recommendations informed by scientific research, Project Seahorse highlighted the huge scale of trade in seahorses and the threat to wild species that unregulated and unsustainable trade was posing. With up to 20 million seahorses traded annually, this listing represented an important step towards sustainability of this trade.

## 17.4  Who campaigns?

Campaigns can be created and delivered by individuals, groups or organisations, whether commercial or charitable. NGOs are particularly associated with campaigning; their fundamental objective is to make the world a better place, and they have members who feel strongly about the issue in hand. NGOs are often very close to their service users and beneficiaries, and can therefore use evidence from their direct experience to highlight changes needed, whether to attitudes, legislation or budgets. The examples in this chapter are drawn from the conservation NGO sector.

A common cause can bring together disparate voices to create a collective campaign that is louder, more wide-reaching and more effective than could be achieved by any single organisation. The campaign to create a marine reserve around the Pitcairn Islands began in 2011, when the Pew Environment Group's Global Ocean Legacy project first discussed with Pitcairn islanders the idea of establishing a large-scale marine reserve within their waters. A number of organisations and celebrities then became involved in the campaign, including the Great British Oceans Coalition, National Geographic, the Zoological Society of London, Hugh Fearnley-Whittingstall, Gillian Anderson, Julie Christie and Helena Bonham-Carter; the Pitcairn Island Marine Reserve was eventually legally designated in September 2016.

## 17.5  Planning a campaign

A well-designed campaign cycle will begin by analysing and selecting the issue, followed by developing the strategy, planning the campaign, delivering it, monitoring progress, evaluating impact and drawing out learning. More complex campaigns may research and develop different strategies and pilot them before conducting monitoring and evaluation on the different groups to determine the most effective strategy. They may begin by establishing an evidence base, developing a theory of change, and embedding within that the system of monitoring and evaluation, to include targets, indicators and means of verification.

Campaigns usually employ a call to action, which will differ depending on the target audience and the chosen goal. Such calls to action need to consider their target audiences. For example, a campaign to conserve water in Europe and the USA may ask people to turn off the tap while brushing their teeth, whereas a water conservation campaign in sub-Saharan Africa may ask farmers to introduce night-time drip irrigation for their crops to minimise evaporation.

If there is no budget previously set aside for the campaign, then funds need to be raised. Communications staff need to work on how to articulate the campaign's concepts and frame the debate. Finally, the organisation needs to be ready to implement the change, perhaps in partnership with others, with all the resources required, and to be able to manage that implementation without detracting from its ongoing work.

## 17.6  Types of campaigns

Campaigns generally fall into three categories: bringing about behaviour change, bringing about policy change, or raising funds. We consider each of these in turn and, for each category, we give an example of a successful campaign, seeking to highlight the aspects that, in our view, contributed to that success.

### 17.6.1  Campaigning to change behaviour

Many campaigns aim to change human behaviour, to reduce the incidence of behaviour that is in some way harmful to wildlife or ecosystems, or promote positive behaviour. Changing behaviour is different to raising awareness of an issue, which involves simply communicating the nature of a threat or conservation problem in the hope that the public or policy-makers will take action. Increasingly, the effectiveness of raising awareness in changing a person's behaviour is being questioned (Christiano & Neimand, 2017).

Greenpeace's palm oil campaign of 2010 (Greenpeace, 2010) targeted both the people buying Kit Kats and Nestlé, the manufacturer. A one-minute video shows a bored office worker shredding documents while watching the clock until 11:00 and his break. He tears open the wrapper of a Kit Kat. We, the viewer, see that the wafer finger is actually an orangutan's finger, complete with furry knuckle and nail. The chocolate bar drips into his keyboard; oblivious, he wipes his mouth and spreads a smear of blood. The video ends with a call to 'Stop Nestlé buying palm oil from companies that destroy the rainforests'. A link to Greenpeace's website, with suggestions for how concerned viewers could take action, was provided. Greenpeace reported 1.5 million views of the advert, more than 200,000 emails and phone calls to Nestlé HQ and countless comments posted on Facebook. This, combined with protests at Nestlé AGM and its headquarters all over the world, and meetings between Greenpeace campaigners and Nestlé executives, resulted in swift action. Nestlé developed a plan to identify and remove any companies in their supply chain with links to deforestation so their products would have 'no deforestation footprint', although it has been reported that they have since backtracked on these commitments (Neslen, 2017).

In a contrasting example, campaigns to increase consumer awareness of the impact of their purchases on overfishing, including labels for certified sustainable products, have been found to have little effect on purchasing choice or consumer demand (Jacquet & Pauly, 2007). Therefore, it is essential that behaviour-change campaigns go beyond simple awareness-raising and base their messages on sound research into when, where, how, why and by whom the behaviour is occurring.

Lynn Johnson has developed a useful pyramid (Figure 17.1) to show the difference between behaviour-change and awareness-raising campaigns. However, the majority of so-called behaviour-change campaigns actually operate at the awareness-raising level, rather than that at the demand-reduction level. Programme managers dealing with the direct consequences of poaching understandably must feel frustrated when they see substantial funds being invested in ineffective efforts to change consumer behaviour in the main consumer countries for illegal wildlife products.

Doug Mackenzie-Mohr (2011) has written extensively about fostering sustainable behaviours and has broken down the steps involved. The process starts by identifying which behaviour you want to change and in whom, while also considering when and where they exhibit this behaviour. The next step involves identifying what might be stopping people from changing

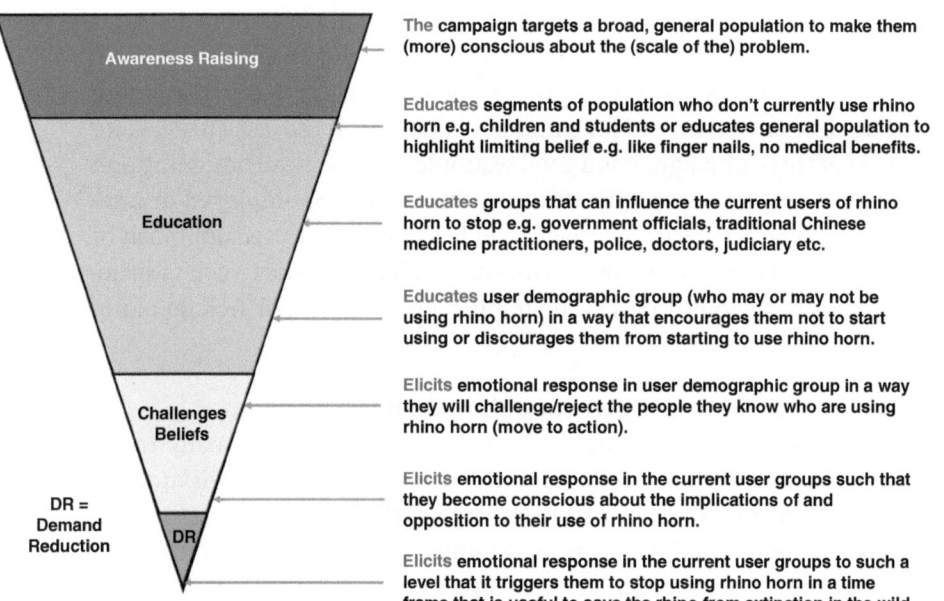

**Figure 17.1** Model showing differences between behaviour-change and awareness-raising campaigns developed by Nature Needs More Ltd for its Breaking The Brand RhiNo Campaign (Breaking The Brand, 2016). (A black and white version of this figure will appear in some formats. For the colour version, please refer to the plate section.)

behaviour, and what the incentives might be for doing so. This allows informed strategies to be developed that consider the design of the messaging but also other factors, such as how social norms can be used to reinforce the desired behaviour. These strategies should then be fully tested in a pilot phase before full-scale implementation, with monitoring and evaluation throughout.

Although behaviour-change campaigns focused on illegal products often suffer from a lack of available data on consumers, there are examples of targeted campaigns that have carefully planned their messages based on evidence. In 2014, TRAFFIC in Vietnam launched the Chi campaign, a behaviour-change campaign based on consumer research into the groups most likely to buy illegal rhino horn. This research established that the key driver for the consumption of rhino horn was its 'emotional' value rather than its 'functional' (i.e. medicinal) value and that the main users were wealthy businessmen aged between 35 and 50 living in Hanoi and Ho Chi Minh City (TRAFFIC, 2013). They valued the strength and power of the animal that had been killed to obtain it, but also the scarcity and high cost of rhino horn and the difficulty of obtaining it; being able to do so demonstrated the extent of the buyer's networks. Having segmented the consumer market, and with the information on the motivations of the prime target audience and the drivers of consumption, there was little point in launching a campaign that relied on photographs of traumatically dehorned rhinos, or on debunking beliefs that rhino horn could cleanse the body of toxins following chemotherapy. Instead, the campaign focused solely on the importance of 'Chi', an inner power and strength that negated the need for rhino horn. While it is too early to evaluate the success of this campaign, it is a good example of the careful designing and tailoring of messages to a specific situation that should be employed in campaigns of this type. Audience segmentation is a commonly used approach of subdividing populations into groups with shared characteristics, such as sociodemographic, behavioural or psychographic profiles (Wedel & Kamakura, 2000).

## 17.6.2 Campaigning to bring about policy change

When it comes to bringing about a change in policy, NGOs usually try to both influence and inform the target audience, who may be legislators or Members of Parliament. They may employ methods that include media campaigns, public speaking, commissioning and publishing research, online petitions (change.org and avaaz.org are two of the most popular English-language online petition websites), organising protest marches or demonstrations, recruiting advice from experts, or making direct approaches to legislators or Members of Parliament on the issue concerned.

In 2017, a group called Two Million Tusks was concerned about the plight of African elephants and the UK's role in the global ivory trade. They researched the quantity of ivory being sold through UK auction houses and whether those auctioneers were compliant with the UK's rules on ivory trade. The resulting report, published in October 2017, exposed weaknesses in auction houses' compliance and called upon the Department for Environment, Food and Rural Affairs to ban all trade in ivory within the UK (Two Million Tusks, 2017). While the debate about ivory sales has been long-fought, a linked television programme, presented by Hugh Fearnley-Whittingstall, revealed new concerns. He arranged for eight ivory items on sale in UK antiques shops to be radiocarbon-dated, and found that three of the pieces were from modern, i.e. post-1947, ivory, and as such could not be legally sold in the UK. During a televised press briefing on this finding, the then Environment Minister, Andrea Leadsom, came under sustained pressure to address the UK's role in laundering ivory from poached African elephants; the eventual result was a Bill to restrict severely the conditions under which ivory can be sold in the UK.

### 17.6.3 Campaigning to raise funds

Fundraising wisdom says that the most effective calls for donations are ones that engage the audience(s) on an emotional level (Hill, 2010). Handling such messaging can be challenging: whether to use images that provoke negative (horror, disgust) or positive (empathy, inspired) emotions; whether to hold donors to ransom ('Unless we act now, this species will go extinct') or focus on success stories; whether to focus on a single, named animal as an ambassador for its species, while being clear that donations will be spent on a wide range of activities, or on a species or habitat as a whole.

In the UK, the Fundraising Regulator, formerly known as the Fundraising Standards Board, sets and maintains the standards for charitable fundraising in England, Wales and Northern Ireland, and aims to ensure that fundraising is respectful, open, honest and accountable to the public, and regulates fundraising practice via *The Code of Fundraising Practice* (Fundraising Regulator, 2016). Its guidance on 'Content of Fundraising Communications' says that organisations: must not imply that donations will be used for a specific purpose if they will be allocated to general funds; must be legal, decent, honest and truthful; must make it clear if they alter any elements of real-life case studies; and must give warnings about and be able to justify the use of any shocking images.

In October 2014, Save the Rhino International (SRI) began planning its annual fundraising appeal for 2015. The decision was made to focus on Kenya, which had not benefited from previous appeals and which had suffered a spike in rhino poaching in 2013, when 59 rhinos were killed, as compared to

29 the previous year. SRI had a long history of supporting rhino conservation efforts with its in-country partners. It was suggested that a focus on the canine units employed by Lewa Wildlife Conservancy, Borana Conservancy, Ol Jogi Conservancy and Ol Pejeta Conservancy, as part of their anti-poaching and community engagement strategies, would provide an interesting and engaging angle for a public fundraising appeal. These units use Belgian Malinois and bloodhounds for tracking (i.e. following poachers' scent trails) and/or detection (i.e. dogs are trained on specific scents to be able to carry out, for example, vehicle searches at road blocks). A name for the appeal, 'Rhino Dog Squad', was chosen as being descriptive, punchy and memorable.

Based on results from previous appeals, SRI's primary objective for the appeal was to raise a total £40,000 for the three canine units in Kenya by February 2016, of which £30,000 would come from a campaign marketed to the general public and £10,000 from zoos via spin-off campaigns.

Three distinct target audiences were identified: the general public/animal lovers, particularly those with pet dogs, living in the UK, continental Europe or the USA, across a broad age range, with some but not detailed knowledge of the rhino poaching crisis; high–net-worth individuals who have visited or have links with Kenya; and zoo visitors. Save the Rhino applied successfully to BBC Radio 4 to have the Rhino Dog Squad featured as one of the station's charity appeals: this greatly increased the charity's 'reach' to the first audience.

SRI's appeal planning team realised early on that the choice of presenter would influence the script, and considered the merits of having a celebrity record the appeal versus one of the Kenyan field programme staff. In the event, SRI recruited Sam Taylor, Chief Conservation Officer at Borana, to read the script, giving SRI an opportunity to personalise the script. Furthermore, knowing that the appeal would be broadcast just before Christmas 2015 (twice on the last Sunday before Christmas and once on Christmas Eve) meant that the SRI team had to consider where radio listeners would be, and how to engage their emotions at such a time.

The BBC Radio 4 appeal alone raised more than £22,000, with the Rhino Dog Squad in total realising about £60,000 by 31 March 2016; some donors set up standing orders and funds are still being received for the canine units at the time of writing (June 2018). The BBC said that the appeal was one of the most successful of its type, and attributed this to:

- a knowledgeable presenter: having someone who worked at one of the beneficiary conservancies read the appeal meant that it could be written in a way that was highly personal and credible;
- an unusual script: the first words of the appeal were 'Sausage bonus! Now there's an image to conjure with. I'm guessing you don't often see the

words "Sausage bonus" in a budget. I do, in my work as Conservation Officer in a wildlife sanctuary in northern Kenya'. The first two words caught and held the attention, as Sam went on to explain how the canine units help the rangers with their work;

- making the most of the timing: SRI knew that listeners would likely be at home with their families, wrapping presents, decorating the tree or beginning to cook Christmas meals. Contrasting listeners' lives at Christmas with that of the rangers in Africa would be powerful. 'This Christmas, as you enjoy time with your families, friends and your pets, please remember our dogs and rangers. They'll be at work, protecting Africa's wildlife. Please help the Rhino Dog Squad';

- the famous British love of dogs: 'We use bloodhounds and Belgian Malinois, and they're awesome. They can track scent for up to three days. They're better than a bullet – they can go around trees and hold poachers until our rangers can safely make an arrest. The dogs work at roadblocks, detecting rhino horn, ivory, and weapons. We also use them to help find lost children or recover stolen property. Our dogs are part of our team';

- the wider appeal held by SRI: in addition to the BBC 4 appeal, SRI had planned a strong social media campaign with many assets: ezines, blogs written in advance ready to be posted, lots of high-quality images (including photographs taken during a visit in March of dogs tearing into parcels wrapped in Christmas paper containing bones and toys), and a main 4-minute film supported by four supplementary 2-minute films.

## 17.7 Potential pitfalls for campaigns

### 17.7.1 Lack of a strong evidence base

While reports of incredible successes offer good news stories for conservation and boost the reputation of the organisations that carry out the campaign, there is the risk that once the evidence base (where it exists) is questioned, the outcomes turn out to be not quite the success story that they initially appeared. Although in the majority of cases this may just lead to wasted donor funds and NGO time, there are also examples of where this has created a conservation problem in itself.

A good example is the 'Save the Bay, Eat a Ray' campaign that followed all of the rules for a good campaign. It used clear messaging to communicate a simple evidence-based action that members of the public could take to help restore Chesapeake Bay: eating more cownose rays (*Rhinoptera bonasus*) (National Aquarium Baltimore, 2016). The evidence said that a huge population increase of cownose rays was decimating the Bay's oyster populations, and some also claimed that the species was invasive. However, further analysis of the science found that the models used were flawed and, not only was the ray a native species that was not responsible for the decline, it was itself extremely vulnerable to overfishing (Grubbs et al., 2016; National Aquarium Baltimore, 2016). In this case, a lack of robust scientific evidence relating to the ecology of the system led to negative conservation consequences, even if these outcomes were intended in the first place.

Behaviour-change campaigns can become particularly complex when they are based around reducing the use of illegal wildlife trade products. Communicating messages to the consumers of an illegal product is difficult because, if admitting to using the product could result in some kind of punishment, even identifying the consumers of it will be a challenge (see Chapter 5 for a discussion of approaches to gathering information about sensitive topics, including illegal resource use). Often, in-depth research focusing on consumer preferences and behaviour is needed to understand motivations for consumption (e.g. Nuno & St John, 2015; Hinsley et al., 2015). However, behaviour-change campaigns are often carried out by NGOs without the time, expertise, resources or capacity to do this kind academic research. This has resulted in several campaigns based on very little knowledge of who the target audience should be, often using high-profile celebrities or eye-catching graphics to get the message out to as many people as possible, with the hope that this will include the actual consumers of the product. Unfortunately, it is not possible to say whether this works: a recent review found that almost no behaviour-change campaigns focused on wildlife consumers report evidence of impact, and very few carry out any kind of robust evaluation at all (Veríssimo & Wan, 2018). One way to address this could be greater collaboration between NGOs that do not have in-house scientists and academics, to ensure that campaigns are based on good scientific evidence, and that results are analysed in depth to evaluate the impact.

## 17.7.2 Over-stated claims of success

Some NGOs have focused their behaviour-change campaigns at children, banking on the 'pester-power' factor (cf. Figure 17.1, activity that 'Educates segments of the population who don't currently use rhino horn, e.g. children'). Humane Society International, for example, launched a campaign aimed at stopping the use of illegal rhino horn in Vietnam via a book called *I'm a little Rhino* that was used in schools to help teach children about rhino poaching concerns and conservation efforts. No information is available on how the campaign was designed, targeted or evaluated, but claims that demand for rhino horn had fallen by 77% in Hanoi following the campaign have been heavily criticised by conservation practitioners (Roberton, 2014).

### 17.7.3  Bias in campaigns

One of the dangers of advocacy/campaigning is that it may not be sufficiently inclusive or consultative. For example, the NGO leading the campaign may have a particular stance on a controversial issue, or an NGO with a direct line to a Member of Parliament or Minister may be able to exert undue influence.

For example, IFAW, Lion Aid and the Born Free Foundation, among others, have worked closely with a group called 'MEPs for Wildlife' (MEPs are Members of the European Parliament). While there was an initial focus on banning the hunting of canned lions (canned hunts are trophy hunts in which an animal is kept in a confined area, such as in a fenced-in area, increasing the likelihood of the hunter obtaining a kill), MEPs for Wildlife expanded its efforts to call for an EU-wide ban on the import of lion trophies, in keeping with decisions made by the Netherlands, French and Australian governments.

However, as an IUCN Briefing Paper for European decision-makers explains (with reference to the then recent and still notorious case of 'Cecil the Lion', shot in July 2015):

Intense scrutiny of hunting due to these bad examples has been associated with many confusions (and sometimes misinformation) about the nature of hunting, including:

- trophy hunting is the same as 'canned' hunting;
- trophy hunting is illegal;
- trophy hunting is driving declines of iconic species, particularly large African mammals like elephant, rhino and lion;
- trophy hunting could readily be replaced by photographic tourism.

None of these statements is correct. (IUCN, 2016)

The Briefing Paper goes on to conclude that 'legal, well-regulated trophy hunting programmes can – and do – play an important role in delivering benefits for both wildlife conservation and for the livelihoods and wellbeing of indigenous and local communities living with wildlife' (IUCN, 2016).

Making the case for positions, particularly 'unpopular' ones such as advocating for well-run trophy hunting, is extremely difficult to do. The IUCN Briefing Paper includes two graphs on rhinos and trophy hunting: the first showing the change in estimated numbers of Southern white rhino in South Africa before and after limited trophy hunting was introduced in 1968; and the second showing growth in estimated total numbers of black rhino in South Africa and Namibia before and after CITES approval of limited hunting quotas (a maximum of five animals per country per year, and even then only if suitable candidate animals can be identified) in 2004. Both graphs show populations increasing exponentially until the current poaching crisis began (IUCN, 2016).

Numerically speaking, the evidence in the Briefing Paper is conclusive: trophy hunting of rhinos, while fatal for the individuals concerned, has not adversely affected the species' meta-population growth. Simultaneously, it has generated incentives for landowners (government, private individuals or communities) to conserve or restore rhinos on their land; and generated revenue for wildlife management and conservation, including anti-poaching activities. This does not hold sway, however, with NGOs that are ideologically opposed to trophy hunting.

### 17.7.4  Conflicting views

It would be wrong to assume that all conservation NGOs speak with a common voice. The Global March for Elephants and Rhinos (GMFER) has become a worldwide campaign, taking place in more than 160 cities in 2016, and thus enabling people from many different countries to take part. In the beginning, the march was about 'raising awareness, generating global media attention on the crisis, and keeping political pressure on world leaders to protect our endangered wildlife'. Such broad aims made it possible for a broad church of elephant- and rhino-focused conservation organisations to take part in the march.

However, in more recent years, the GMFER has focused on banning trade in ivory and rhino horn, including applying pressure on South Africa to maintain a ban on domestic rhino horn trade (the ban was eventually overturned in early 2017) and on Japan and Hong Kong to ban online and domestic sales of ivory. A number of NGOs that are working to tackle the rhino and elephant poaching crises are actually pro-sustainable use, and have taken the decision not to participate in GMFER's annual event, because its aims were incompatible with their own.

### 17.7.5  Inappropriate use of emotion

Conservation or animal welfare/animal rights NGOs must tread a fine line when campaigning about emotive subjects. Some of the most difficult images to view are those showing animal abuse or suffering, bushmeat and the impact of poaching. A photograph that is too upsetting will result in the viewer turning the page quickly without taking in the call to action.

There are ways around this challenge. Photographs of dead elephants with their tusks hacked out certainly tell the story behind the poaching crisis, but so too does Nick Brandt's monochrome image, *Line of rangers holding tusks killed at the hands of man, Amboseli 2011.* As the photographer writes (Brandt, 2015),

I wish that I had never had to take this photo. I wish that it had never been possible to take this photo. The photo was taken as a deliberate visual echo of *Elephants Walking Through Grass*, a very different world – a vision of paradise and plenty – taken only a couple of miles away three years earlier. But instead of a herd of elephants striding across the grassy plains of Africa, we see only their remains: the tusks of 22 elephants killed at the hands of man within the Amboseli/Tsavo Ecosystem.

Brandt's post goes on to hold out hope in the form of the work being done by Big Life Foundation's rangers; a good example of a strong image, which does not in itself provoke feelings of disgust or revolt in the viewer (Fundraising Regulator, 2016), but which explains the catastrophe that has occurred and offers a way of helping to solve the problem.

### 17.7.6 Risk of unintended consequences

Ensuring that communications are well-designed and that the campaign's main messages are evidence-based can make achieving the ultimate aim more likely, but it does not always protect against unintended, often negative, consequences of the campaign.

To date in conservation there has not been enough robust evaluation of campaigns to measure the occurrence of unintended consequences, but evidence from other fields demonstrates the risk. In the field of health, the risk of unintended consequences is well-recognised. For example, multiple studies have found that campaigns aimed at reducing drug and alcohol consumption frequently create so-called 'boomerang effects', where the result is an increase in consumption rather than a decrease (Ringold, 2002). This extent to which this phenomenon may be occurring in response to demand-reduction campaigns for high-profile wildlife products is unknown, but the complexity of these markets and the use of conflicting messages by different groups may increase the risk. For example, the legal bear bile trade in China has been the focus of extensive campaigns by animal welfare organisations, with the ultimate aim of closing down all bear farms. While some campaigns use the ineffectiveness of bear bile as a medicine as the key message, others instead focus on the cruelty of the farms, or the health risks to consumers of using farmed bile, such as the 2012 *Healing without Harm* campaign (Watts, 2012). While these messages may be intended to close down the market for bear bile, and with it the farms themselves, little is known about how regular consumers of bile – who believe that it is an effective treatment for a serious condition, such as liver cirrhosis – will react. For example, will these consumers switch to wild-sourced bear bile instead where it is available, or will they start using another product? Currently there is little evidence either way, making this a risky strategy for conservation. To mitigate this, campaigns should fully consider all potential consequences of their messaging and evaluate the risks of carrying out the campaign before it starts, drawing on existing evidence from other fields.

Another problem area lies in the way that illegal wildlife trade products are described by some NGOs, which is then repeated in the media. Products such as orchids, pangolin scales and rhino horns are often described as rare and hard to obtain by well-meaning organisations or researchers. However, in markets that often prize rarity, such messages can increase consumers' desire for the forbidden item, the acquisition of which will demonstrate both their wealth and their ability to use their networks to obtain it. For example, specialist consumers of slipper orchids, all species which are on CITES Appendix I, have been found to be willing to pay more for a rare species (Hinsley et al., 2015). Although several of these species have already been collected to near extinction for trade (e.g. *Paphiopedilum canhii*: Rankou & Averyanov, 2015), highlighting their rarity is likely to be counter-productive. Similarly, mentioning high prices for wildlife products can raise awareness of their value among both consumers and traders, and organisations like TRAFFIC and Wildlife Conservation Society have drawn up clear internal guidelines for their staff, explaining why they should never discuss the black-market price of an illegal wildlife product.

## 17.8  Future directions for campaigns in conservation

Campaigning to bring about change is central to much of conservation action, and it is essential that the importance of a well-designed campaign is recognised and appreciated. There are numerous examples of campaigns that have brought about change, many that did not achieve their intended goals, and even more that have never been carefully evaluated. As described in this chapter, the most successful campaigns will undertake careful planning and tailor their messages to the specific aim and context to ensure that they engage the target audience effectively. Other important steps include clear goal-setting, development of indicators and means of verification; monitoring, and a comprehensive evaluation of outcomes.

Competition for donor funds or the support of the public can sometimes mean that collaboration and open dialogue between different conservation actors is not always a priority. However, partnerships between different NGOs can extend the reach of a campaign and provide new perspectives, and collaboration with academics can provide a strong scientific research base for its design. Possibly the most important action should be to share lessons learned from successes and failures, as this is an important way that campaigns can continue to improve and avoid the pitfalls described here. These steps are essential, as a good campaign cannot only prevent the waste of donor funds, but increase the likelihood of conservation delivering change for the common good.

# References

Brandt, N. 2015. *Behind the photo*. Facebook, 7 July 2015. Available from www .facebook.com/NickBrandtPhotography/ photos/a .10150167609326087.301158. 161975326086/10152917589861087/? type=3

Breaking the Brand. 2016. *How Much Is Spent On Rhino Horn Demand Reduction Campaigns?* Available from https://breakingthebrand .org/how-much-is-spent-on-rhino-horn-demand-reduction-campaigns/?doing_ wp_cron=1534087399 .6144099235534667968750

Christiano, A. & Neimand, A. 2017. *Stop Raising Awareness Already*. Stanford Social Innovation Review. Available from https:// ssir.org/articles/entry/ stop_raising_awareness_already

Fundraising Regulator. 2016. *The Code of Fundraising Practice*. Available from www .fundraisingregulator.org.uk/wp-content /uploads/2018/05/Code-of-Fundraising-Practice-v1.7–25052018.pdf accessed May 2018

Greenpeace. 2010. *Success! You made Nestlé drop dodgy palm oil!* Available from https://green peace.org.uk/success-nestle-palm-oil/

Grubbs, R. D., Carlson, J. K., Romine, J. G., et al. 2016. Critical assessment and ramifications of a purported marine trophic cascade. *Scientific Reports*, 6, 20970.

Hill, D. 2010. *Emotionomics: Leveraging Emotions for Business Success*. London: Kogan Page.

Hinsley, A., Verissimo, D. & Roberts, D. L. 2015. Heterogeneity in consumer preferences for orchids in international trade and the potential for the use of market research methods to study demand for wildlife. *Biological Conservation*, 190, 80–86.

IUCN. 2016. Informing decisions on trophy hunting. A Briefing Paper for European Union Decision-makers regarding potential plans for restriction of imports of hunting trophies. April 2016. Available from www .iucn.org/downloads/iucn_informingdeci sionsontrophyhuntingv1.pdf

Jacquet, J. & Pauly, D. 2007. The rise of seafood awareness campaigns in an era of collapsing fisheries. *Marine Policy*, 31, 308–313.

Mackenzie-Mohr, D. 2011. *Fostering Sustainable Behavior: An Introduction to Community-Based Social Marketing*. Toronto: New Society Publishers.

National Aquarium Baltimore. 2016. Cownose Rays in the Chesapeake Bay: What do we know? Final Report of a workshop hosted on October 22nd 2015 by the Chesapeake Bay Program's Sustainable Fisheries Goal Implementation Team. National Aquarium Baltimore, Maryland. Available from www .chesapeakebay.net/channel_files/23141/ cnr_workshop_report_final_1–29-16.pdf

Neslen, A. 2017. Nestlé, Hershey and Mars 'breaking promises over palm oil use'. *The Guardian*. Available from www .theguardian.com/environment/2017/oct/ 27/nestle-mars-and-hershey-breaking-promises-over-palm-oil-use-say-campaigners (accessed June 2018).

Nuno, A. & St John, F. A. V. 2015. How to ask sensitive questions in conservation: a review of specialized questioning techniques. *Biological Conservation*, 189, 5–15.

Project Seahorse. 2018. Conservation Programmes: Trade and Policy. Available from www.projectseahorse.org/action-programs#trade

Rankou, H. & Averyanov, L. 2015. *Paphiopedilum canhii*. The IUCN Red List of Threatened Species 2015: e.T191858A2009477. Available from http://dx.doi.org/10.2305 /IUCN.UK.2015–2 .RLTS.T191858A2009477.en

Rawlinson, K. 2017. Michael Gove 'haunted' by plastic pollution seen in Blue Planet II. Available from www.theguardian.com /environment/2017/dec/19/michael-gove-haunted-by-plastic-pollution-seen-in-blue-planet-ii

Ringold, D. J. 2002. Boomerang effects in response to public health interventions: some unintended consequences in the alcoholic beverage market. *Journal of Consumer Policy*, 25, 27–63.

Roberton, S. I. 2014. Has demand for rhino horn truly dropped in Vietnam? *National Geographic*, 3 November 2014. Available from https://blog.nationalgeographic.org/2014/11/03/has-demand-for-rhino-horn-truly-dropped-in-vietnam/

TRAFFIC. 2013. Rhino horn consumers: who are they? Available from www.traffic.org/home/2013/9/17/pioneering-research-reveals-new-insights-into-the-consumers.html

Two Million Tusks. 2017. Ivory: the grey areas. A study of UK auction house ivory sales – the missing evidence. Available from www.twomilliontusks.org/

Veríssimo, D. & Wan, A. K. 2018. Characterising the efforts to reduce consumer demand for wildlife products. Available from https://osf.io/preprints/socarxiv/642pb

Watts, J. 2012. End Chinese bear-bile farming, says UK animal rights activist. The Guardian, 9 January 2012. Available from www.theguardian.com/environment/2012/jan/09/china-bear-bile-farming-animal-rights

Wedel, M. & Kamakura, W. A. 2000. *Market Segmentation: Conceptual and Methodological Foundations*. New York, NY: Springer US.

# CHAPTER EIGHTEEN

# Behavioural insights for conservation and sustainability

TOBY PARK

*The Behavioural Insights Team*

## 18.1 Introduction

Many of society's ailments and ambitions, from obesity and corruption to economic growth and conflict, are ultimately about human behaviour. Sustainability and conservation challenges are no different, and although legal, economic and engineering solutions will be key, so will a shift in individual actions around resource use and waste, diet, fishing and agricultural practices, wildlife consumption, tourism and beyond (Rowson & Corner, 2015). Policy-makers, educators and conservation NGOs are therefore unavoidably in the business of behaviour change, but the conventional toolkit of regulation, incentives and information provision is increasingly being recognised as incomplete, and too rooted in a rudimentary model of human behaviour (Shafir, 2013).

On the rise is a more realistic understanding of behaviour, drawing on the latest insights from behavioural economics, social marketing and cognitive and social psychology. By harnessing these new tools we can radically improve policy and campaign outcomes and achieve greater social impact (Halpern, 2015). The field is rapidly growing in some parts of the sustainability community, as well as in public health, international development and consumer finance, but conservationists have so far been slow to embrace the behavioural perspective (Reddy et al., 2017). This is now beginning to change, particularly among NGOs faced with explicitly human challenges such as poaching, corruption, the illegal consumption of wildlife and common pool resource depletion, including water and coastal fisheries.

In this chapter I provide an overview of behavioural insights for sustainability and conservation, aimed at readers with little prior expertise in the subject. I do this by first reviewing a conventional understanding of behaviour change, discussing its shortcomings and then presenting some additional strategies.

## 18.2  A flawed starting point – rational choice

In both economics and psychology the dominant models of behaviour have historically been rooted in the concept of *subjective expected utility*, describing individuals as making rational choices that maximise the benefits to themselves (Scott, 2000) (see also Darnton, 2008, for a review of behaviour-change models). The axioms underlying these models are first that behaviour is cognisant and deliberate; second, that we are self-interested in the sense that we maximise our own utility as defined by our preferences, typically construed as wealth, enjoyment or subjective well-being; and finally that the locus of decision-making is the individual, implying a degree of indifference to context (Becker, 1976).

In economics, this account of behaviour is formalised in standard micro and macro models, and has long provided the dominant intellectual framework for policy, regulation and law, business and finance, international development, public health and natural resource management. Indeed, the economic concept of cost–benefit analysis is highly analogous to this understanding of behaviour, implying we make choices by carefully trading off pros and cons. Among environmental campaigners and educators the language draws more from the field of psychology, speaking of *values* and *attitudes* rather than *preferences* and *utility*, but the assumptions of intentional, reasoned and individual choice are usually still implicit.

With this conventional model of behaviour in mind, a suite of tools for behaviour-change emerge, and capture the bulk of government and NGO activity.

1. **Regulation.** Influencing our behaviour through the threat of sanction via bans, quotas or standards.
2. **Economic levers.** Self-interest is harnessed by making pro-environmental behaviours the more appealing option, typically through the provision of economic incentives including taxes, subsidies, fines, grants, or payments for eco-services.
3. **Social marketing and attitudinal campaigns.** An attempt to alter our preferences, values or attitudes by promoting greater environmental concern.
4. **Information provision.** Assuming pro-environmental values to be present, people cannot act on them if they have flawed beliefs or lack awareness of the environmental impact of their choices. This information deficit may be overcome through education, awareness-raising, guidance, or product labels and kite marks. In practice, the line between 'merely' providing information and attempting to influence our attitudes is often blurred.

## 18.3 Going beyond conventional wisdom

A great deal has been achieved through the above approaches. In particular, regulation and economic incentives can be highly effective, reflecting the fact that self-interest is a powerful driver of behaviour. Information provision can also be effective *if* information deficit is a major barrier – product labels can have powerful effects in otherwise shrouded markets, for example. Raised awareness is also often a critical step towards building public consent for big-ticket policy initiatives, such as a carbon tax or the banning of wildlife products (Marteau, 2017; Portney et al., 2018). In and of itself, however, awareness is often not enough to shift individual behaviour due to the dominance of other factors, such as competing motivations or practical and psychological barriers to action (Barr, 2004; Olander & Thøgersen, 2014).

The wider criticism is that these tools, and the behavioural assumptions underpinning them, overlook important aspects of human nature. I highlight three insights below as particularly in need of greater focus, before outlining some additional tools that emerge from these insights.

### 18.3.1 The importance of context

By focusing on the *individual* as the locus of behaviour, rational accounts of behaviour fail to recognise the extent to which our actions are shaped by the social, physical, economic, political and cultural context (Shove, 2009). Indeed, evidence suggests that interventions that alter the setting in which choices are made, by making the desired behaviour cheap, convenient, politically cultivated and socially normative, are often more effective than those which focus solely on individual beliefs, attitudes and choices (Thøgersen, 2014). They do, however, require fundamentally different levers than conventional information-provision approaches often relied upon by conservation NGOs, targeting not the individual's unsustainable choice, but the socio-technical structures which encourage unsustainable practices to flourish.

### 18.3.2 The importance of non-conscious processes

This sensitivity to context is best explained by dual-process models of cognition, which define two parallel systems of mental activity. One is slow, reflective, cognisant and deliberative. This system most resembles rational choice, although more accurately is *boundedly rational*, operating under limited information and cognitive bandwidth, and usually aiming to *satisfice* (find a good enough solution) rather than to optimise (Simon, 1972). The second system, which dominates more of our decision-making than we tend to realise, is fast, largely automatic and driven by intuitive processes such as ingrained habit, emotion and *heuristics* (mental shortcuts) (Kahneman, 2011).

These fast-and-frugal processes are mostly unreflective responses to cues in our social and physical environment, and hence our great susceptibility to external influence. They also leave us susceptible to predictable errors of judgement, or *cognitive biases*, as we trade-off accuracy for cognitive efficiency. For example 'choose the middle option', 'stick with the default and the familiar unless there is a strong reason to risk the unknown' and 'do what most people like me appear to be doing' are all heuristics we instinctively adopt – serving us well enough most of the time without demanding much mental resource, but often leading us to err from optimal decisions (Kahneman, 2011). Designing environments, and campaigns, which reflect these more automatic processes can be an effective strategy for enabling and encouraging more sustainable behaviour (Thaler & Sunstein, 2008).

### 18.3.3 The importance of behaviour over values, attitudes and beliefs

Conservation campaigns typically attempt to raise awareness and elevate pro-environmental values, on the premise that greater concern for the planet, or a species or habitat, will drive financial support or more sustainable behaviour. However, it can be difficult to engage citizens in these issues. Research shows that pro-environmental information often has the intended impact only on those already sympathetic to the message, as we update our views asymmetrically, skewed towards the direction of our prior convictions (Sunstein et al., 2016). This observation is rooted in *confirmation bias* – our tendency to gravitate towards information which corroborates our existing views, while we discount, ignore or distort information which challenges us (Nickerson, 1998).

That said, encouragingly, the battle for hearts and minds is slowly being won: pro-environmental attitudes are now common across much of Europe, for instance (Steentjes et al., 2017). This is helping to raise the policy agenda (Carrington, 2019). However, few citizens are independently giving up their cars, overseas holidays or beef burgers. It would also be naïve to expect fishers, farmers, poachers and loggers to compromise their livelihoods so willingly. Clearly, there is more to behaviour change than awareness and attitudes, highlighting the problem of a widely observed *value–action gap* (Kollmuss & Agyeman, 2002). The reasons for this gap are myriad and complex, although two broad categories are worth highlighting: insincerity of our values and barriers to acting on them.

First, pro-environmental values are frequently in tension with self-interest, creating cognitive dissonance and guilt for habits we are unwilling to forego. Guilt can be a powerful motivator for action, but we also have a tendency to resolve this dissonance not by curbing our unsustainable behaviour, but by ignoring the issue (wilful ignorance), or employing various acts of psychological fudging, including motivated reasoning (rationalising towards a convenient and ego-serving, rather than logical, conclusion), moral licensing (excusing our-selves the flight because we recycled) and biased social comparisons (inflated convictions that 'I do more than most' and deferring responsibility to govern-ment/industry/other countries) (Barkan et al., 2015). In other words, our beha-viour reveals that our concern for cost, profit, convenience and enjoyment frequently outranks our concern for the planet, despite our ability to maintain sincere environmental values and a sense of integrity – the psychological equivalent of having our cake and eating it (Shalvi et al., 2015; Gino et al., 2016).

Second, even where intentions are sincere, we may fail to act due to various psychological and practical barriers. These include hassle, a lack of options, lack of know-how, upfront cost barriers, lack of willpower, lack of self-efficacy (belief that we can make a worthwhile difference), procrastination, forgetful-ness, ineffective planning, ingrained habit and various cognitive biases that favour a 'do-nothing' strategy, including loss aversion, present bias, uncer-tainty-aversion, inertia and risk-aversion. These factors constitute the second major element of the value–action gap (Webb & Sheeran, 2006), and although they often seem trivial, they can be disproportionately impactful. They there-fore deserve disproportionate attention when designing interventions and campaigns to help bridge the divide between good intentions and action. For example, helping people plan better to reduce food waste, removing the hassle of switching to a green energy tariff, providing easy substitutes to medicinal wildlife products, or providing timely reminders and tips for redu-cing water consumption are all strategies which can help turn green aspira-tions into green actions.

## 18.4 Effective strategies for promoting conservation behaviours

With the above points in mind, the most effective route to change, whether tackling wildlife crime, energy conservation, or the protection of common pool resources, is often a 'twin-track' approach (Burgess, 2016). The aim is to target both the individual (motivations, decision processes, habits, emotional engagement, attitudes and awareness) *and* the enabling environment (ensuring that policy, the built environment, social norms and incentives promote and facilitate the sustainable beha-viour). These are often two sides of the same coin: the choices we make as individuals are often inseparable from the enabling environments in which we make them.

Below, I briefly outline four levers for change that span individual forces on behaviour and three key environmental dimensions: social, material and economic. Many of the examples given are drawn from other contexts where the behavioural mechanisms are relevant, acknowledging that the use of behavioural interventions is nascent within the field of wildlife con-servation (Reddy et al., 2017).

### 18.4.1  Inner motives: ego, emotion and meaning

Two fundamental motivations influence our adoption of beliefs and attitudes: to construe our lives in a positive (ego-enhancing) fashion, and to construe them in a way which makes sense and is consistent (Chater & Loewenstein, 2016). Thus, we are rarely convinced by mere truth, but by *narrative fidelity* and *self-enhancement*: the extent to which something concords with our prior world-view and with the flattering autobiography we curate in our minds. We also tend to think automatically first (with emotion, intuition, gut instinct) and rationalise second. Thus, our reflective and deliberative faculties often act more as interpreters of our instincts than as executives guiding our judgement (Haidt, 2001).

Therefore, successful campaigns rarely pose cerebral facts or logical arguments, but cater to deeper emotional triggers, operating at the 'human level' we have evolved to think at, and are rooted in meaning, plot and personal relevance (Schiff, 2012). This largely explains the *identifiable victim effect* – our greater tendency to donate or make efforts to save an individual animal/ecosystem/community member than a statistical one (Jenni & Loewenstein, 1997). It also explains why campaigns evoking guilt or anxiety can lead to disengagement, because these emotions undermine the ego and present uncomfortable truths, inviting psychological defence rather than engagement (particularly if giving up the unsustainable behaviours is difficult or unappealing). In contrast, research suggests that harnessing emotions with positive valence (intrinsic attractiveness), in particular anticipated pride from acting sustainably, can be more effective (Schneider et al., 2017).

Recognising our tendency to find the 'wiggle room' to rationalise our self-interested actions also sheds light on wildlife crime. Evidence shows interventions that reduce the ease of rationalisation can be effective. For instance, we can highlight the prevalence of good behaviour to correct self-serving misconceptions that 'everyone does it' (see discussion on social norms below). We might also create less-malleable boundaries between acceptable and unacceptable behaviours to constrain our ability to re-frame dishonest actions as acceptable (e.g. we are less likely to steal money than do something which indirectly equates to us acquiring money dishonestly, such as by paying a lower price in cash to avoid taxes). Drawing people's attention to their moral standards, through religious reminders or honesty commitments, can also be effective by reducing the level of dishonesty we are able to reconcile with our self-concept of integrity (Mazar et al., 2008). Such strategies offer attractive alternatives to fines and punishment, particularly in remote situations where monitoring and enforcement are difficult.

Our social identity (the portion of self-concept expressed by membership of social groups or categories such as gender, race, or political beliefs) is also vital in shaping our beliefs, values and actions. We listen to and mimic people we identify with, like and perceive as credible, but may do the opposite of people in our perceived 'out-group' simply as a way of dis-identifying with them (Turner, 1991). This is partly why the politicised nature of the environmental debate is so damaging, but also means certain messengers can be disproportionately effective – in the UK, the so-called 'David Attenborough effect', for instance (Haynes-Worthington, 2018). In China, the efforts of Jackie Chan and other celebrities to campaign against shark-fin soup have led to dramatic reductions in consumption (WildAid, 2014).

The broader point through all of this is that we need to understand and cater to the underlying motivations of individuals involved in the depletion or consumption of wildlife and natural resources. We would be naïve to presume a message of sustainability will, for many people, prevail over potent drivers of self-advancement (e.g. in wildlife corruption or over-fishing), convenience (e.g. in air travel or disregard for environmental protections), pleasure and hedonism (e.g. in eating beef or hunting), status (e.g. in ivory ownership), self-expression (e.g. in car ownership) and so on. This demands pragmatism: sustainable outcomes need not be fought only on the basis of sustainability, restraint or moral duty if these more powerful motivations can be harnessed to good effect. Sometimes this is about choosing the right *framing*. For example, public health researchers have found that food explicitly sold as indulgent out-sells identical food sold as healthy, and although a niche market for health food surely exists, on average healthy connotations may harm sales even compared to entirely neutral, non-descriptive labels (Turnwald et al., 2017). Similar findings are now emerging in the promotion of sustainable diets (Vennard et al., 2019). Other approaches include finding compelling ways to displace or supplant these competing motivations, for example by making sustainable travel significantly cheaper and more convenient than air travel; by tapping into identity and offering more sustainable avenues for self-expression; or by attempting to substitute ivory products with an alternative market for high-status jade carvings (e.g. Burgess, 2016).

### 18.4.2  Social dimension: peer-influence

The social dimension of our behaviour is particularly relevant to conservation issues because the protection of public resources, including fish stocks, rainforests, freshwater or clean air, frequently depends on collective action and the restraint of personal self-interests for communal benefit. Rational choice theory, painting us broadly as self-serving individuals, highlights the risk of a tragedy of the commons in such circumstances, and suggests taxes on externalities (Pigouvian taxes) or privatisation of resources are necessary to realign individual and collective interests (Ostrom, 2000).

However, in reality we are deeply social creatures: we have the capacity to cooperate, a tendency to reciprocate and conform to social norms and to shun freeloaders and deviants (Trivers, 1971; Ostrom, 2000). These are processes of evolutionarily ingrained peer pressure: feelings of social obligation, guilt and desire for public acceptance are the proximate drivers for deeper benefits of group cohesion and collaboration. For example, evidence suggests adherence to social norms, and the taboo of breaking them, has traditionally been enough to ensure sustainable harvesting practices in Madagascar (Jones et al., 2008). Our objective is to harness and further strengthen these traits.

One effective approach is to highlight the prevalence of a desirable behaviour, harnessing our tendency for conformity, but also for reciprocity: the knowledge that others are contributing to a public good encourages us to do the same. For example, comparing householders' energy use to their more efficient neighbours reliably reduces consumption by a few percent (Allcott, 2011), and telling hotel guests that most others re-used their towel led to 44% doing so, significantly more than with a conventional environmental message (Goldstein et al., 2008). Another strategy is to imply reciprocity more directly. For instance, the conservation charity Rare brokered agreements between downstream and upstream water users. Downstream users financed payments and materials for upstream users in the hope they would reciprocate and be stewards of upstream ecosystems, protecting 16,000 hectares of land in the watershed (Rodríguez-Dowdell et al., 2014).

The corollary is that advertising undesirable norms, often done inadvertently in an attempt to highlight the severity of a problem, can unintentionally license the undesirable behaviour. For example, a US national park suffering the theft of fossilised wood found that thefts increased in response to a sign which read 'thousands of visitors are taking fossilised wood and deteriorating the natural environment' (Cialdini, 2003). Note the connection to an earlier point: we often rationalise selfish behaviours through convenient social comparisons, a form of *social licensing* through which freeloading can become normalised and resource extraction risks competitively escalating (Dimant, 2017).

Our tendency to adhere to norms is often strengthened by peer observation, because being watched adds real social cost to deviance (Argyle, 1957). Hence, we can promote cooperation in conservation contexts by making behaviours less anonymous and improving the mechanisms for communication, peer monitoring and self-governance within fishing and farming communities (Ostrom et al., 1994). Public league tables are one way of achieving this: taking an example from a different setting, UK government departments' energy consumption dropped by up to 22% after publication in a ranked league table. Operating through similar principles of observability, a national park in Costa Rica found that donations were more likely to be made when they were public (Alpizar et al., 2008). Making behaviour more observable doesn't only leverage peer pressure to act pro-socially, but also helps build the perception of a social norm. For instance, solar panels installations have been shown to be 'contagious' – neighbours are more likely to install them if other houses nearby have them visibly installed (Plumer, 2015).

### 18.4.3 Material dimension: choice architecture, nudging and effort

The term 'choice architecture' refers to the presentation, setting, or framing of choices. This might include the manner in which ethical investments are presented to pension customers, the design of a plastic bottle return scheme, or the layout of a supermarket, restaurant menu or canteen. We can be greatly influenced by the minutiae of these choice environments, which can therefore be designed to gently promote more sustainable outcomes without precluding freedom of choice or relying on conventional incentives – this is the basis of *nudging* (Thaler & Sunstein, 2008). Nudges can take many forms, from the provision of timely prompts to the design of information and choice environments. Often they aim to address or directly harness a particular cognitive bias or trait, for example putting sustainable options first on menus in canteens. Such examples barely scratch the surface of the opportunities to use choice architecture, which are well-reviewed elsewhere (Johnson et al., 2012).

Two particular aspects of our choice environment are in particular worth highlighting: effort and timing. First, we are consummate effort minimisers, and in the words of Nobel prize-winning behavioural economist Richard Thaler, if you want to encourage a particular behaviour, 'make it easy' (Halpern, 2015). This has major implications for sustainable behaviours and conservation efforts which, even with good intentions, are often thwarted by minor hassle. Importantly, this goes beyond what might be considered rational, with small *friction costs* (seemingly trivial points of hassle) having a disproportionate impact on our behaviour and often leading us to act against our best interests or intentions (Behavioural Insights Team, 2014).

Removing or introducing small frictions is therefore a powerful and widely applicable intervention. For example, shaped bin lids that remove the friction of recycling, making it easy to see where to put bottles, cans and paper, have been shown to significantly increase recycling and reduce contamination (Duffy & Verges, 2009). Similarly, multiple studies have shown that removing the tray from canteens (but allowing plate refills) makes it slightly harder to take too much food, significantly reducing food waste by up to 40% (e.g. Thiagarajah & Getty, 2013).

One of the most powerful ways to make something easy is to make it the default, in part because we often fail to make an active choice, and in part because defaults are often taken as implicit recommendations or safe/standard options. For instance, one study found a 10-fold increase in the uptake of a renewable energy tariff by making it the default (Ebeling & Lotz, 2015).

Similarly, in 2012, UK auto-enrolment legislation changed private pensions from 'opt in' to 'opt out', leading to a dramatic 42% increase in the number of people saving for retirement, more effective than billions of pounds in subsidies (Department for Work and Pensions (DWP), 2017). Perhaps a natural progression from this success is for pension providers to make the default portfolio an ethical investment – a policy idea surely capable of pushing trillions into the green economy, considering the great majority of us never change our investment portfolio.

Second, timing really matters. We find some behaviours much easier at certain moments, and policies and campaigns should be targeted to harness this fact; for instance, promoting uptake of loft insulation among new home movers while the loft is empty. Similarly, evidence shows that we are more likely to adopt new transport behaviours after disruptions such as a house move or train strike, having been forced to break our usual habits and explore new options (Larcom et al., 2017). This so-called 'fresh-start effect' was evidenced by the Behavioural Insights Team in the City of Portland, finding that promotions to use a bike-sharing scheme were nearly four times as effective among people who had just moved home (unpublished data, 2017).

Although some of these examples may seem removed from the field of conservation, the broader point is that it pays to understand the relevant micro-behaviours and processes, as there are invariably points at which default outcomes can be set, timely moments identified and frictions introduced or removed, often with surprisingly large impacts. This might include, for example, making it easier to accurately record fish-take, to apply for licences or land stewardship schemes, or to whistle-blow on poachers and ivory sellers. This approach embraces the concept of *radical incrementalism*, noting that multiple incremental changes, each targeting a small part of the problem, can sum to dramatic improvements in outcomes.

### 18.4.4  Economic dimension – incentive design

Incentives are often effective, and there is a large literature in economics devoted to this which I do not cover here. However, they can also have more subtle psychological consequences, and these factors should be considered to maximise their effect and minimise their risk of backfiring.

A key insight is that payments and fines embody meaning beyond their economic value, signalling the desirability of the behaviour and altering its social acceptability and thus interacting with our intrinsic motivation to do something. For example, Swiss residents were found to be less likely to support the construction of a nearby nuclear facility when offered compensation, as the payment implied risk (Frey & Oberholzer-Gee, 1997). Under such circumstances, common advice is to 'pay enough, or don't pay at all' (Gneezy &

Rustichini, 2000). Similarly, pro-social activities such as volunteering are valuable to those who do them partly because they satisfy a feeling of virtue or duty, which payment can undermine (Ariely et al., 2009). In such cases, non-financial rewards such as public recognition, which can amplify the value of virtue rather than crowd it out, can be more effective (Ashraf et al., 2014; Gallus, 2016).

Several studies have similarly highlighted the risk that individual or community payments for conservation outcomes can backfire, crowding out intrinsic motivations. By creating the option of foregoing the payment, these incentives can unintentionally create a guilt-free route to ignoring the conservation agenda, as this is now an option you can 'pay for' (e.g. Vollan, 2008). In other words, the punishment becomes more tolerable and morally acceptable, compared to the guilt of breaking local norms, community trust and social obligation – these intrinsic motivators can be a potent form of enforcement.

The importance of self-governance and local norms must therefore be reflected in any outside regulation or incentive scheme, which should aim to support and augment (*crowd-in*) these intrinsic motivations, not supplant or undermine them (Vollan, 2008). This is not always easy to achieve by design or to predict. One good example from a disparate context is the UK's £0.05 plastic bag charge, which has led to an 83% reduction in use (HMG, 2017). Such a large impact is implausible through price elasticity alone, but occurs because it reinforces intrinsic motivations by altering the choice architecture: the payment acts as a reminder; the default is now to forego a bag and social expectation of not using one is strengthened – no longer can we unthinkingly use a bag in wilful ignorance, but must proactively and publicly request to harm the planet.

A second cluster of research focuses on designing incentives to harness the heuristics and biases through which we think about costs and rewards. For instance, our tendency to steeply discount the future and bias our attention towards the present (Laibson, 1997) implies effective incentives should be front-loaded and costs delayed. Finance solutions may achieve this, for example to encourage home energy improvements where the reverse (high upfront costs and long-term benefits) is ordinarily a barrier. Similarly, simply redesigning product labels to highlight lifetime cost rather than only the price tag can nudge us towards more energy-efficient purchases (DECC, 2014).

Prospect Theory, an empirical account of our perception of gains, losses and risks (probabilistic outcomes), shows us to be loss-averse, i.e. more motivated to avoid a loss than receive an equivalent gain (Kahneman & Tversky, 2013). Implementing this in a literal fashion may be contentious in some contexts but effective: giving teachers a bonus at the beginning of a year and then taking it back if they fail to meet certain performance standards has been found to be more effective than conventional payment on performance (Fryer et al., 2012). Topical at the time of writing, this bias may prove useful in the UK if we transition from EU agricultural subsidies to a system of payments for conservation outcomes – farmers' historic receipt of these payments will likely drive a stronger motivation not to lose them compared to new incentives being introduced.

Lotteries can also be a powerful tool. Despite being equivalent in expected utility, we tend to value a 1-in-a-million chance of £1m more than a guaranteed £1, while a guaranteed loss of £1 is preferable to a 1-in-a-million chance of losing £1m. Through the lens of rational choice, this equates to a biased over-weighting of small probabilities. A more intuitive psychological explanation is that we are willing to pay for the hope of winning, or the peace of mind of having no risk rather than some risk. Regardless, lotteries offer creative policy options and are widely usable in many contexts, although they have not yet been tested in a conservation context. In another context, one compelling example comes from China, where authorities introduced state lottery tickets on the back of retail receipts to reduce tax avoidance. The expected value was tiny due to very long odds, meaning an equivalent fixed incentive would be ineffective. However, the disproportionate value customers put on the lottery meant they asked for their receipt, putting the sale on record and making it harder for retailers to evade tax (Wan, 2010).

To translate these insights to a sustainability context, imagine a plastic bottle deposit scheme which, rather than returning £0.10 per bottle, entered you into a lottery where every millionth returned bottle won £100,000 (this would yield 35 news-worthy winners per day based on current UK bottle use; House of Commons, 2017). Or – quite hypothetically to illustrate the point – would anyone dare use a plastic bag if rather than being charged £0.05, a spot fine of £1000 was levied on every 20,000th bag-user? Clearly, not all incentive designs are equal through the lens of behavioural science.

In this chapter I have only scratched the surface of what behavioural science can offer the field of conservation, but the key lessons are this: there are myriad influences on our behaviour, many of them contextual and operating through subtle, non-conscious processes. Effective interventions must consider these forces alongside a conventional understanding of regulation, incentives, information and awareness. In doing this, entirely novel approaches are often revealed. Other times, conventional tools can be made more effective. Ultimately, however, the most effective intervention will not be the one which draws upon the most novel finding from behavioural science, but the one which addresses the relevant barriers and motivations. As such, none of these strategies are 'one-size-fits-all', but should be brought to bear through a grounded and empirical understanding of the nature of the problem among the population of interest. Sometimes, this may be as simple as making things a bit easier or a bit cheaper.

# References

Allcott, H. 2011. Social norms and energy conservation. *Journal of Public Economics*, 95, 1082–1095.

Alpizar, F., Carlsson, F. & Johansson-Stenman, O. 2008. Anonymity, reciprocity, and conformity: evidence from voluntary contributions to a national park in Costa Rica. *Journal of Public Economics*, 92, 1047–1060.

Argyle, M. 1957. Social pressure in public and private situations. *The Journal of Abnormal and Social Psychology*, 54, 172.

Ariely, D., Bracha, A. & Meier, S. 2009. Doing good or doing well? Image motivation and monetary incentives in behaving prosocially. *American Economic Review*, 99, 544–555.

Ashraf, N., Bandiera, O. & Jack, B. K. 2014. No margin, no mission? A field experiment on incentives for public service delivery. *Journal of Public Economics*, 120, 1–17.

Barkan, R., Ayal, S. & Ariely, D. 2015. Ethical dissonance, justifications, and moral behavior. *Journal of Personality and Social Psychology*, 74, 63–79.

Barr, S. 2004. Are we all environmentalists now? Rhetoric and reality in environmental action. *Geoforum*, 35, 231–249.

Becker, G. S. 1976. *The Economic Approach to Human Behavior*. Chicago, IL: University of Chicago Press.

Behavioural Insights Team. 2014. EAST: four simple ways to apply behavioural insights. Available from www.bi.team/wp-content /uploads/2015/07/BIT-Publication-EAST _FA_WEB.pdf

Burgess, G. 2016. *Powers of persuasion? Conservation Communications, Behavioural Change and Reducing Demand for Illegal Wildlife Products*. Paris: OECD.

Carrington, D. 2019. Public concern over environment reaches record high in UK. *The Guardian*. Available from www .theguardian.com/environment/2019/jun/ 05/greta-thunberg-effect-public-concern-over-environment-reaches-record-high (accessed 7 June 2019).

Chater, N. & Loewenstein, G. 2016. The under-appreciated drive for sense-making. *Journal of Economic Behavior & Organization*, 126, 137–154.

Cialdini, R. B. 2003. Crafting normative messages to protect the environment. *Current Directions in Psychological Science*, 12, 105–109.

Darnton, A. 2008. *An Overview of Behaviour-change Models and Their Uses*. Government Social Research, Knowledge Review.

DECC (UK Department of Energy and Climate Change). 2014. Evaluation of the DECC/John Lewis energy labelling trial.

Dimant, E. 2017. On peer effects: contagion of pro-and anti-social behavior and the role of social cohesion. Discussion Papers 2017–06, The Centre for Decision Research and Experimental Economics, School of Economics, University of Nottingham.

Duffy, S. & Verges, M. 2009. It matters a hole lot: perceptual affordances of waste containers influence recycling compliance. *Environment and Behavior*, 41, 741–749.

(DWP) Department for Work and Pensions. 2017. Automatic Enrolment Review 2017: Analytical Report.

Ebeling, F. & Lotz, S. 2015. Domestic uptake of green energy promoted by opt-out tariffs. *Nature Climate Change*, 5, 868.

Frey, B. S. & Oberholzer-Gee, F. 1997. The cost of price incentives: an empirical analysis of motivation crowding-out. *The American Economic Review*, 87, 746–755.

Fryer Jr, R. G., Levitt, S. D., List, J., et al. 2012. *Enhancing the Efficacy of Teacher Incentives through Loss Aversion: A Field Experiment* (No. w18237). Cambridge, MA: National Bureau of Economic Research.

Gallus, J. 2016. Fostering public good contributions with symbolic awards: a large-scale natural field experiment at wikipedia. *Management Science*, 63, 3999–4015.

Gino, F., Norton, M. I. & Weber, R. A. 2016. Motivated Bayesians: feeling moral while acting egoistically. *Journal of Economic Perspectives*, 30, 189–212.

Gneezy, U. & Rustichini, A. 2000. Pay enough or don't pay at all. *The Quarterly Journal of Economics*, 115, 791–810.

Goldstein, N. J., Cialdini, R. B. & Griskevicius, V. 2008. A room with a viewpoint: using social norms to motivate environmental conservation in hotels. *Journal of Consumer Research*, 35, 472–482.

Haidt, J. 2001. The emotional dog and its rational tail: a social intuitionist approach to moral judgment. *Psychological Review*, 108, 814.

Halpern, D. 2015. *Inside the Nudge Unit*. London: Random House.

Haynes-Worthington, S. 2018. The Attenborough effect: searches for plastic recycling rocket after *Blue Planet II*. Resource. Available from https://resource.co/article/attenborough-effect-searches-plastic-recycling-rocket-after-blue-planet-ii-12334 (accessed 24 July 2018).

HMG. 2017. Data available from www.gov.uk/government/publications/carrier-bag-charge-summary-of-data-in-england/single-use-plastic-carrier-bags-charge-data-in-england-for-2016-to-2017 (accessed May 2018).

House of Commons Environmental Audit Committee. 2017. Plastic bottles: Turning Back the Plastic Tide. First Report of Session 2017–19.

Jenni, K. & Loewenstein, G. 1997. Explaining the identifiable victim effect. *Journal of Risk and Uncertainty*, 14, 235–257.

Johnson, E. J., Shu, S. B., Dellaert, B. G., et al. 2012. Beyond nudges: tools of a choice architecture. *Marketing Letters*, 23, 487–504.

Jones, J. P., Andriamarovololona, M. M. & Hockley, N. 2008. The importance of taboos and social norms to conservation in Madagascar. *Conservation Biology*, 22, 976–986.

Kahneman, D. 2011. *Thinking, Fast and Slow*. London: Macmillan.

Kahneman, D. & Tversky, A. 2013. Prospect theory: an analysis of decision under risk. In MacLean, L. C. & Ziemba, W. T., editors, *Handbook of the Fundamentals of Financial Decision-Making* (pp. 99–127). Singapore: World Scientific.

Kollmuss, A. & Agyeman, J. 2002. Mind the gap: why do people act environmentally and what are the barriers to pro-environmental behavior? *Environmental Education Research*, 8, 239–260.

Laibson, D. 1997. Golden eggs and hyperbolic discounting. *The Quarterly Journal of Economics*, 112, 443–478.

Larcom, S., Rauch, F. & Willems, T. 2017. The benefits of forced experimentation: striking evidence from the London underground network. *The Quarterly Journal of Economics*, 132, 2019–2055.

Marteau, T. M. 2017. Towards environmentally sustainable human behaviour: targeting non-conscious and conscious processes for effective and acceptable policies. *Philosophical Transactions of the Royal Society A*, 375(2095), 20160371.

Mazar, N., Amir, O. & Ariely, D. 2008. The dishonesty of honest people: a theory of self-concept maintenance. *Journal of Marketing Research*, 45, 633–644.

Nickerson, R. S. 1998. Confirmation bias: a ubiquitous phenomenon in many guises. *Review of General Psychology*, 2, 175.

Ölander, F. & Thøgersen, J. 2014. Informing versus nudging in environmental policy. *Journal of Consumer Policy*, 37, 341–356.

Ostrom, E. 2000. Collective action and the evolution of social norms. *Journal of Economic Perspectives*, 14, 137–158.

Ostrom, E., Gardner, R. & Walker, J. 1994. *Rules, Games, and Common-Pool Resources*. Ann Arbor, MI: University of Michigan Press,

Plumer, B. 2015. Solar power is contagious: installing panels often means your neighbors will too. Available from www .vox.com/2014/10/24/7059995/solar-power-is-contagious-neighbor-effects-panels-installation

Portney, K. E., Hannibal, B., Goldsmith, C., et al. 2018. Awareness of the food–energy–water nexus and public policy support in the United States: public attitudes among the American people. Environment and Behavior, 50, 375–400.

Reddy, S. M., Montambault, J., Masuda, Y. J., et al. 2017. Advancing conservation by understanding and influencing human behavior. Conservation Letters, 10, 248–256.

Rodríguez-Dowdell, N., Yépez-Zabala, I., Green, K., et al., editors. 2014. Pride for ARAs: A Guide to Reciprocal Water Agreements for People and Nature. Arlington, VA: Rare.

Rowson, J. & Corner, A. 2015. The Seven Dimensions of Climate Change: Introducing a New Way to Think and Talk and Act. London: RSA & Climate Outreach and Information (COIN).

Schiff, B. 2012. The function of narrative: towards a narrative psychology of meaning. Narrative Works: Issues, Investigations and Interventions, 2, 33–47.

Schneider, C. R., Zaval, L., Weber, E. U., et al. 2017. The influence of anticipated pride and guilt on pro-environmental decision making. PLoS ONE, 12(11), e0188781.

Scott, J. 2000. Rational choice theory. In Browning, G., Halcli, A. & Webster, F., editors, Understanding Contemporary Society: Theories of the Present. Browning (pp. 126–138). London: SAGE.

Shafir, E., editor. 2013. The Behavioral Foundations of Public Policy. Princeton, NJ: Princeton University Press.

Shalvi, S., Gino, F., Barkan, R., et al. 2015. Self-serving justifications: doing wrong and feeling moral. Current Directions in Psychological Science, 24, 125–130.

Shove, E. 2009. Beyond ABC: climate change policies and theories of social change. Environmental and Planning, 42, 1273–1285.

Simon, H. A. 1972. Theories of bounded rationality. Decision and Organization, 1, 161–176.

Steentjes, K., Pidgeon, N. F., Poortinga, W., et al. 2017. European Perceptions of Climate Change (EPCC): Topline Findings of a Survey Conducted in Four European Countries in 2016. Cardiff: Cardiff University.

Sunstein, C. R., Bobadilla-Suarez, S., Lazzaro, S. C., et al. 2016. How people update beliefs about climate change: good news and bad news. Cornell Law Review, 102, 1431.

Thaler, R. & Sunstein, C. 2008. Nudge. New Haven, CT: Yale University Press.

Thiagarajah, K. & Getty, V. M. 2013. Impact on plate waste of switching from a tray to a trayless delivery system in a university dining hall and employee response to the switch. Journal of the Academy of Nutrition and Dietetics, 113, 141–145.

Thøgersen J. 2014. Unsustainable consumption: basic causes and implications for policy. European Psychologist, 19, 84–95.

Trivers, R. L. 1971. The evolution of reciprocal altruism. Quarterly Review of Biology, 46, 35–57.

Turner, J. C. 1991. Social Influence. Milton Keynes: Open University Press.

Turnwald, B., Boles, D. & Crum, A. 2017. Association between indulgent descriptions and vegetable consumption: twisted carrots and dynamite beets. Journal of the American Medical Association, 177, 1216–1218.

Vennard, D., Park, T. & Attwood, S. 2019. Encouraging Sustainable Food Consumption By Using More-Appetizing Language. Washington, DC: World Resources Institute.

Vollan, B. 2008. Socio-ecological explanations for crowding-out effects from economic field experiments in southern Africa. Ecological Economics, 67, 560–573.

Wan, J. 2010. The incentive to declare taxes and tax revenue: the lottery receipt experiment in China. *Review of Development Economics*, 14, 611–624.

Webb, T. L. & Sheeran, P. 2006. Does changing behavioral intentions engender behavior change? A meta-analysis of the experimental evidence. *Psychological Bulletin*, 132, 249–268.

WildAid. 2014. *Evidence of Declines in Shark Fin Demand, China*. San Francisco, CA: WildAid.

# CHAPTER NINETEEN

# Social marketing and conservation

ROBERT J. SMITH
*Durrell Institute of Conservation and Ecology*
GABBY SALAZAR
*Imperial College London*
JOSEPH STARINCHAK
*US Fish and Wildlife Service*
LAURA A. THOMAS-WALTERS
*Durrell Institute of Conservation and Ecology*
and
DIOGO VERÍSSIMO
*University of Oxford and San Diego Zoo Global*

## 19.1 Introduction

Most conservation issues stem from people's actions and choices, so halting biodiversity loss depends on changing human behaviour (Schultz, 2011). The two main approaches traditionally used to achieve such behaviour change are based on education, where people are encouraged to understand and appreciate the natural world, and legislation, where people are punished for breaking rules and laws designed to protect nature (Rothschild, 2000). Both approaches have advantages, but evidence suggests they are often ineffective because increasing awareness is rarely sufficient to change behaviour (Waylen et al., 2009; Chapter 18) and effective conservation legislation in the face of opposing social norms depends on costly enforcement (Cooney et al., 2017). This is why conservation scientists and practitioners increasingly recognise the value of approaches based on social marketing, which seeks to change people's behaviour for the benefit of wider society by using techniques originally developed in the business world to sell products and services (Smith et al., 2010; Wright et al., 2015). This link to commercialism makes many conservationists queasy. However, the current extinction crisis shows we need to move outside our comfort zone and consider new techniques with proven success. In this chapter we discuss the use of social marketing in conservation, beginning with definitions of the terms and an explanation of how it differs from conservation education. We then briefly review how social marketing has been used in community-based natural resource management, demand reduction and flagship species fundraising, and end by discussing lessons that relate more broadly to conservation.

## 19.2  Defining marketing and social marketing

Marketing is widely used in the private sector and is defined as 'the process of planning and executing the development, value, promotion and distribution of products, services, and ideas to create exchanges that are mutually beneficial' (Silk, 2006). It is an important component of most successful businesses, so it was probably inevitable that other sectors would apply marketing techniques to their work. In particular, this led to the development of social marketing, defined as 'the systematic application of marketing along with other concepts and techniques to achieve specific behavioural goals for a social good' (French et al., 2006). It should be noted that while social media is often used in social marketing, they are not the same thing. Instead, social media is just one type of communication channel, with other examples including radio, billboards and street theatre.

In the behaviour change field, social marketing is seen as one of four approaches (Rothschild, 2000; Santos et al., 2011). Two of the others, education and law, are widely recognised in conservation. The fourth is technical intervention, which is defined as those aspects of technology, infrastructure or equipment that are critical to enable behaviour change to take place. The appropriateness of these four approaches in a particular context can then be defined based on three components: a person's ability, opportunity and motivation to change their behaviour. These three components determine whether a person is prone, unable or resistant to behaviour change (Figure 19.1a), and hence which combinations of approaches should be used in response (Figure 19.1b). Law-based approaches should be used when people lack motivation, education-based approaches when they lack the ability and technical intervention-based approaches when they lack opportunity (Figure 19.1b). In contrast, marketing-based approaches are useful in a much wider range of circumstances, because they are designed to overcome a lack of all three components. Moreover, while social marketing and education campaigns are often confused, there are other fundamental differences between the two approaches. In particular, social marketing focuses on exchange, with both sides willing to engage in the transaction and happy with the outcome, whereas conservation education depends on people changing their behaviour for the greater good. In addition, while both approaches are designed with a target audience in mind, this is fundamental in social marketing, and involves identifying and defining the target audience based on factors that relate to their relevant values and interests (Wright et al., 2015).

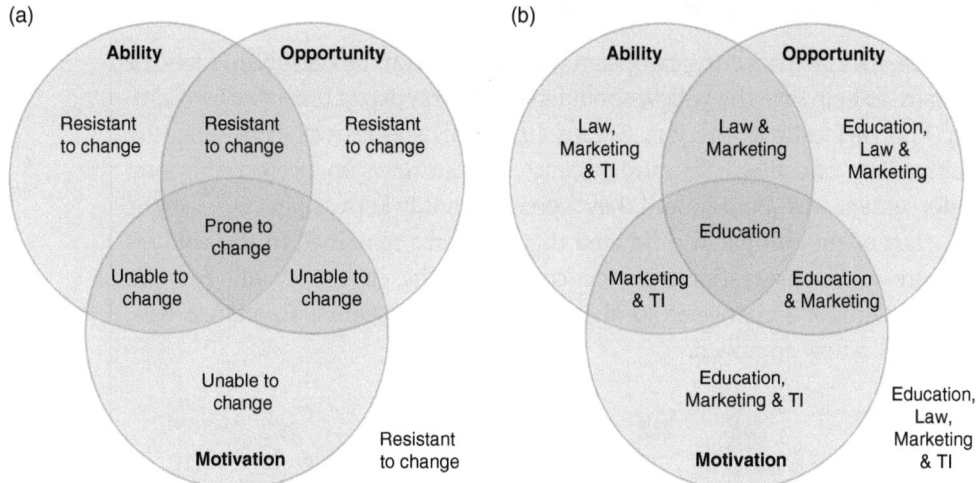

**Figure 19.1** Diagram showing how a person's ability, opportunity and/or motivation determines (a) whether they are prone, unable or resistant to change and (b) the appropriateness of the four different behaviour change approaches of education, law, marketing and technical intervention (TI) under these different conditions (adapted from Rothschild, 2000; Santos et al., 2011). (A black and white version of this figure will appear in some formats. For the colour version, please refer to the plate section.)

Social marketing has been used for over 50 years in areas such as health, development, financial literacy and transportation (Lefebvre, 2013) and is now represented by a number of practitioners and professional bodies. These groups came together to develop a broader definition, stating: 'Social Marketing practice is guided by ethical principles. It seeks to develop and integrate marketing concepts with other approaches to influence behaviours that benefit individuals and communities for the greater social good. It seeks to integrate research, best practice, theory, audience and partnership insight, to inform the delivery of competition sensitive and segmented social change programmes that are effective, efficient, equitable and sustainable' (iSMA et al., 2013). A key component is the application of a systematic, step-by-step process that is described and illustrated in Box 19.1.

## 19.3 Social marketing in conservation

The application of social marketing in conservation is relatively new compared to sectors like health and development, although its role in fundraising goes back decades (Nicholls, 2011). More recently, a number of conservation scientists and practitioners have recognised the approach's value, and social marketing is becoming a more common component of the conservation

## Box 19.1. Bonaire parrot campaign

In 1998, a social marketing campaign was launched on the Caribbean island of Bonaire to help save the yellow-shouldered Amazon parrot (*Amazona barbadensis*), known locally as the lora (Figure 19.2). This species was threatened by habitat loss and illegal capture because, despite laws to protect the lora, enforcement was sporadic and they were commonly kept as pets. A survey at the start of the campaign estimated that 300 loras remained in the wild on Bonaire and conservationists were concerned the species would become extinct without a change in local attitudes and behaviours. To address this, they took a new approach.

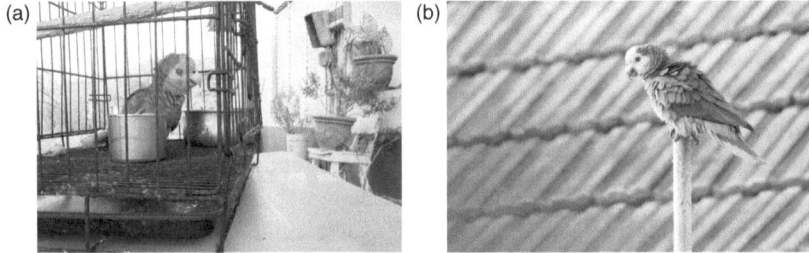

(a)  (b)

**Figure 19.2** The lora or yellow-shouldered Amazon parrot (*Amazona barbadensis*) that was the focus of a social marketing campaign on the Caribbean island of Bonaire. (A black and white version of this figure will appear in some formats. For the colour version, please refer to the plate section.)

The conservation organisation Rare had already run social marketing campaigns on other Caribbean islands, which were based on creating national pride in a target species to shift attitudes and behaviours towards that species (Scholtens & Butler, 1999). Conservationists on Bonaire approached Rare and together formed a committee of local organisations to plan a year-long social marketing campaign to 'Save the Lora'. Following social marketing theory, the campaign included the following six core concepts (ESMA, 2017):

1. **Setting of explicit social goals**. The first step is identifying the behaviour the campaign is trying to influence and setting clear, quantifiable goals related to that behaviour. On Bonaire, the goal was to reduce the number of people purchasing loras as pets and so, ultimately, reduce the number of these parrots removed from the wild.
2. **Citizen orientation and focus**. In social marketing programmes, citizens should be engaged in the process of identifying issues and developing solutions. On Bonaire, a consortium of environmental organisations, government departments, media companies and volunteers was created to plan and implement the campaign. Before the campaign, the committee

**Box 19.1. (cont.)**

conducted a formative evaluation to understand citizens' knowledge of and attitudes towards the lora. To do this, they distributed a questionnaire to approximately 4% of the island's population. The data they collected helped inform the campaign message and provided baseline information for measuring the campaign's impact.

3. **Highlighting target audience benefits via a mix of marketing interventions**. Social marketing campaigns ask people to exchange a detrimental behaviour or value for a more desirable one. On Bonaire, the campaign asked people to exchange the opportunity cost of having a pet lora for a new symbol of collective national identity. The campaign sought to reframe the lora, which was traditionally seen as a pet, as a symbol of national pride. This 'product' was sold using a mix of marketing interventions, including radio broadcasts, songs and pamphlets. Volunteers also dressed in a giant lora costume to emphasise the species' role as a national mascot.

4. **Theory, insight, data and evidence informed audience segmentation**. Social marketing is based on the idea that a one-size-fits-all approach rarely works. Instead, it is important to spend time and resources identifying, understanding and selecting which parts of the population (known as 'segments') should be the focus of subsequent campaigns. Following the Rare Pride Campaign model, the group on Bonaire developed campaign materials to target different audiences. They used formative research to help identify the most popular news sources on the island and produce radio shows, music videos and articles to reach different segments of the population. For example, they created a song about the lora to reach school children and a religious sermon to reach church congregations.

5. **Competition/barrier and asset analysis**. Social marketing programmes also seek to identify and remove barriers that could keep their target audiences from adopting or sustaining positive behaviours. On Bonaire, keeping loras as pets had become a social norm, so the Rare campaign focused on reframing this species as a wild animal that should stay in the wild. The campaign also worked with local newspapers and radio stations to inform citizens of the illegality and consequences of keeping this parrot in captivity.

6. **Critical thinking, reflexivity and being ethical**. To be effective and ethical, social marketing campaigns require flexibility and an understanding of the local context. When the campaign started on Bonaire, hundreds of loras were estimated to be in captivity, making it impossible to confiscate all illegal pets. Instead of confiscating the birds, which would have resulted in animal welfare issues, the campaign focused on creating a context in which no new pet loras would be acquired.

> **Box 19.1. (cont.)**
>
> Following this campaign, there has been a long-term increase in the lora population on Bonaire. Recent research suggests the campaign played a role in this conservation success by helping shift social norms around keeping loras as pets and increasing support for the enforcement of existing laws and regulations (Salazar, 2017).

toolbox (Wright et al., 2015). To illustrate this, we briefly outline how social marketing has been used in three different aspects of conservation practice.

### 19.3.1 Community-based natural resource management

Increasing the sustainability of natural resource management by local communities is perhaps the most widespread use of social marketing in conservation (DeWan et al., 2013; Green et al., 2013). For example, the US Fish and Wildlife Service created the 'Stop Aquatic Hitchhikers' campaign to empower recreational waterway users to help prevent the spread of aquatic invasive species (Larson et al., 2011; Figure 19.3). These species are a major threat to global biodiversity and have important economic impacts (Gallardo et al., 2016), but often remain forgotten because they are underwater and thus out of mind. The campaign used social marketing to make the issue more salient among groups such as boaters, anglers, rafters, kayakers, sailors and waterfowl hunters who inadvertently transport aquatic invasive species across waterways on their equipment. As most of these activities require licensing or registration, the Fish and Wildlife Service represented not only an important source of information about the profile of its target audience, but also active partners to promote the appropriate cleaning of recreational equipment. Using branding, the campaign leveraged the links between natural resources and the identity of communities who live on or near the water. They instilled a sense of stewardship in recreational users, so that the target audience was willing to exchange old behaviours for new ones to keep the rivers clean for the benefit of themselves and others (Ries & Trout, 1982). The support of local businesses and other government agencies was vital, as they not only acted as key influencers but also created additional visibility and salience for the message around the need for more thorough cleaning of equipment.

### 19.3.2 Demand reduction

One of the earliest uses of social marketing in conservation was to reduce demand for wildlife and wildlife products, based on campaigns to discourage

**Figure 19.3** Promotional material encouraging boat owners in the Greater Yellowstone Area to adopt practices that will reduce the spread of invasive species. (A black and white version of this figure will appear in some formats. For the colour version, please refer to the plate section.)

people from buying selected species as pets (as detailed in Box 19.1). More recently, increases in the illegal wildlife trade has created wider recognition of the value of social marketing for demand reduction, as a way to tackle the resultant threats to biodiversity, public health, local livelihoods and effective governance (Veríssimo et al., 2012). One example is the Chi Initiative, which was launched in 2014 and seeks to reduce rhino horn consumption in Vietnam by targeting wealthy businessmen (Offord-Woolley, 2017). The campaign messaging built on the Vietnamese concept of Chi, or 'strength of will', and emphasised that masculinity and good fortune come from an individual's character, not from products purchased on the market. Thus, they sought to create conditions in which taking rhino horn is seen as a sign of weakness, so that business men are willing to exchange this behaviour for one that does not support the illegal wildlife trade. This campaign pioneered the use of social marketing techniques to tackle the illegal wildlife trade, but also illustrates some of the difficulties. In particular, it shows how hard it can be to measure campaign impacts in the context of dynamic rhino horn demand (TRAFFIC, 2017) and multiple ongoing demand reduction efforts in Vietnam. This should become easier in the future, though, as the number of demand-reduction interventions has grown in the last decade (Veríssimo & Wan, 2018), increasing the amount of research and monitoring of market trends and interventions.

### 19.3.3 Conservation flagships

There is a long history of organisations using particular species for fundraising and awareness-raising. Traditionally, flagship status was seen as an intrinsic characteristic, failing to recognise that flagship species are actually marketing tools. This has changed, with a new definition of a flagship as 'a species used as the focus of a broader conservation marketing campaign based on its possession of one or more traits that appeal to the target audience' (Veríssimo et al., 2011). Viewing flagship species through this lens implies that these campaigns should adopt core social marketing concepts, including setting explicit social goals at the beginning of the process (ESMA, 2017). This is important because people generally prefer species that are large, brightly coloured and/or have human-like traits (Gunnthorsdottir, 2001; Barua et al., 2012; Borgi & Cirulli, 2015). Thus, setting goals helps guide actions towards the species most needing conservation (Veríssimo et al., 2017), rather than those that are most popular with the target audience (Smith et al., 2012). Emphasising that it is the species' traits that are important, rather than the species itself, also suggests the flagship approach can be applied to broader aspects of biodiversity. For example, Conservation International's biodiversity hotspots (Mittermeier et al., 2004) have been described as a new type of flagship, designed to appeal to a target audience of international donors by emphasising traits based on endemic biodiversity, return on investment and scientific credibility (Smith et al., 2010). The main aim of this campaign was to raise funds rather than change people's behaviours, meaning it cannot be defined as social marketing. However, the creation of this new type of flagship did have wider social marketing impacts, by building local pride in countries containing these hotspots, leading to new conservation policies and wider civil society engagement (Visseren-Hamakers et al., 2012).

## 19.4 Broader lessons from social marketing

Social marketing is a structured and systematic approach for achieving positive conservation outcomes and so many of its fundamental principles are shared with other aspects of conservation decision science and implementation.

However, it provides a number of specific insights that have broader relevance for conservation, which we highlight below.

## 1. Acknowledging ethical issues

Some critics are uneasy about the ethical issues underpinning social marketing, partly because of its links with capitalism and consumerism (Smith et al., 2010). One accusation is that campaigns are a form of 'brainwashing', so it should be stressed that social marketing is always based on choice and mutually beneficial exchange. A more fundamental issue comes from campaign development, as while the social marketing definition states the approach 'is guided by ethical principles' (iSMA et al., 2013), it does not specify whose ethics should do the guiding. This is a key concern, because marketing need is often identified by external actors with world views and priorities that differ from those of the target audience (Adams & Mulligan, 2003). Obviously, this issue applies to all behaviour change initiatives and, by focusing on choice and beneficial exchange, social marketing might be better at producing locally supported solutions than approaches based on education and legislation. However, social marketers should always be mindful of the power imbalances involved and be open to outside scrutiny and criticism.

## 2. The importance of evaluation

It is almost universally agreed that monitoring and evaluation should be core parts of any conservation activity (Sutherland et al., 2004), although their relative rarity shows that conservationists often fail to dedicate the necessary time and resources (Lindenmayer & Likens, 2010). This is less of a problem in business, where learning how to increase effectiveness pays for itself, and helps explain why evaluation is a fundamental component of marketing. This focus on effectiveness is probably why social marketing campaigns were some of the first behaviour-change projects to systematically evaluate their work (Jenks et al., 2010), as an important way to understand their target audience and adapt their campaigns to increase impact. Just as importantly, social marketers recognise that behaviour change projects can have a range of unintended consequences, including negative impacts. For example, a campaign based in Dominica, similar to that used in the lora project, raised the profile of the flagship species but created a negative association with another parrot species (Douglas & Winkel, 2014). Examples such as this illustrate why social marketers are obliged to learn from their actions and improve.

## 3. Changing behaviour is not easy

While social marketing offers many valuable opportunities for achieving conservation goals, behaviour change can often be slow and expensive. This is illustrated by campaigns from other sectors, such as public health, which have been working on behaviour changes for decades with varying success. Many of these campaigns failed to make any impact, or even had the opposite of the intended effect. For example, one of the US government's flagship programmes to reduce teen substance abuse actually led to an increase in adolescent drug use in certain contexts (Rosenbaum & Hanson, 1998). Such findings have contributed to the results of a recent systematic review on the effectiveness of global health programmes, which found the majority had no positive behavioural results, although success increased with the quality of the campaign (Firestone et al., 2017). Thus, caution is needed when describing the potential gains from social marketing in conservation, especially because funding for such work is likely to be relatively small compared to the health sector. However, evidence from interventions like 'Save the Lora' suggests behaviour change is possible, especially when campaigns influence societal norms and allow governments to improve regulation and enforcement (Salazar et al., 2019).

## 4. The myth of 'the general public'
A fundamental insight from marketing is that the 'general public' is an empty concept when communicating with people. This is why audience segmentation is a core concept in social marketing (Box 19.1), based on categorising people into relatively homogeneous subgroups, so that the resultant campaigns can be tailored for maximum impact. Demographic factors like age and income can play a role in defining these groups, although psychographic factors like attitudes, interests and beliefs are often more important (Wright et al., 2015). More broadly, conservationists should recognise the audience-specific nature of their messages, rather than broadcasting them to as many people as possible. For example, while messages based on 'ecosystem services' have been successful at highlighting the financial value of nature to government bureaucrats, they have created possibly avoidable tension when aimed at people who value nature for other reasons (Jones, 2018).

## 5. Value is more than a financial metric
The huge profits of some companies can be viewed as illustrations of all that is wrong with marketing, where advertising campaigns lead people into buying over-priced goods and services. However, it also reveals a fundamental marketing premise: a product's value is neither fixed nor dependent on its manufacturing costs (Sutherland, 2019). This insight also underpins social marketing in conservation, where people change their behaviour because campaigns foster stronger, more positive links with specific species, ecosystems and actions. Thus, for example, the 'Stop Aquatic Hitchhikers' campaign empowered people to reduce their negative impacts on the places they love and the 'Save the Lora' campaign built local pride in an endemic species. In each case this increase in

value was not measured financially, although the target audience may now be more willing to fund and support activities to conserve these species and habitats.

In conclusion, in this chapter we have discussed how social marketing has been used in conservation and highlighted its strengths and weaknesses. However, benefiting from these strengths involves accepting uncomfortable truths: many conservationists are uneasy about learning from the corporate world or accepting that their reasons for loving nature are not universally shared. However, we can only stem biodiversity loss by engaging with the widest possible range of people, and social marketing is one of the better ways of understanding these multiple audiences and working with them to increase how they value nature.

## References

Adams, W. M. & Mulligan, M. 2003. *Decolonizing Nature: Strategies for Conservation in a Post-colonial Era*. Abingdon: Earthscan.

Barua, M., Gurdak, D. J., Ahmed, R. A., et al. 2012. Selecting flagships for invertebrate conservation. *Biodiversity and Conservation*, 21, 1457–1476.

Borgi, M. & Cirulli, F. 2015. Attitudes toward animals among kindergarten children: species preferences. *Anthrozoös*, 28, 45–59.

Cooney, R., Roe, D., Dublin, H., et al. 2017. From poachers to protectors: engaging local communities in solutions to illegal wildlife trade. *Conservation Letters*, 10, 367–374.

DeWan, A., Green, K., Xiaohong, L., et al. 2013. Using social marketing tools to increase fuel-efficient stove adoption for conservation of the golden snub-nosed monkey, Gansu Province, China. *Conservation Evidence*, 10, 32–36.

Douglas, L. R. & Winkel, G. 2014. The flipside of the flagship. *Biodiversity and Conservation*, 23, 979–997.

ESMA. 2017. *Global Consensus on Social Marketing Principles, Concepts and Techniques*. Monksferry: European Social Marketing Association.

Firestone, R., Rowe, C. J., Modi, S. N., et al. 2017. The effectiveness of social marketing in global health: a systematic review. *Health Policy and Planning*, 32, 110–124.

French, J., Dip, H. E. & Blair-Stevens, C. 2006. From snake oil salesmen to trusted policy advisors: the development of a strategic approach to the application of social marketing in England. *Social Marketing Quarterly*, 12, 29–40.

Gallardo, B., Clavero, M., Sánchez, M. I., et al. 2016. Global ecological impacts of invasive species in aquatic ecosystems. *Global Change Biology*, 22, 151–163.

Green, K. M., DeWan, A., Balcázar Arias, A., et al. 2013. Driving adoption of payments for ecosystem services through social marketing, Veracruz, Mexico. *Conservation Evidence*, 10, 48–52.

Gunnthorsdottir, A. 2001. Physical attractiveness of an animal species as a decision factor for its preservation. *Anthrozoös*, 14, 204–215.

iSMA, ESMA, AASM. 2013. *Consensus Definition of Social Marketing*. International Social Marketing Association, European Social Marketing Association and Australian Association of Social Marketing.

Jenks, B., Vaughan, P. W. & Butler, P. J. 2010. The evolution of Rare Pride: using evaluation to drive adaptive management in a biodiversity conservation organization. *Evaluation and Program Planning*, 33, 186–190.

Jones, J. P. G. 2018. How can we communicate all that nature does for us? Available from http://theconversation.com/how-can-we-communicate-all-that-nature-does-for-us-94761 (accessed 17 June 2018).

Larson, D. L., Phillips-Mao, L., Quiram, G., et al. 2011. A framework for sustainable invasive species management: environmental, social, and economic objectives. *Journal of Environmental Management*, 92, 14–22.

Lefebvre, R. C. 2013. *Social Marketing and Social Change: Strategies and Tools For Improving Health, Well-Being, and the Environment*. San Francisco, CA: Jossey-Bass.

Lindenmayer, D. B. & Likens, G. E. 2010. The science and application of ecological monitoring. *Biological Conservation*, 143, 1317–1328.

Mittermeier, R. A., Robles-Gil, P., Hoffmann, M., et al. 2004. *Hotspots Revisited: Earth's Biologically Richest and Most Endangered Ecoregions*. Mexico City: Cemex.

Nicholls. H. 2011. The art of conservation. Available from www.nature.com/articles/472287a (accessed 16 June 2018).

Offord-Woolley, S. 2017. The Chi Initiative: a behaviour change initiative to reduce the demand for rhino horn in Viet Nam. *Pachyderm*, 58, 144–147.

Ries, A. & Trout, J. 1982. *Positioning: The Battle For Your Mind*. New York, NY: McGraw-Hill.

Rosenbaum, D. P. & Hanson, G. S. 1998. Assessing the effects of school-based drug education: a six-year multilevel analysis of project D.A.R.E. *Journal of Research in Crime and Delinquency*, 35, 381–412.

Rothschild, M. L. 2000. Carrots, sticks, and promises: a conceptual framework for the management of public health and social issue behaviors. *Social Marketing Quarterly*, 6, 86–114.

Salazar, G. R. 2017. A qualitative impact evaluation of a conservation social marketing campaign. MSc thesis, Imperial College, London.

Salazar, G., Mills, M. & Veríssimo, D. 2019. Qualitative impact evaluation of a social marketing campaign for conservation.

*Conservation Biology*, 33, 634–644. https://doi.org/10.1111/cobi.13218

Santos, C., Simões, A. M., Atalaia, J., et al. 2011. *Melhorar a Vida – Um Guia de Marketing Social*. Alverca: Fundação CEBI.

Scholtens, E. & Butler, P. 1999. *Learning About the Lora: A Campaign to Promote Pride in Bonaire's Yellow Shouldered Parrot*. Arlington, VA: Rare.

Schultz, P. W. 2011. Conservation means behavior. *Conservation Biology*, 25, 1080–1083.

Silk, A. 2006. *What Is Marketing?* Harvard, MA: Harvard Business Press.

Smith, R. J., Veríssimo, D., Isaac, N. J. B., et al. 2012. Identifying Cinderella species: uncovering mammals with conservation flagship appeal. *Conservation Letters*, 5, 205–212.

Smith, R. J., Veríssimo, D. & MacMillan, D. C. 2010. Marketing and conservation: how to lose friends and influence people. In Leader-Williams, N., Adams, W. M. & Smith, R. J., editors, *Trade-offs in Conservation: Deciding What to Save* (pp. 215–232). Oxford: Wiley-Blackwell.

Sutherland, R. 2019. *Alchemy: The Surprising Power of Ideas That Don't Make Sense*. London: WH Allen.

Sutherland, W. J., Pullin, A. S., Dolman, P. M., et al. 2004. The need for evidence-based conservation. *Trends in Ecology & Evolution*, 19, 305–308.

TRAFFIC. 2017. Chi Briefing Paper. World Rhino Day 2017. Available from www.trafficj.org/publication/17_Briefing_CHI-World_Rhino_Day.pdf.

Veríssimo, D., Challender, D. & Nijman, V. 2012. Wildlife trade in Asia: start with the consumer. *Asian Journal of Conservation Biology*, 1, 49–50.

Veríssimo, D., MacMillan, D. C. & Smith, R. J. 2011. Towards a systematic approach for identifying conservation flagships. *Conservation Letters*, 4, 1–8.

Veríssimo, D., Vaughan, G., Ridout, M., et al. 2017. Increased conservation marketing effort has major fundraising benefits for even the least popular species. *Biological Conservation*, 211, 95–101.

Veríssimo, D. & Wan, A. K. Y. 2018. Characterising the efforts to reduce consumer demand for wildlife products. Open Science Framework. Available from https://osf.io/tu4jy/ (accessed 27 May 2018).

Visseren-Hamakers, I. J., Leroy, P. & Glasbergen, P. 2012. Conservation partnerships and biodiversity governance: fulfilling governance functions through interaction. *Sustainable Development*, 20, 264–275.

Waylen, K. A., McGowan, P. J. K., Group, P. S., et al. 2009. Ecotourism positively affects awareness and attitudes but not conservation behaviours: a case study at Grande Riviere, Trinidad. *Oryx*, 43, 343–351.

Wright, A. J., Veríssimo, D., Pilfold, K., et al. 2015. Competitive outreach in the 21st century: why we need conservation marketing. *Ocean & Coastal Management*, 115, 41–48.

# Conclusion

# CHAPTER TWENTY

# Successfully translating conservation research into practice and policy: concluding thoughts

NATHALIE PETTORELLI
*Zoological Society of London*
PETER N. M. BROTHERTON
*Natural England*
ZOE G. DAVIES
*Durrell Institute of Conservation and Ecology (DICE)*
NANCY OCKENDON, WILLIAM J. SUTHERLAND
*University of Cambridge*
and
JULIET A. VICKERY
*RSPB Centre for Conservation Science*

In the Anthropocene, when our environment is changing rapidly and the windows of opportunity for action to prevent further biodiversity loss are narrow, conservation researchers are increasingly encouraged to think and operate beyond the traditional approaches of producing peer-reviewed papers and presenting results to other members of the research community. Indeed, the perception that researchers belong in their ivory tower, from which they deliver evidence for others to interpret, disseminate and use in decision-making, is thankfully now widely recognised as outdated. The rise of fake news, a deliberate lack of consideration for scientific evidence, and changes to the ways of assessing the value of researchers' work probably all play a role in supporting this shift in perception. Moreover, for many researchers, the prospect of their work 'making a difference' and having an impact on wider society is at least as great a motivation for doing research as generating new knowledge, however interesting that may be. In addition, researchers and research institutions are nowadays not only required to contribute to advancing knowledge, but also play a part in societal development. Impact thus matters to a growing number of researchers and funders, and it increasingly shapes the functioning of research institutions worldwide.

Research impact can come in a plethora of forms, but the pathway to delivery will typically involve negotiating the interface between research and policy/practice successfully. One recurring theme that emerges in this book is the need for close working relationships between those generating evidence and the practitioners and policy-makers that apply it. The nature, quality and regularity of these interactions are instrumental in ensuring that pertinent evidence underpins solid decision-making. For too long, one of the biggest misconceptions about the interface between research and policy/practice has been that it follows a linear model, whereby decision-makers pose questions, and researchers generate appropriate evidence, which is then used by the decision-makers to make well-informed choices. We hope that this book helps to further dispel this myth.

Instead, we lay out some of the potentially more complex models by which researchers, policy-makers and practitioners can be brought together, across all stages of the knowledge generation, exchange and application process. These models range from collaborative efforts to identify future research priorities through to co-producing projects that can provide outputs that address clear and topical policy or practice needs. Within a policy arena, such relationships require time to establish, and this can be challenging when political leaders change position relatively frequently. Researchers therefore need to consider being 'in it for the long haul', possibly well after the papers have been published and the novelty factor has worn off. Moreover, balancing the sometimes slow pace of change in policy with a research world always looking for new and exciting opportunities can be challenging. Close collaboration with policy staff within non-governmental organisations can be one way to overcome this problem, as they are often well networked within the decision-making communities and have a good understanding of how the 'system' works. Equally, developing meaningful relationships with those holding non-political appointments in government (such as in government agencies or the civil service) can prove fruitful. Ultimately, while the methods adopted, for example to synthesise and present different types of information or assess the cost-effectiveness of a range of policy options, are fundamental steps to enable the use of research in decision-making, building mutual trust and respect between individual researchers and decision-makers can make the difference to whether the available evidence is used.

Another key relationship is that between scientists and the practitioners or project managers making conservation decisions on the ground. Building relationships with, and learning from, practitioners can provide a unique opportunity to gain detailed insights into how research supports (or fails to support) management interventions. Much work has been done to better connect policy-makers and researchers in various countries, but we are yet to provide similar national platforms for researchers to better connect with practitioners. This step matters, as the consequences of not using scientific evidence when making decisions about conservation interventions can be damaging, in terms of both wasting limited human and financial resources and failing to meet biodiversity objectives.

An important point we hope to have conveyed with this contribution is that societal change, and thus impact, can in some cases be secured without direct engagement with policy-makers and practitioners. The internet and social media have considerably changed the modus and speed with which researchers can communicate with the public, in effect making campaigning an accessible tool for everyone. However, such a strategy can come at a cost, and lead to unintended consequences. Importantly, the choices we make as individuals (such as avoiding products with palm oil or buying organic food) are often inseparable from the enabling environments (including social norms, the political and economic situation, and incentives to promote certain behaviours) in which we make them. Because of this, the most effective route to change often means targeting both the individual (for example by working with approaches that help change motivations, habits, emotional engagement and awareness) and the enabling environment.

While encouraging the improved use of evidence by practitioners and policy-makers, it is important to remain mindful of the intricacies of the multitude of factors that influence decision-making in both these domains. These may include layers of advice from colleagues and personal experience, as well as a myriad of multi-faceted social, economic and cultural factors. This can be a frustration to researchers, and may influence the nature and content of their communications with decision-makers and the type of relationships that are built. This deeply human dimension to working at the research–policy/practice interface remains underappreciated by many in conservation. Researchers that are inexperienced with the research–policy/practice interface may arguably achieve more by collaborating with communication or behavioural change experts from the initial stages of project through to completion, rather than by going it alone or only considering communication as an add-on extra once the results of a project are complete.

This book would have not seen the light of day if researchers around the world were not increasingly recognising the need to engage and collaborate with decision-makers from the outset, to ensure the value and timeliness of their work for conservation policy and practice. Our aim here was to ask a diversity of experts to respond to a single question, namely how best to ensure impact is realised. Most of our contributors agree on one thing: researchers need to use a variety of approaches and invest in a range of different relationships to make sure high-quality evidence is co-developed and co-produced with relevant stakeholders.

Some also point to the importance of providing increased formal or informal training to current and future conservation researchers in the skills that are needed to work productively at the research–policy/practice interface. Teaching the next generation about effective knowledge exchange and how to interact with governmental and parliamentary procedures within the confined walls of universities can only go so far. Early-career researchers will likely model their behaviour and approaches on the senior researchers that they are exposed to, and learning in a classroom setting will never replace first-hand experience. Moreover, while establishing mutual trust and understanding between researchers and decision-makers is vital, it may be challenging to find the space to meet and develop relationships when both communities are subject to different work priorities, constraints and cultures. To address these issues, a number of initiatives have surfaced to increase and enhance opportunities for direct interactions between, and in-situ training for, researchers and decision-makers at all levels of seniority, many supported by research councils and learned societies.

Ultimately, our compilation of case studies and opinion pieces clearly demonstrates how engagement with policy and practice ultimately challenges us as researchers to individually confront our fears and impostor syndrome relating to our ability to generate good, useful and 'as certain as it gets' knowledge that may be appreciated and valued by society as a whole. It also challenges our egos, forcing us to realise that scientific evidence is, at the end of the day, only one of the many considerations shaping decisions. Finally, these case studies highlight how the research–policy/practice interface ultimately consists of a collection of individual research–policy/practice interfaces, shaped over the years by those who appreciate how rewarding collaborating together can be, but also understand that it requires long-term commitment. Within this book, we have tried to bring together collective wisdom of how each of us can best build our own interface, with the aim of equipping current and future generations of conservation researchers with the tools and knowledge to help them to decide how to best navigate the specific policy/practice context within which they work. We hope that this broad diversity of experience and advice will provide a valuable resource, enabling people interested in translating their research to bring about real-world change for the benefit of biodiversity.

# Index